Regulatory Issues in Organic Food Safety in the Asia Pacific

GOH Bee Chen • Rohan Price
Editors

Regulatory Issues in Organic Food Safety in the Asia Pacific

Springer

Editors
GOH Bee Chen
School of Law and Justice
Southern Cross University
Gold Coast, QLD, Australia

Faculty of Law
Chiang Mai University
Chiang Mai, Thailand

Rohan Price
School of Law and Justice
Southern Cross University
Gold Coast, QLD, Australia

ISBN 978-981-15-3579-6 ISBN 978-981-15-3580-2 (eBook)
https://doi.org/10.1007/978-981-15-3580-2

This Springer imprint is published by the registered company Springer Nature Singapore Pte Ltd.
The registered company address is: 152 Beach Road, #21-01/04 Gateway East, Singapore 189721, Singapore

Preface

In late 2017, Southern Cross University established the newly formed Centre for Organics Research. And, in early 2018, the coeditors of this book were awarded an internal seed grant for a collaborative industry research project with the Van Horne Institute in Canada to investigate organic food security in China. Beyond this research project, the idea of a book was conceived to canvass regulatory issues in organic food safety in the Asia Pacific.

We are grateful to all the chapter contributors from across the various select jurisdictions in the Asia Pacific for coming on board in this meaningful book project. In an era where health and well-being are key and climate change an urgent agenda, organic food safety on a global scale has become all the more important. Law plays a significant regulatory role in organics. It is hoped that this book is able to make a major contribution to an understanding of regulatory issues affecting organics in our region.

We would like to thank Southern Cross University Centre for Organics Research for seeding this research project and the School of Law and Justice for extending collegial support.

Bee Chen would especially like to thank the Faculty of Law at Chiang Mai University, Thailand, for her Visiting Professorship in late 2019 and, prior to that, for arranging beneficial field trips in 2018 to organic farms and markets to obtain an understanding of Thai organic agriculture and the regulatory framework surrounding certification and distribution of organic foods. During the course of the preparation of this edited volume, in particular, she is grateful to her husband, Chamkaur, for his abiding support and to their daughter, Mindy, for shining her own light in her poetic calling. They remain always her pillars of strength and inspiration in Bee Chen's academic endeavours.

The ongoing support of Rohan's wife, Tameeka Stewart, is noted with heartfelt gratitude.

Gold Coast, Australia
31 December 2019

GOH Bee Chen
Rohan Price

Contents

Contributors

Mohammad Firdaus Abdul Aziz is a senior lecturer in the Faculty of Law at the University of Malaya (UM) in Kuala Lumpur, Malaysia. He is a coordinator for a number of courses namely Medical Law and Ethics, Biosafety Law, Biodiversity Law, Law and Society, and Ethical Issues of Emerging Sciences. He holds BSc (Hons) in Genetics and Molecular Biology from the University of Malaya, MA in Biotechnological Law and Ethics from the University of Sheffield and DPhil in Public Health from the University of Oxford. For his doctoral research, he examined the regulatory techniques adopted by selected Commonwealth countries to regulate human stem cell research and proposed an appropriate regulatory framework for Malaysia. He is also currently a researcher and production editor at Centre for Law and Ethics in Science and Technology (CELEST) at the University of Malaya; an academic affiliate at the Centre for Health, Law and Emerging Technologies (HeLEX) at Oxford; and an affiliate member of Young Scientists Network – Academy of Sciences Malaysia working on policy and governance. His research interests are in the legal and ethical issues of emerging technologies. He is also actively involved with UNESCO Bioethics as a trainer for the UNESCO Ethics Teachers Training Course (for Higher Education) in the region.

GOH Bee Chen A former Malaysian Rhodes Scholar, GOH Bee Chen is Professor of Law and Director of the Judge-in-Residence Program, School of Law and Justice, Southern Cross University, Australia. Bee Chen has recently been a Visiting Professor at the Chiang Mai University Faculty of Law in Thailand. She is a Director and Fellow of the Australian Academy of Law, Fellow of Cambridge Commonwealth Society and Fellow of the Society for Advanced Legal Studies in London. Besides Organics regulatory framework, Bee Chen's current collaborative research project deals with Law and Theatre. Her scholarly interests include Mediation and ADR, especially on Cross-Cultural (Sino-Western) Dispute Resolution and International Law of Peace. Her publications include *Negotiating with the Chinese* (Dartmouth/Routledge, 1996), *Law Without Lawyers, Justice Without Courts: On Traditional Chinese Mediation* (Ashgate/Routledge, 2002), Goh, Offord and Garbutt (eds) *Activating Human Rights and Peace: Theories, Practices and Contexts* (Ashgate/

Routledge, 2012); Farrar, Lo and Goh (eds) *Scholarship, Practice and Education in Comparative Law: A Festschrift in Honour of Mary Hiscock* (Springer, 2019).

Thararat Chitov has had teaching and research experience in the fields of food microbiology and food safety for more than 20 years. Her passion in food safety began when she studied at the University of Strathclyde, Glasgow, United Kingdom, and continued to grow during her years of academic service at the Department of Food Technology, Maejo University, and at the Department of Biology, Chiang Mai University, Thailand, where she currently works. Some of her other experiences include science education, consultancy services for food industry and analytical laboratory, working with dairy farms and mixed crop-livestock farms to assist transitioning from non-organic to organic farming, and, together with her family, running the family's own organic pomelo orchard.

Febri Doni is a research fellow at the Institute of Biological Sciences, Faculty of Science, University of Malaya. He obtained his PhD in microbiology from the School of Biosciences and Biotechnology, Faculty of Science and Technology, The National University of Malaysia. His research interest includes plant–microbe interactions, agroecology and plant pathology. He published dozen of papers focusing on microbe–rice plant interactions, System of Rice Intensification (SRI), organic agriculture and agroecology. He is the executive secretary of the Malaysian Agroecology Society (SRI-Mas) and is involved in various social activities in South East Asia in promoting agroecology and sustainable agriculture.

Tracey Epps is a Trade Law Consultant at Chapman Tripp in Wellington, New Zealand. Previously, she was a senior advisor in the New Zealand Ministry of Foreign Affairs (MFAT) Legal Division. At Chapman Tripp, she advises government and private sector clients on all aspects of international trade and investment law. Tracey has authored and edited a number of books and articles on topics related to international economic law, including climate change and international trade, subsidies, sanitary and phytosanitary measures, and technical barriers to trade. She teaches International Trade Regulation and International Investment Law in the Faculty of Law at the University of Otago. She is Chair of the International Trade Committee in the Inter-Pacific Bar Association, and regularly attends international conferences in the field of international economic law.

Satoshi Kodera is a professor of the Department of Law in Seinan Gakuin University, Fukuoka, Japan. His work focuses specifically on public international law, international economic law and international law of development. He is a co-author of *The Introduction to WTO and FTA Law* (Horitsu Bunkasha, 2016, in Japanese).

Hui Li PhD, Associate Professor, School of Law, Dalian University, China. Projects presided: Research on Allocation and Coordination Mechanism of Authorities and Functions Relating to Banking Supervision – based on Deposit

Insurance Ordinance of China, project supported by the National Social Science Foundation of China, 2015. (Grant No. 15BFX174); Research on Banking Deposit Insurance Legislation, Project Supported by Humanities and Social Science Foundation for Young Scholars, Ministry of Education of China, 2014. (Grant No. 14YJC820029).

Muhamad Shakirin Mispan is currently a senior lecturer at the Institute of Biological Sciences, University of Malaya, Malaysia, teaching Bachelor degree courses on Ecology and Management of Weeds, Soil Ecology and Management, and Quantitative Ecology and Applied Weed Science and Pest Management in Organic Crops for Master of Science in Crop Protection. He is also the coordinator of Glami Lemi Biotechnology Research Centre, University of Malaya, where part of the Centre's responsibilities is to promote sustainable agriculture and develop new agriculture technologies for local farmers. He is a PhD holder from South Dakota State University working on QTL mapping for weedy adaptive traits in rice, haplotype analysis for a major seed dormancy QTL region in rice, and seed longevity and photoperiod sensitivity in weedy rice. His current research focuses are mainly on ecological genetics of weedy rice focusing on adaptation to various ecological systems, seedbank dynamics and longevity, environmental impact from various weed management, integration of sustainable agriculture for rice production, and precision agriculture in rice farming.

Mehfuz Mollah is a lawyer with a background in economics. At Economic Laws Practice, he regularly advises clients on regulatory and policy issues. He has a degree in law from Delhi University, a Master's Degree in Economics from the Delhi School of Economics and an LLM from the London School of Economics. He is also a visiting Professor at NMIMS University, India.

Suhail A. Nathani is the Managing Partner of Economic Laws Practice and has over 25 years of experience spanning corporate and commercial laws, international trade and customs, and competition law and policy. He has been widely recognised as a leading lawyer and has been named as a top 30 International Trade practitioners in the world by the Best of the Best Expert Guides 2016 and India Business Law Journal's A List as India's Top 100 Lawyers in 2017 and 2018. He has appeared before the WTO Panel and Appellate Body on behalf of the Government of India. Suhail earned his Master's Degree at Cambridge University, England, and has also received an LLM from Duke University, USA. Apart from India, he is also admitted to the State Bar of New York. He is an Adjunct Professor at the Jindal Global Law School in India.

Van Kien Nguyen grew up in a small-scale rice-based farming family in Soc Trang province of Mekong Delta, Vietnam. Kien is a current honorary senior lecturer in agroecology at Fenner School of Environment & Society and the Australian National University as well as the founder of the company Mekong Organics Pty Ltd. Kien received a BSc in Land Management in 2000 (Cantho University), a master's degree in Environmental Management and Development in 2006, a PhD in Sociology in 2013 from the Australian National University (ANU), and a post-doctoral fellowship (2014–2016) from Luc Hoffmann Research Institute (LHI) at ANU Fenner School of Environment & Society. Kien has been leading the Research Centre for Rural Development in An Giang University, Vietnam, for 5 years (2013–2018). His research focuses on agroecology, sustainable agriculture and livelihoods, food security and water governance, and climate change adaptation of rural households in the Mekong Region. Since 2017, Kien has been involved in linking Australia and Southeast Asia in promoting organic farming.

Alex Phillips is the President and CEO the Van Horne Institute and a trade and transportation lawyer based in Calgary, Alberta, Canada. The Van Horne Institute is a national 'think tank' focused on growth issues related to trade, transportation, infrastructure and sustainable resource development. In his 35-year career, Alex has provided law, economics and regulatory policy advice to carriers, shippers, forwarders, airports, ports, electricity grids and related supply chain stakeholders. He holds a BComm(Hons) and LLB degree from the University of Manitoba; a Master of Laws degree in aviation deregulation from the London School of Economics, the United Kingdom; and a Doctorate (SJD degree) in infrastructure privatisation, trade access rights and dispute resolution from Bond University in Australia.

Rohan Price A lecturer in the Southern Cross University School of Law and Justice, Dr Rohan Price has established a reputation as a foremost historian of the role of nationalism in the British/Chinese colonial encounter. Rohan was awarded a Doctor of Philosophy from the University of New England (Australia) for his thesis on the use of property law to encourage civic loyalty to colonial Hong Kong between the world wars. His books including *Reading Colonies: Property and Control of the British Far East* and *Resistance in Colonial and Communist China* (1950–1963) are based on extensive archival and digital repository research. His books have been described in reviews as 'passionate, containing 'argumentative strength and forthright originality' and 'enormous attention to historic, theoretical and political detail'. Rohan has enjoyed lengthy stints as a visiting professor in three Chinese universities over the last decade, teaching in fields including common law history, maritime law and the law of trusts. His interest on Chinese food safety issues was prompted by a friend in Hong Kong who, in 2008, casually mentioned he should not buy the same brand of noodles every time he went to the supermarket. Rohan has finished the year 2019 with a new monograph published by City University Press Hong Kong, *Violence and Emancipation in Colonial Ideology: Hong Kong and British Malaya.* It has been described in an academic review as 'a major contribution to the literature'.

Ying Shen is an early career researcher with multijurisdictional educational qualifications, having earned her LLB, LLM and PhD from China, Germany and Australia, respectively. Her expertise lies in the field of Environmental Law, with a special focus on climate change issues. She was an Endeavour Research Fellow and the first Australia-APEC Women in Research Fellow representing China in 2015 and has since held different research positions in several universities and institutions in both Australia and Germany. She was awarded the German Chancellor Fellowship by the Alexander von Humboldt Foundation in 2019. Dr Shen is the author of *China's Way to Carbon Emissions Reduction: The Choice of Regulatory Instruments and Its Legal Challenges*, a monograph published by Wolters Kluwer in 2015.

Francis Snyder C.V. Starr Professor of Law, EU Jean Monnet Chair ad personam and Director, Centre for Research on Transnational Law, Peking University School of Transnational Law, Peking University Shenzhen Graduate School, China; Adjunct Professor, University of Macau; Emeritus Professor, CERIC/Faculty of Law and Political Science, Aix-Marseille University; and Visiting Professor, College of Europe, Bruges. Previously Fellow, Wissenschaftskolleg (Institute of Advanced Study), Berlin; Centennial Professor of Law, London School of Economics; Professor of EU Law, European University Institute (Florence); Professor of Public Law Aix-Marseille University; and Professor of European Law University College London. Publications include more than 40 books and 200 articles, including *Food Safety Law in China: Making Transnational Law* (E.J. Brill, Leiden, 2016) and *The EU, The WTO and China: Legal Pluralism and International Trade Regulation* (Hart, 2010). Invited by Chinese Central Government in 2013 to serve as lead foreign expert on reform of food safety system in China, leading to 2015 Chinese Food Safety Law. Educated at Yale University, Sciences Po Paris, Harvard Law School, and Université de Paris I (Panthéon-Sorbonne). In 2018, he was awarded the People's Republic of China Friendship Award and also the Peking University Friendship Award.

Anastasia Telesetsky is a tenured professor at the University of Idaho College of Law, United States of America, where she is a core faculty member for the Natural Resources and Environmental Law Program. She has published numerous article and chapters on matters involving marine living resources and food law and co-authored and edited books on ecological restoration and disaster relief and planning law. She is a former co-chair of the International Environmental Law Committee for the American Bar Association's Section of International Law and a current member of the IUCN World Commission on Environmental Law. She is a former Fulbright Scholar (Philippines) and Axford Fellow (New Zealand). She has taught food law at the University of Auckland in New Zealand and The Chinese University of Political Science and Law in Beijing, China.

Danae Wheeler is a solicitor at Chapman Tripp in New Zealand. Danae has experience advising clients on arbitration and litigation proceedings and international trade and investment law. Having lived and studied in 6 countries over the past 8 years, she has a strong interest in how organisations operate in different cultural and legal settings. Before joining Chapman Tripp, she completed honours degrees at the Chinese University of Hong Kong (LLM in Chinese Business Law), Melbourne Law School (JD) and New York University (BA). In 2018, she was awarded the AMINZ-AUT Arbitration Young Practitioner Scholarship. She wrote her Master's thesis on the enforceability of arbitral awards under the New York Convention in the People's Republic of China. Her undergraduate thesis explored the relationship between the rule of law and economic development, for which she was awarded the European and Mediterranean Studies Award at New York University. Danae speaks Spanish, Mandarin and Cantonese.

Natalie W. M. Wong currently is Visiting Fellow at Department of Public Policy, City University of Hong Kong. She completed her PhD in Politics at The University of York, United Kingdom, and her dissertation mainly focuses on incinerator and waste policy in China. Despite her working experiences in academia, Natalie was a researcher at a Hong Kong–based labour NGO, which concerns the labour conditions in Pearl River Delta of China, and an international environmental NGO to investigate the air quality of Beijing before commencement of the Olympic Games in 2008. She has published extensively on topics on environmental governance and environmental activism in *China Information*, *Voluntas*, and the *Journal of Environmental Planning and Management*.

Chapter 1
Introduction: Issues Affecting the Regulatory Environment for Organic Food Safety in Asia Pacific

GOH Bee Chen

Abstract As organic production and consumption continue to be on the rise globally, and specifically in the Asia Pacific region for the purpose of this book, the regulatory environment for organics has become all the more relevant and significant. Concerns for food safety, animal welfare, and climate emergency propel a drive for organics worldwide. Yet, as contributors of this edited volume have observed, there is no consistency across jurisdictions in the Asia Pacific for the regulation of organics. For instance, the regulatory framework that exists for organic certification, and for organic labelling, may be mandatory or voluntary. International trade, whether bilateral or multilateral, can be enhanced by way of regulatory harmonization or through mutual recognition regimes in organics. Historical accounts leading to an organic regulatory regime for organic agriculture and organic food products have also demonstrated the challenging issues involved in complying with standards and market distribution. It is hoped that this book will make a major contribution to the reader's understanding of the regulatory issues involved in organic food safety in the Asia Pacific.

Keywords Organic agriculture · Organic food safety · Regulatory framework · Asia Pacific

GOH Bee Chen (✉)
School of Law and Justice, Southern Cross University, Gold Coast, QLD, Australia

Faculty of Law, Chiang Mai University, Chiang Mai, Thailand
e-mail: beechen.goh@scu.edu.au

GOH Bee Chen, R. Price (eds.), *Regulatory Issues in Organic Food Safety in the Asia Pacific*, https://doi.org/10.1007/978-981-15-3580-2_1

1.1 Introduction

In a world of organics, images of good health and well-being immediately come to mind. The contributing factors that have brought forth the surging trend in organics include the world's concern for food safety, strive for a cleaner planet, and a compassionate treatment of animals. With the rampant occurrences of worldwide natural disasters like earthquakes, floods or bushfires, a climate crisis is undoubtedly at hand. The Stockholm Declaration of 1972 which was hailed as a clarion call for humanity to act in aid of environmental protection will soon be celebrating its 50th Anniversary. It then was, and still is, 'Only One Earth'.[1] Such a climate urgency has now been translated into a climate emergency. It can no longer be left to a dire situation of 'too little too late'. In this connection, the phenomenal rise of organic agriculture on a global scale may thus be seen as a positive contributor in our world of rapid climate change.

Food security issues, generally speaking, and organic food safety issues in particular, are a major concern for consumers worldwide. It is not just a matter of eating healthily, or about the consumption of nutritious foods devoid of chemicals or artificial ingredients. Accessibility and affordability are the prime considerations when approaching the matter of food safety. Organic food safety further adds another layer to this consideration. The consumption of organic food, no doubt, contributes to the consumer perception of eating nutritious food and healthy eating. However, although there is a general grasp of what 'organic' entails, there is no universally defined terminology for 'organic'. Added to this is a fringe category of 'wild' collection of food products.[2] It may be said that the underlying understanding about what may be claimed as organic is that there is a refrain from utilizing chemicals or synthetic materials in food production. As a consequence, consumer safeguard against inauthentic claims has to fall within the purview of consumer protection laws where national jurisdictions have extended such consumer protection.[3]

The future of organic food products appears to be bright indeed. Global growth has been exponential and phenomenal. The most recent available report in 2017 has shown that the region covering Asia, Australasia, Latin America and Africa held a

[1] Slogan used at the 1972 Stockholm Declaration at the conclusion of the United Nations Conference on the Human Environment held in Stockholm, 16 June 1972.

[2] See Helga Willer, Julia Lernoud and Laura Kemper, 'The World of Organic Agriculture 2019: Summary' in Helga Willer and Julia Lernoud (eds), *The World of Organic Agriculture: Statistics and Emerging Trends 2019*, Research Institute of Organic Agriculture (FiBL), Frick, and IFOAM Organics International, Bonn, 2019, at page 26; Julia Lernoud and Helga Willer, 'Current Statistics on Organic Agriculture Worldwide: Area, Operators and Market' in Helga Willer and Julia Lernoud (eds), ibid, at page 55; Julia Lernoud, Helga Willer and Bernhard Schlatter, 'Asia: Current Statistics' in Helga Willer and Julia Lernoud (eds), ibid, at page 200.

[3] For instance, in Australia, section 18 of the Australian Consumer Law deals with misleading or deceptive conduct in trade or commerce. The Australian Consumer Law is contained in Schedule 2 of the Competition and Consumer Act 2010.

combined total value of US$8.7 billion dollars of the organic food market.[4] Relevantly, in 2015, all Member States of the United Nations adopted the Transforming Our World: the 2030 Agenda for Sustainable Development containing, *inter alia*, the much celebrated 17 Sustainable Development Goals (SDGs). A future in organic foods plays a critical role in helping to achieve several of the SDGs such as the first three SDGs like SDG 1 (No poverty), SDG 2 (Zero hunger) and SDG 3 (Good health and well-being). To this end, it is of note that, in 2017, the IFOAM Strategy was adopted which seeks to promote sustainable food production systems in aid of the SDGs.[5] It has also been keenly observed that organic agriculture can beneficially contribute to the reduction of poverty and hunger.[6] Again, the latest statistics in 2017 indicate that worldwide, there were 69.8 million hectares of organic agricultural land, with Oceania accounting for the largest share at 35.9 million hectares.[7] Relevantly for the Asia Pacific region covered in this book, a combined total of 45.2 million hectares of organic agricultural land were in countries in Oceania, Asia and North America.[8] Overall, Australia would appear to be the leading country in terms of ownership of organic agricultural land, followed by Argentina and the United States of America.[9]

1.2 The Regulation of Organics

Laws and regulations play an integral role for the protection of the environment, animal welfare, and organic food safety. As an organic claim is regarded as a credence claim,[10] it is highly critical that there be a framework surrounding its regulation. In the domain of organics, it is even more important that the regulatory environment for standards compliance in respect of organic agriculture and organic food production and distribution be stringent in formulation and strict in implementation. After all, we are dealing with the vital matter of food safety.

Currently, the regulatory environment for organic food safety across jurisdictions has been found to be varied.[11] In this connection, mention must be made of the

[4] Amarjit Sahota, 'The Global Market for Organic Food and Drink' in Helga Willer and Julia Lernoud (eds), note 2, at page 147.

[5] The new IFOAM Strategy adopted by IFOAM – Organics International: Helga Willer, Julia Lernoud and Laura Kemper, 'The World of Organic Agriculture 2019: Summary' in Helga Willer and Julia Lernoud (eds), ibid, at page 33.

[6] See China's Poverty Reduction Online: http://p.china.org.cn/2017-08/24/content_50003182.htm

[7] See Julia Lernoud and Helga Willer, note 2, at page 40.

[8] Ibid.

[9] Ibid.

[10] M Canavari, N Cantore, E Pignatti and R Spadoni, 'Role of Certification Bodies in the Organic Production System' in Haas et al (eds), *Looking East, Looking West: Organic and Quality Food Marketing in Asia and Europe*, Wageningen Academic Publishers, 2010, at pages 85–86.

[11] Beate Huber, Otto Schmid, Verena Batlogg and Flavia Moura E Castro, 'Public Standards and Legislation' in Helga Willer and Julia Lernoud (eds), note 2, at pages 152–156.

immense efforts made by the International Federation of Organic Agriculture Movements – Organics International, a non-governmental organization established in the early 1970s, for conscientiously steering the world of organic agriculture in terms of, *inter alia*, certification standards for organic food production and distribution.[12] Alongside domestic or international legislative measures, public standards or applicable organics guidelines, there also exist Participatory Guarantee Systems as a complementary certification tool.[13] According to the definition provided by IFOAM – Organics International, 'Participatory Guarantee Systems are locally focused quality assurance systems. They certify producers based on active participation of stakeholders and are built on a foundation of trust, social networks and knowledge exchange.'[14] Given such a plethora of compliance platforms, it would be desirable, both for the protection of food safety and promotion of international trade in organics that there be a harmonization of applicable laws, rules and regulations for governing standards.

When consumers purchase food labelled 'organic', 'certified organic', or '100% organic', or other descriptions alike, there is the element of trust associated with such purchase. Such trust is based fundamentally on a regulatory system assuring consumers of the authenticity of claims at the risk of legal penalties imposed on the producers or distributors. As stated earlier, an organic claim on food products is a credence claim, owing to the fact that information concerning organics is asymmetrical in that organic food producers are the ones knowledgeable about the inherent quality of the organic foods being distributed for purchase. Since organic food products can generally command premium pricing, organic food trade can thus be perceived as quite a lucrative business enterprise. There is inevitably fraud in global organic trade. Countries such as the United States of America recently set up the Global Organic Supply Chain Integrity Task Force through the Organic Trade Association in an effort to combat fraud.[15] In the interests of safeguarding organic food safety, the regulatory regime ought to be clear and uniform in order to gain effective consumer trust.

As we have seen earlier, globally, there are no universal standards governing the regulation of organic foods. And, across jurisdictions, compliance with organic standards may also either be mandatory or voluntary. Consumers who are willing to pay a premium price for organic food products can be forgiven for mistakenly presuming that the market distribution of certified organic foods or foods labelled as organic have undergone a mandatory compliance regime by the so-called organic food producers and distributors. By way of an example, organic food labelling in Australia for domestic consumption is voluntary. It has keenly been remarked that, in Australia, it is more a case of 'market demand, rather than regulation, [which]

---

[12] For information on IFOAM – Organics International, see https://www.ifoam.bio/

[13] Ibid, at page 152.

[14] https://www.ifoam.bio/en/organic-policy-guarantee/participatory-guarantee-systems-pgs

[15] Barbara Fitch Haumann, 'United States: New Sales Records' in Helga Willer and Julia Lernoud (eds), note 2, at page 279.

drives certification'.[16] And so is the case with New Zealand where a voluntary regime prevails in the organic food domain. Interestingly, both these countries are considered to be mature organic economies. The increased flow of international trade in organic food products and the increased consumption of domestic organic food products definitely call for the timely harmonization of enacted rules and regulations affecting organics. Consumers worldwide need the clarity, certainty and consistency of a regulatory environment in organics.

1.3 Structure of the Book

It may be said that a book canvassing the issues of regulation in relation to organics is timely, given the current significance of worldwide organic food production and consumption, and cross-border trade in organics. Its relevance is particularly compelling due to the rise of emerging organic economies in the Asian region. Hence, the deliberate focus of this edited volume on select jurisdictions in the Asia Pacific.

The book is arranged thematically in three parts. Part I takes an overarching approach on international perspectives in relation to organic foods. In Chap. 2, Snyder comprehensively investigates the scenarios of food safety in China under the brand of Ecological Food, and proceeds to suggest that the Chinese model could usefully be adopted by other countries seeking to improve food safety and food quality. Goh, in Chap. 3, then proceeds to examine the issues of labelling, and consumer trust, in the contexts of emerging organic economies (using China and Thailand as the country examples) and mature organic economies (primarily centred on Australia as the country example). In Chap. 4, Shen examines the various impacts of Chinese organic food law on climate change, and in particular, discusses regulatory impacts on climate change mitigation and adaptation.

Part II traces relevant history and other factors influencing organics, with contributors focusing on Japan and China. In Chap. 5, Kodera takes a twenty-year view over the development of the system of organic Japanese Agricultural Standard, and alerts us to the associated challenges. In Chap. 6, Price casts a historical eye on a near-three decades of the science in relation to food production in China, with a cautionary note that the Organic Guarantee imported from the West may be the answer to China's future food production. In looking also at China, in Chap. 7, Wong uses the case study of gutter oil in food waste management as a rising concern in food safety.

Finally, Part III canvasses organics regulatory framework across select jurisdictions in the Asia Pacific. Continuing with China is Li's Chap. 8 on the regulatory issues on Chinese organic food, raising challenges and deficiencies of the current legal regime. Another major player in Asian organics is India, with its organic food

[16]Andrew Lawson, Amy Cosby, Derek Baker, Shawn Leu, Ed Lefley, Amarjit Sahota and Nick Bez, 'Australia' in Helga Willer and Julia Lernoud (eds), note 2, at page 298.

safety laws examined jointly by Nathani and Mollah in Chap. 9. Moving across to South-east Asia, in Chap. 10, we have Firdaus, Shakirin and Febri jointly investigating the development and challenges of organic food policy and regulation in Malaysia where the organic food economy is still at its nascent stage. In Thailand, in Chap. 11, Chitov employs a scientific approach in investigating Thai organic food safety. In Chap. 12, Nyugen takes us on an evidence-based journey of organics qualitative research in Vietnam examining opportunities for education and research in organics. Part III concludes with an overview of three Pacific nations, i.e. the United States of America, New Zealand and Canada. In Chap. 13, Telesetsky examines food safety law in the United States through the risk management of the organic food supply chain. In Chap. 14, Epps and Wheeler jointly walk us through the New Zealand organics regulatory environment and its voluntary regime, shedding light on the legislative reform proposal. Finally, in Chap. 15, Phillips outlines the scenario of Canadian organics and further explores organic food safety from a collaborative trade approach.

The foregoing Parts I to III are framed by an introductory chapter by Goh and a concluding chapter by Price and Goh.

1.4 Conclusion

Across the world, and the Asia Pacific region in particular, organic agriculture is increasingly making a major contribution to our sense of good health and well-being, environmental protection and animal welfare. However, the regulatory framework for organic foods is neither uniform nor consistent. In some cases, claims by food producers, that their products are organic, may appear also to be intuitive, rather than strictly scientific. In the case of an intuitive claim, it could be based on a belief that food production is natural, and therefore organic, because there is no or little application of artificial ingredients or chemicals or other toxic substances.[17] Agricultural small-holders or small-scale family-owned enterprises involved in organic agriculture are prone to making such organic claims, in the absence of an overarching stringent regulatory framework. In an allied fashion, consumer approach to what is labelled as organic, depending on the national legal framework, may be attributed to a matter of trust, and the fact that there is consumer protection legislation against misleading, deceptive or fraudulent trading conduct.

With the contemporary significance of organic food safety, it is hoped that the regulatory environment for organics can be harmonized across jurisdictions to achieve a more satisfying outcome for organic food producers, distributors and consumers alike in both domestic and international trade arenas.

[17] For instance, the current regulatory regime in New Zealand allows organic food growers to determine for themselves if they believe their production methods adopted are organic.

References

Australian Consumer Law (Schedule 2 of the Competition and Consumer Act 2020), section 18

Canavari M, Cantore N, Pignatti E, Spadoni R (2010) Role of certification bodies in the organic production system. In: Haas R et al (eds) Looking east, looking west: organic and quality food marketing in Asia and Europe. Wageningen Academic Publishers, Wageningen

China's Poverty Reduction Online. http://p.china.org.cn/2017-08/24/content_50003182.htm

Haumann BF (2019) United States: new sales records. In: Willer H, Lernoud J (eds) The world of organic agriculture: statistics and emerging trends 2019. Research Institute of Organic Agriculture (FiBL), Frick, and IFOAM Organics International, Bonn, p 276

Huber B, Schmid O, Verena Batlogg V, Castro FME (2019) Public standards and legislation. In: Willer H, Lernoud J (eds) The world of organic agriculture: statistics and emerging trends 2019. Research Institute of Organic Agriculture (FiBL), Frick, and IFOAM Organics International, Bonn, p 152

IFOAM – Organics International. https://www.ifoam.bio/

Lawson A, Cosby A, Baker D, Leu S, Lefley E, Sahota A, Bez N (2019) Australia. In: Willer H, Lernoud J (eds) The world of organic agriculture: statistics and emerging trends 2019. Research Institute of Organic Agriculture (FiBL), Frick, and IFOAM Organics International, Bonn, p 294

Lernoud J, Willer H (2019) Current statistics on organic agriculture worldwide: area, operators and market. In: Willer H, Lernoud J (eds) The world of organic agriculture: statistics and emerging trends 2019. Research Institute of Organic Agriculture (FiBL), Frick, and IFOAM Organics International, Bonn, p 36

Lernoud J, Willer H, Schlatter B (2019) Asia: current statistics. In: Willer H, Lernoud J (eds) The world of organic agriculture: statistics and emerging trends 2019. Research Institute of Organic Agriculture (FiBL), Frick, and IFOAM Organics International, Bonn, p 199

Participatory Guarantee Systems by IFOAM – Organics International. https://www.ifoam.bio/en/organic-policy-guarantee/participatory-guarantee-systems-pgs

Sahota A (2019) The global market for organic food and drink. In: Willer H, Lernoud J (eds) The world of organic agriculture: statistics and emerging trends 2019. Research Institute of Organic Agriculture (FiBL), Frick, and IFOAM Organics International, Bonn, p 146

Willer H, Lernoud J, Kemper L (2019) The world of organic agriculture 2019: summary. In: Willer H, Lernoud J (eds) The world of organic agriculture: statistics and emerging trends 2019. Research Institute of Organic Agriculture (FiBL), Frick, and IFOAM Organics International, Bonn, p 25

GOH Bee Chen A Former Malaysian Rhodes Scholar, Professor GOH Bee Chen is Professor of Law and Director of the Judge-in-Residence Programme, School of Law and Justice, Southern Cross University, Australia. She has recently been a Visiting Professor in the Faculty of Law, Chiang Mai University, Thailand. She is a Director and Fellow of the Australian Academy of Law and Fellow of Cambridge Commonwealth Society and of the Society for Advanced Legal Studies in London. Besides organic regulatory framework, her current collaborative research project deals with Law and Theatre. Her scholarly interests include mediation and ADR, especially on cross-cultural (Sino-Western) dispute resolution, and international law of peace. Her publications include *Negotiating with the Chinese* (Dartmouth/Routledge, 1996), *Law Without Lawyers, Justice Without Courts: On Traditional Chinese Mediation* (Ashgate/Routledge, 2002), *Activating Human Rights and Peace: Theories, Practices and Contexts* (Ashgate/Routledge, 2012, with Offord and Garbutt); and *Scholarship, Practice and Education in Comparative Law: A Festschrift in Honour of Mary Hiscock* (Springer, 2019, with Farrar and Lo).

Part I
International Perspectives on Organics Regulatory Framework

Chapter 2
Understanding the Regulation of Ecological Food in China: Regulatory Intermediation, Path Dependence and Legal Pluralism

Francis Snyder

Abstract Ensuring food safety and quality for ordinary people in China is a continuing challenge. Among the major responses to this challenge has been ecological food, which can be defined as the product of 'ecological agriculture (*shengtai nongye*). Ecological food in China takes three principal forms: hazard-free food (*wu gonghai*, also known as 'pollution-free' or 'no public harm' food), green food (*lǘse shipin*) and organic food (*youji shipin*). This chapter identifies the major factors which have shaped the regulation of ecological food in China. It first examines how these forms of ecological food have been regulated so far. It then offers a theoretical explanation for the co-existence of these forms by referring to the theories of regulatory intermediaries, path dependence and legal pluralism. The discussion shows that the distinctive Chinese pattern of regulating ecological food tends to perpetuate lack of consumer trust, domestic regulatory competition, tensions between different economic interests, and conflicts among national food policy objectives. It also suggests, however, that the Chinese model may in the short run be a useful template for many other countries seeking to improve food safety and food quality, while in the long run it appears to be consistent with current global developments in the regulation of ecological food.

Keywords China · Food quality · Ecological food · Organic food · Green food · Regulatory intermediary theory (RIT) · Path dependence · Legal pluralism

---

F. Snyder (✉)
Peking University School of Transnational Law, Peking University Shenzhen Graduate School, Shenzhen, China

College of Europe, Bruges, Belgium

GOH Bee Chen, R. Price (eds.), *Regulatory Issues in Organic Food Safety in the Asia Pacific*, https://doi.org/10.1007/978-981-15-3580-2_2

Glossary of Acronyms

AFN	Alternative Food Network
AQSIQ	Administration of Quality Supervision, Inspection and Quarantine
CAA	Certification and Accreditation Administration
CAC	Codex Alimentarius Commission
CCP	Chinese Communist Party
CNAB	China National Accreditation Board for Certifiers
CNAL	National Accreditation Board for Laboratories
CNCA	China National Certification Administration
COFCC	China Organic Food Certification Centre
CSA	Community Supported Agriculture
GFDC	Green Food Development Centre
GTZ	Deutsche Gesellschaft für Technische Zusammenarbeit
IFOAM	International Federation of Organic Agriculture Movements
INR	International normative repertoire
ISO	International Organization for Standardization
MEP	Ministry for Environmental Protection
MOA	Ministry of Agriculture
NDRC	National Development and Reform Commission
NIES	Nanjing Institute of Environmental Science
OBOR	One Belt, One Road
OCIA	Organic Crop Improvement Association
OFDC	Organic Food Development Centre
RIT	Regulatory intermediary theory
SEPA	State Administration for Environmental Protection
SFS	State Farm System
SRI	Silk Road Initiative
TNR	Transnational normative repertoire
WTO	World Trade Organization

2.1 Introduction

Ensuring safe food safety for ordinary people in China is a continuing challenge.[1] Among the major responses to this challenge has been ecological food, which can be defined as the product of 'ecological agriculture (*shengtai nongye*). Ecological food in China takes three principal forms: hazard-free food (*wu gonghai*, also

[1] I am grateful to GOH Bee Chen, Yi Seul Kim, Ni Lili, Phil McConnaughay, Anne-Lise Strahtmann, Tian He and Zhang Xuan for their contributions to this chapter. I thank also Peking University School of Transnational Law and Peking University Shenzhen Graduate School for their support. Any shortcomings in the chapter are my own.

known as 'pollution-free' or 'no public harm' food), green food (*lǘse shipin*) and organic food (*youji shipin*). This chapter identifies the major factors which have shaped the regulation of these forms of ecological food in China. It first shows how these forms of ecological food have been governed so far and then it provides a theoretical explanation for the co-existence of these multiple regulatory schemes. The chapter focuses on ecological food from China and sold on the domestic Chinese market. It does not deal with food from China that is exported to foreign markets, or food from outside China that is imported into the domestic Chinese market, or food from outside China that is sold on markets outside China. This focus results from the principal objective of the chapter, which is to contribute to the improvement of food quality for ordinary people in China.

2.2 Materials, Methods and Argument

This chapter is based on theory and practice. It involved a thorough search on the law and academic literature in English and Chinese on food safety, food quality and eco-food in China. It gives special weight to recent empirical research. It also takes account of the author's research and practical experience in China, Europe and Africa concerning law, administrative regulations and standards, both international and domestic, on food safety and food quality. This data makes it possible to construct, in the first part of the chapter, a succinct account of the development of the regulation of eco-food in China. The chapter seeks also to offer a theoretical perspective on this development. Its second part draws on theories concerning standards, inter-institutional competition, and legal pluralism to give a convincing theoretical explanation for the Chinese pattern of regulatory pluralism.

The chapter suggests that the distinctive Chinese pattern of regulating eco-food, based on conflicting standards promoted by different, competing domestic and international interests, tends to perpetuate domestic regulatory competition, lack of consumer trust, tensions between different economic interests, and conflicts among national food policy objectives. Against these shortcomings, however, it is important to balance the very significant facts that, in the context of globalization and China's Silk Road Initiative (SRI, or One Belt, One Road, OBOR), the Chinese model may be in the short-to-medium terms a useful template for many other countries which seek to improve domestic food safety and food quality, and that in the longer term the Chinese pattern of regulatory pluralism appears to be consistent with current global developments in the regulation of ecological food.

2.3 Evolution of Eco-Food Regulation in China

Two key benchmarks, despite certain precursors (Schmalzer 2016; Eisenman 2018) marked the historical development of concepts, standards and certification for organic food, green food and hazard-free food: the beginning of China's reform and opening in December 1978; and Chinese accession to the World Trade Organization (WTO) on 11 December 2001. Shortly after the December 1978 beginning of Reform and Opening, the Chinese State Council advanced the idea of building up eco-agriculture in the country' (CCIC 1996, see also Sanders 2000). However, no concrete action on eco-food was taken at the time, despite the existence of serious environmental pollution. For the next decade, the main priority of national agricultural policy was to guarantee food security by increasing food production, with greater use of agro-chemicals.

In 1990, the Nanjing Institute of Environmental Science (NIES), together with the Dutch organic certification body Skal (now Skal Biocontrol (2019)) issued China's first organic certification (Sternfeld 2009: 8). 'Organic' meant the products of organic agriculture, which is

> a production system that sustains the health of soils, ecosystems and people. It relies on ecological processes, biodiversity and cycles adapted to local conditions, rather than the use of inputs with adverse effects[, and] combines tradition, innovation and science to benefit the shared environment and promote fair relationships and a good quality of life for all involved. (IFOAM 2019a, b)

NIES was a small state-sponsored *shiye danwei*, 'government-funded, not-for-profit' institution (Yang 2005: 26), 'service unit' (Foster 2015: 93) or 'service organisation….' (Lam and Perry 2001), founded in 1978 as a research institute by the newly established State Environmental Protection Agency (SEPA). SEPA then did not yet have ministerial status and, in a period of severe environmental disasters, it sought to develop environmental policy outside the control of the Ministry of Agriculture (MOA). In 1994, 'SEPA set up the Organic Food Development Centre (OFDC), attached to Nanjing Institute of Agricultural Science' (Sternfeld 2009: 5). NIES established close links with the international (but American-based) NGO, the Organic Crop Improvement Association (OCIA) (OCIA International 2019; Wikipedia 2019). Between 1997 and 2003 the project was supported by the German international cooperation agency GTZ (Deutsche Gesellschaft für Technische Zusammenarbeit). In 2002 OFDC with German support 'qualified to become the first Chinese organic certifier accredited by the International Federation of Organic Agricultural Movements (IFOAM) and the International Organization for Standardization (ISO-65)' (Sternfeld 2009: 8). As early as 1995, SEPA promulgated 'Technical Norms on Organic Food and an Approach to Management of Organic Certification (Sternfeld 2009: 8). In 2001 SEPA's OFDC published 'some guidelines and regulations for organic certification and production which were closely following' IFOAM guidelines (Sternfeld 2009: 8); in 2002 IFOAM accredited OFDC. Academic research reported in 2005 that an 'OFDC staff serves on the important IFOAM standards board' (Thiers 2005: 7). Consequently, 'OFDC was simultaneously a subunit of a Chinese state ministry and a chapter of an international NGO' (Thiers 2009: 154).

Partly for domestic reasons (Pauli 2007: 11), and partly in response to state-supported NIES success in creating international links and obtaining international financial support, for example from the Rockefeller Brothers Fund (RBF) (Thiers 2009: 151–154), the Ministry of Agriculture (MOA), in the early 1990s, introduced a Green Food label and then in 1992 established a Green Food Development Centre (GFDC) (Sternfeld 2009: 8; Chen et al. 2018: 41) as part of the existing State Farm System (SFS) (Thiers 2006: 198–199). Described as broadly comparable to Integrated Crop Management in western countries (Sternfeld 2009: 6; CABI 2019), Green Food was never intended to be the same as organic food (Zhu 2008). Green Food certification was more likely to be within the capacity of domestic producers. The conditions for qualifying for the Green Food logo were:

> (1) The environment in which the product or product raw material is produced meets the environmental quality standards of the green food producing area; (2) The use of pesticides, fertilizers, feedstuffs, veterinary drugs and other inputs meets the guidelines for the use of green food inputs; (3) The product quality meets the quality standards of green food products; and (4) Packaging, storage and transportation are in line with the storage and transportation standards for green food packaging. (MOA Order No. 6, Article 9)

As with organic food, products must comply with China's Food Safety Law and with the Law on Agricultural Product Quality and Safety. However, Green Food, unlike organic food, allows limited use of chemical fertilisers and pesticides and thus is based on product standards rather than process standards (Denmark 2015: 32).

The Green Food logo is a registered trademark of the GFDC (MOA 2012). According to the 2012 Administrative Measures on Green Food Marks, it designates 'safe, high-quality edible agricultural products and related products that are produced from an excellent ecological environment, produced in accordance with green food standards, and implemented full quality control and access to green food signs' (Article 2). Green Food was regulated first through the Green Food A Standard and then in 1995 by the Green Food AA Standard 'for premium products mainly designed for export and rarely found in the local markets' (Sternfeld 2009: 5). Originally intended mainly for exports (Thiers 2005, 2006; Yasuda 2018: 60–77), the development of Green Food represented a typical strategy for developing countries, such as China or Brazil (Snyder and Kim 2018: 24) which give priority to exports because they wish to break into more expensive, higher-quality foreign markets, as well as a feasible policy in a highly polluted agricultural environment, populated by many small farms. Green Food joined IFOAM in 1993 (Thiers 2006: 205). However, Green Food was not accepted internationally as organic food, but, as Paul Thiers (2006: 199) points out, 'legitimacy, access and subsidies provided by association with the central government made it unnecessary for Green Food officials to consider market access and the possibility that external standards might contest Green Food authority'. The AA Standard was abandoned in 2002 when the GFDC 'set up its own organic certification body, the China Organic Food Certification Centre (COFCC)' (Sternfeld 2009: 6; Thiers 2006: 200–201; COFCC 2019) GFDC became China's leading organic certifier with about 30% of all organic farms and enterprises. Local governments subsidised certification if done by certifiers attached to government institutions; this included both OFDC and COFCC (Sternfeld 2009: 6).

In 2001 the MOA launched the Hazard-free Food Action Plan, which it had bruited in the 1980s (AQSIQ 2001; MARA 2012; Marchesini et al. 2010: 156). Hazard-free food consists of 'agricultural produce that is grown using chemical pesticides and fertilisers that are applied in quantities that meet or fall below maximum accepted limits' (Cody 2019: 89). In some settings the term 'hazard free food' might seem to be an oxymoron, but in China the level of agricultural pollution, overuse of toxic pesticides, lack of knowledge of and respect for standards, and food fraud resulted in the creation of a food quality category that, while suggesting perhaps that ordinary food was not hazard-free, urged producers to meet national standards. The Hazard-free Action Plan focused on control of highly toxic pesticides and on use of clenbuterol hydrochloride in meat production. It was intended to encompass stricter supervision of agricultural production and marketing, consumer education, and greater standardisation of product quality and safety (MOA 2012). It was based on a catalogue promulgated by the MOA and the Certification and Accreditation Administration (CAA) which included products for which producers could apply for certification (PRC 2003a) Initially voluntary, it reportedly became mandatory in 2006 (Berti 2015: 50; but see Scott et al. 2013; 7, 20). Since 2007 the MOA has abolished more than 100 industrial standards for hazard-free agricultural products, which meant that the MOA no longer reviews and supervises this category of products, which now are governed only by national standards (Sun 2018: 110). Recent research indicates increased regulation and considerable improvement in control of pesticide residues but nevertheless a continuing high level of use of toxic pesticides (Snyder and Ni 2017a, b).

Despite a continuing fragmentation of administrative authority (*zhengchu duomen*) for supervising food safety, which was consolidated by the State Council's 2004 Decision about Strengthening Food Safety (PRC 2004), these developments concerning food quality occurred within a gradual centralisation of national rules about standardisation and certification, which directly affected eco-food production and marketing. They were influenced also by the WTO Agreements and the WTO Trade Policy Review Mechanism (Snyder 2015). The 1988 Standardization Law provided for the formulation of standards and supervision of implementation of standards and sought to 'encourage active adoption of international standards' (PRC 1988: Article 4). Implementing Measures adopted in 1990 opened the Chinese standardization world further by providing that 'The State encourages the adoption of international standards and advanced standards abroad and takes an active part in the formulation of international standards' (PRC 1990).

State Council 2001 administrative reforms resulted in the ministerial-level Administration of Quality Supervision, Inspection and Quarantine (AQSIQ) (for organisational history, see Mertha 2005: 173–194), [now part of the State Administration for Market Regulation], attached directly to the State Council. In 2003 the State Council adopted Regulations on Certification and Accreditation (PRC 2003b), which established a unitary system for certification and accreditation, led by the newly created China National Certification Administration (CNCA) as a vice-ministerial-level (http://english.aqsiq.gov.cn/AboutAQSIQ/) department attached to AQSIQ. The Regulations stated that only certification bodies registered at the CNCA and duly accredited by the relevant State Council department (then the China National Accreditation Board for Certifiers, CNAB) could 'engage in certification activities' (PRC 2003b: Article 9). They also required certification bodies to be independent of government departments (PRC 2003b: Article 9). The former requirement fitted MOA's GFDC/COFCC but not SEPA's OFDC, while the latter corresponded to the interests of the OFDC but was directly contrary to the interests of the GFDC/COFCC; both agencies remained within the jurisdiction of the MOA, though some researchers state that formerly governmental certification agencies are now third-party certification agencies (Chen et al. 2018: 45). COFCC was the first to be registered (Sternfeld 2009: 6). It became the main Chinese partner for BioFach, 'the world's leading organic trade fair' (CNAS 2019). Then in 2006 the China National Accreditation Service for Standardization (CNAS) was founded to merge the CNAB and the National Accreditation Board for Laboratories (CNAL); the CNAB itself resulted from an earlier merger of four administrative bodies (CNAS 2019). In principle, this spelled the end of a period of administration fragmentation concerning certification and accreditation. In practice, it marked an attempt to restore the role of central government in the regulation of food quality.

In 2015 AQSIQ revised the Administrative Measures on Certification of Organic Products (AQSIQ 2013; PRC 2015), which are still in force. The Measures are mandatory for all '[p]ersons engaging in *certification* of organic products and manufacturing, processing, importation and sale of *certified* organic products in the People's Republic of China' (PRC 2015: Article 2, emphasis added). 'Organic products referred to in these Measures shall mean products manufactured, processed, and sold for human and animal consumption which *comply with the Chinese national standards* for organic products' (PRC 2015: Article 3, emphasis added). Reflecting the interests of central government, the Measures bind very closely together three terms: 'organic products', 'certification' and 'manufacturing, processing, importation and selling'. As a result, if we take the words seriously, the Measures do not apply to organic products which are not certified as organic or to products which do not comply with Chinese national standards for organic products.

2.4 Explaining the Chinese Regulatory Model

2.4.1 Introduction

We can understand the principal features of the Chinese regulatory model for eco-food by focusing on standard-setting, institutional competition and normative diversity. For this purpose the chapter draws on three bodies of theory: regulatory intermediation, path dependence and legal pluralism.

2.4.2 Standards

Standardisation is the subject of a vast literature, which deals with rules, rule-making and rule-makers (eg Braithwaite and Drahos 2000; Grindley 2002; Delimatsis 2015a, b). The theories of international normative repertoires (INR) or transnational normative repertoires (TNR) focus on the creation of international standards and their links to domestic law and politics. Research on anti-dumping law (Snyder 2001), competition law (Wu 2013) and food safety regulation (Snyder and Kim 2018) reveals the existence in each field of 'a handful of basic principles [involving] the constitution of a conceptual and normative repertoire, drawing on [international or transnational law], national legislation, multilateral negotiations, and administrative [and judicial] decisions in several jurisdictions within a specific historical context' (Snyder 2001: 419–420). This complex framework may be called an 'normative repertoire', using an analogy to musicians or theatre companies, which typically have a repertoire from which they select elements for specific performances. In some contexts, an INR or TNR may be implemented more or less directly in a national legal system, to the extent that we can speak of a 'transnational legal order', in which international, regional and national sets of substantive norms are essentially identical (Halliday and Shaffer 2015); a good example is regulation of pesticides in China (Snyder et al. 2019) In other circumstances, the connections between the different sites of governance which contribute to or are linked to the INR/TNR may be more tenuous or more indirect (Snyder and Kim 2018)

Theories of regulatory intermediation help us understand how international standards are made. Instead of a command-and-control model of regulation, they posit relationships between regulator, intermediary and regulatory target (R-I-T) (Abbott et al. 2017a, b). In the field of food quality, the regulator is usually assumed to be an international organization. For example, the Codex Alimentarius Commission (CAC, or Codex) is the principal source of international food safety standards, according to the WTO Agreement on Sanitary and Phytosanitary Measures (SPS Agreement) (Scott 2009; Arcuri 2015). Codex now has highly developed standards for organic food (WHO and FAO 2007). Codex standards are proposed by committees composed of national government delegations, which tend to be dominated by 'international food manufacturers and agrochemical corporations' (Braithwaite and Drahos 2000: 408; see also Smythe 2009) which together with trade associations in fact played a key role as both regulators and intermediaries even before the establishment of Codex in 1963 (Snyder 2016: 217–218). In principle, Codex standards are not legally binding, but in practice the WTO Appellate Body decision in the *EC-Hormones* case gave them legal effect (Snyder 2006).

The striking fact about the eco-food sector, however, is that international standards stem essentially from an international non-governmental network, the International Federation of Organic Agriculture Movements (IFOAM), which was founded in 1972 and which amalgamated standards from numerous national associations (Lockeretz 2007). It acted (and acts) as both regulator and intermediary in developing an international normative repertoire for organic agriculture and organic food. It published its first organic standards in 1980, and it participated as an observer in the development of Codex standards starting in 1992 and culminating in 1991 and 2001 (Geier 2007; Pacific Organic Policy Toolkit 2016: 3–5).

In formulating and developing a TNR for organic products, IFOAM and Codex thus represent a 'governance partnership': 'a formal or informal relational arrangement between a founder partner (any actor that institutes a partnership) and one or more enlisted partners (frequently transnational actors that, having certain common goals and objectives, while remaining organisationally independent, decide to share in the exercise of regulatory authority over a common set of target actors or issues' (Mattli and Seddon 2015: 17). IFOAM is the rule maker, endowed with 'discursive power' (Clapp and Fuchs 2009: 10), because of its expertise, standing in the field and established network, while Codex plays a supporting role, which nevertheless is essential because of its position as a legitimate, internationally accepted standard-setter under WTO public international law. It is no exaggeration to assert, borrowing Bűthe and Mattli (2011)'s expression, that in the organic food sector, IFOAM is 'the new global ruler'. Such partnerships 'exist in virtually every area of global governance' (Mattli and Seddon 2015: 171), but usually they are invisible. They contribute to the development of private TNRs and to the privatisation of global regulation, as other research has shown for global retailing (Marques 2019) and building regulation (van der Heijden 2017). Whether this generalisation applies to IFOAM as an organisation, as distinct from international market harmonisation through its standards, lies outside the scope of this chapter.

IFOAM now has affiliates in more than 120 countries, including China. Its original focus was on agricultural production, and its standards still concern only farmers and processors, but today IFOAM Principles also address sellers and consumers (Freyer et al. 2015: 92). In principle, IFOAM standards, just as Codex standards, are not legally binding. Almost invariably, however, they are adopted by WTO Members, including China, either alone or as Codex standards to regulate production, processing and marketing of organic food. Chinese Food Safety Law (both 2009 and 2015) require food safety standards and food safety-related standards to be legally binding (Snyder and Kim 2018; Lepeintre and Sun 2018). This may not apply however to organic standards as such, but these are expressed in legal form by certification and accreditation measures.

Foreign cooperation and foreign investment were central in importing IFOAM standards into China. They helped create links between specific Chinese institutions and foreign export markets (Scorzon et al. 2014: 182–183) Recourse to transnational standards enabled SEPA to carve out a legal and policy space separate from the MOA, even though environment was always lower than agriculture in the bureaucratic hierarchy (Thiers 2006: 215, note 19; Li 2018). After the MOA established Green Food, foreign investment helped steer the MOA toward a negotiated compromise with SEPA, following a typical Chinese pattern of bargaining among governmental units (Lieberthal and Oksenberg 1989; Lieberthal and Lampton 1992) within the framework of IFOAM standards and export market opportunities. SEPA's [now MEP's] OFDC remains 'the only Chinese organic certification body which is accredited by IFOAM' (OFDC 2019), while the MOA's COFCC only became an

IFOAM affiliate in 2002 (IFOAM 2019a, b). All these developments occurred alongside the evolution of government-sponsored Green Food.

2.4.3 Path Dependence

In addition to standards, a second key factor shaping the Chinese regulatory regime for eco-food concerns path dependence: institution stability and change. Path dependence refers to the persistence and social reproduction of institutions, and more precisely to 'those historical sequences in which contingent events set into institutional patterns or event chains that have deterministic properties' (Mahoney 2000: 507). As long ago as 1985, Paul David used the term 'path dependence' to refer to the lasting use of the QWERTY keyboard, which though inefficient benefitted from the advantage of being first on the market (David 2001). Brian Arthur developed the concept further in research on the economics of high technology and then on increasing returns in the economy (Arthur et al. 1987, Arthur 1989, 1994). Subsequently, the political scientist Paul Pierson (2000) and the sociologists Mark Granovetter (1985, 1992) and Neil Fligstein (2013) have been among the leaders in importing this originally economic concept into other social sciences. Building on this work, a recent study by Daniel Klutzz argues that that path dependence is not 'fully deterministic' and instead is compatible with social change; in fact, the two may be mutually constitutive (Klutzz 2019: 268). The concept of path dependence helps us to understand the current regulation of ecological food in China.

Relations with international and foreign institutions fundamentally altered inter-institutional relations in China. Their standards served as new resources for Chinese institutions and indeed became embedded in domestic administrative turf battles (see also Thiers 2006; 193). SEPA benefited from foreign and international support and recognition in staking out new eco-food territory before the MOA. According to Paul Thiers, this was 'an attempt by one state agency to create a policy outpost in the established turf of another' (Thiers 2009: 150; see also Thiers 2006: 203–207). With support from the American-based Organic Crop Improvement Association, the German development agency GTZ, the international network IFOAM and the international standardization body ISO, it gained a crucial 'first-mover' advantage. These ties gave SEPA both domestic and international legitimacy in the social field of eco-food. In addition, its place in the web of international organic food organisations enabled it to avoid being displaced by the hierarchically superior MOA, despite the latter's greater resources, even after the MOA in 1992 established the GFDC.

Sequencing is a crucial element in constructing path dependence. Daniel Klutzz identifies the importance of a 'critical juncture', 'a period during institutional emergence or formation that triggers a process of self-reinforcement' (Klutzz 2019: 46; Collier and Collier 1991). In the development of the Chinese eco-food regime, this critical juncture occurred in 2002–2003. Following SEPA's creation of OFDC, the MOA *qua* State Council department established the COFCC in 2002, followed closely by the State Council Certification and Accreditation Regulations in 2003. The Regulations conferred on CNCA, a vice-ministerial department attached to AQSIQ, a monopoly on accreditation of organic products. MOA's COFCC was the first certifier which was accredited by CNCA, and the MOA definitively entered the value chain for organic food. The 2003 State Council Regulations applied to all products, not only organic food, and they did not give MOA a monopoly on certification of products as organic. Nevertheless they reinforced the position of MOA/COFCC at the centre of the eco-food sector by tying it closely to the social field of accreditation, an indispensable step in organic food marketing, as distinct from agricultural production. It is likely that mediation of this institutional configuration was decided at the highest level of China's party state, particularly by the National Development and Reform Commission (NDRC) (Heilmann and Shih 2017: 77–78; Conrad 2017: 356–361)

By 2003 the Chinese eco-food regulatory regime was basically in place, encompassing both Green Food and organic food. Concerning organic food, regulatory intermediaries provided new resources in the form of transnational standards, shaping domestic bureaucratic politics which over time resulted in a new configuration of governmental institutions; national administrative measures decisively consolidated these various elements. These 'core components' of the Chinese eco-food regulatory regime were not displaced by subsequent legal reforms, which amounted to what 'institutional layering' (Klutzz 2019: 246). The 2015 AQSIQ Administrative Measures on Certification of Organic Products sought to unify organic product certification, an authentification directory, standards and certification mark (Article 5). In practice, they simply reinforced the MOA's position concerning certification of organic products. In addition to the two government certification agencies, COFCC and OFDC, there are now many other other certification organisations, all of which must now be accredited by the CNCA (Mamas 2019).

Path dependence is taken usually to refer only to institutional arrangements, but it almost inevitably also includes normative configurations. International and domestic standards were used as resources and anchored in the pattern of 'fragmented authoritarianism' (Lieberthal and Oksenberg 1989), which was the dominant pattern of administrative organisation at least until the creation of the State Administration for Market Regulation (SAMR) in 2018. Rules, including international standards, and domestic procedures, such as certification, gained in importance as the Chinese Party-State increasingly emphasised 'rule by law'. If we consider only SEPA/EPA and MOA, the high level of path dependence with regard to organic standards was strengthened by regulatory capture, in which standardisation was dominated by a small group, which 'advocat[d] for a particular, well-established method of behaving, coordinating and decision-making' (Delimatsis 2015a, b: 8) concerning the production, processing and marketing of eco-food. Path dependence and institutional change were mutually constitutive (Klutzz 2019: 268), but both were deeply conditioned by institutions which cut across all social fields, notably the Chinese Communist Party (CCP) (Zhu 2007; Lam 2018). Again, they occurred alongside the development of Green Food.

This chronology shows, first, that a tension between Chinese and foreign conceptions, norms and institutions was introduced at the beginning of the eco-food sector in China. Second, it illustrates the institutional fragmentation and competition which long have been features of Chinese food safety and food quality policies. Third, it reveals the uneasy fit between agricultural productivism, environmental protection and food safety which remains a central part of Chinese rural and food policy. These features converge in China because of the size and complexity of the country, even though recent research (Lockeretz 2007; Freyer and Bingen 2015) suggests that none of them, taken individually, are limited to China.

2.4.4 Legal Pluralism

Standards and path dependence have shaped Chinese regulation of eco-food as a form of legal pluralism. In 1936 Eugen Ehrlich defined legal pluralism as 'the presence in a social field of more than one legal order' (Ehrlich 1936). The basic ideas in this definition have undergone three transformations during their long intellectual life. First, from (state) 'law' to 'governance'. 'Law' has been broadened to embrace private norms as well as state norms; reshaped to include non-legally-binding norms, or soft law (Snyder 1993, 1994; Luo and Song 2013); and thus redefined so as frequently to blur previously sharp distinctions between law, governance and regulation (Scott 2004; Bavinck 2018). Second, from a nationally bounded social field to transnational, semi-autonomous social fields. The concept of social field has been defined more precisely as 'the totality of coexisting facts that are conceived of as mutually interdependent' (Lewin 1951: 240). Then, it has been differentiated in more specific terms such as 'legal field' (Bourdieu 1987), disaggregated as 'semi-autonomous social fields' (Falk Moore 1973) and, with increasing globalization, applied in studies of transnational legal fields (Trubek et al. 1994) the development of markets (Fligstein 2013) and the creation of food safety law in China (Snyder 2016). Third, the concept of legal pluralism, articulated initially by sociologists and anthropologists, has been used in numerous national, regional and international settings (Griffiths 1986; Merry 1988; von Benda-Beckmann 1988; von Benda-Beckmann and von Benda-Beckmann 2008) by scholars from many disciplines, and further elaborated as global legal pluralism (Teubner 1997; Snyder 1999; Berman 2012), legal hybridity (Sousa Santos 1995; Dolan and Heirbaut 2015) and varieties of and perspectives on transnational law (Zumbansen 2019). The Chinese regulatory regime for eco-food exemplifies a complex, evolving form of legal pluralism.

Using standards, certification and accreditation as criteria, we can distinguish the following normative configurations of eco-food, leaving aside hazard-free food which requires simply following minimum national standards: (1) products produced according to international and/or national organic agriculture standards and certified as organic products by OFDC (SEPA), which is accredited by CNCA; (2) products produced to according to international and/or national according to organic agriculture standards and certified as organic by COFCC (MOA), which is accredited by CNCA; (3) products produced according to international and/or national organic agriculture standards and certified as organic by other organizations accredited by CNCA (Sternfeld 2009: 6–7; Denmark 2015: 34; Snyder and Ni 2019); (4) organic products produced for the political and economic elite according to special supply arrangements (*tegong*) (Demick 2011; Yan 2012: 723; Zhou 2017: 27, 58–61, 128-129, 175–176, 215–219; see also Ho 2019) (5) organic products produced for export according to the 'one pattern, 10 systems' arrangement under the supervision of AQSIQ (now part of SAMR) to meet the standards of export markets or, perhaps more frequently, standards provided by contract (Zhou 2017: 28, 57–58, 152–160 171–173; Snyder and Ni 2017b); (6) products which are called organic, or which consumers consider to be organic, but which are not officially certified and not labeled as organic; this includes many forms of Community Supported Agriculture (CSA) (Yasuda 78–92, 150–163), such as Little Donkey Farm (Beijing), which rely not on formal certification but on 'informal, participatory "certification"' (Shi et al. 2011: 553): 'a good reputation gradually established through word-of-mouth among CSA members, other local farmers and regular consumers to ensure consumer trust'; and (7) farmers' markets, such as the Beijing Organic Farmers' Market, which also bring together farmers, retailers (often the same) and consumers and which rely on 'networking certification' *(qinggan renzheng)* (Si 2018) or 'ethical inspections' (*lianxin renzheng*) (Si et al. 2018: 89); (8) other forms of Alternative Food Networks (AFNs) (Si and Scott 2016) such as the so-called 'exemplary' agriculture of 'chemical-free and small-scale farming' in rural areas near Shanghai, which are farmed by urban-to-rural migrants and which sell their products to farmers' markets (Cody 2019), by box sales or online (e.g. Organic Farm 2019), without certification or labeling; and (9) Green Food, certified and labeled as such.

If the key issue is ensuring consumer trust in food quality (Ilcan and Philips 2006; Karpik 2010: 174–175; Augustin-Jean 2014; Wang et al. 2015), the plurality of normative configurations indicates that governmental standards, certification and labelling are not the only ways to do so. Some configurations focus on farmers and organic agriculture; others emphasise consumers and markets for organic food. Some rely on government and law, some partly on closed value chains, and others on face-to-face transactions. Lack of awareness or confusion about organic food is common (Sternfeld 2009: 4–5; Liu et al. 2013; Klein 2015: 241; Denmark 2015: 14–17; Bekele et al. 2017: 65). However, China is not unique (Hughner et al. 2007; Aschemann et al. 2007): a 2014 U.S. study found that about 20% of American and Canadians thought organic food meant 'locally grown', and more than 20% thought that 'locally grown food [was] automatically organic' (Foodtank 2015). Recent research suggests that Chinese consumers evince little interest in farmers or rural areas and are concerned mainly if not only with food safety and quality (Scott et al. 2018a, b). In addition, the governance of eco-food in China, as elsewhere (Schmid 2007), is informed by the nineteenth century European model of 'food standards as a central state responsibility (Braithwaite and Drahos 2000: 409). While still embodied in governmental legal principles, standard-setting and certification for food quality are not in fact governmental monopolies (Cody 2019: 90), except to enable producers or sellers to create political alliances, improve their reputation or enlarge their export or domestic markets (Li 2014: 99–100). Some consumers view organic products as an undesirable foreign fashion (Bekele et al. 2017: 68). Much research reports that Chinese producers, retailers and consumers lack trust in government or third-party certification or labeling (e.g. Garnett and Wilkes 2014: 66; Li 2014: 81–82; Berti 2015: 54), in addition to wishing to avoid the complexity, cost or poor reputation of official certification (Yasuda 2018: 145–148; Cody 2019: 185; Si 2018; Chen et al. 2018: 52–53). This is especially true in CSA farms and other alternative food networks (AFNs)(Si et al. 2018: 89; Schumilas 2018).

The term 'multiple organics' (Li 2014: 5) suggests that there is no single, generally accepted conception of 'organic food' in China, even though this is not unique to China (see Loconto and Van der Kamp 2015; Pacific Organic Policy Toolkit 2016: 4), and even though it does not describe accurately the entire eco-food sector in China because it neglects Green Food. It proposes that the meanings of 'organic' are not defined solely by government, law or international standards, but rather are socially constructed, with the multiple normative aspects being captured by the term 'legal pluralism', thus reflecting fundamental disagreements about the meaning of 'organic'. Concepts are expressed in language and shaped by context. According to the anthropologist Jakob Klein (2015: 241), the Chinese term *youji shipin* and English term 'organic' have different connotations: 'the Chinese term carries none of the positive connotations of its English-language equivalent'. Research in Kunming and elsewhere in China found that the term 'organic food' was not widely used and not always distinguished from Green Food (Klein 2009, 2015: 241). An unusual exception is a study of Little Donkey Farm near Beijing, which found that '[t]he term 'organic' is often used in describing the farm as shorthand for a larger set of values about the relationship between humans and the environment at the site of agricultural production….' (Shi et al. 2011: 584). According to the recently published results of a large-scale Canadian research project, 'in Chinese government and research circles, there is a narrow understanding of organic farming and a strong "technological managerialism", linked to the broader scientism and its manifestations in government policies' (Si et al. 2018: 98). These factors, in turn, have long been informed by Party and government concerns for food security, export markets and multi-faceted reforms of China's rural areas.

The Chinese context thus differs fundamentally from that of the origins of organic farming in Germany and the United Kingdom (Vogt 2007), where organic farming and organic food were partly a reaction against the intensive use of agrochemicals and the industrialization of the food sector, and which are reflected to some extent in international organic standards, both legally binding and soft law. The development of eco-food as the basis of a market-driven social movement, as sometimes claimed for western countries (Scott et al. 2013), is unlikely in the very different Chinese context (Ho and Edmonds 2008; Unger 2008), except for limited initiatives (Wen et al. 2012; Si and Scott 2016). Recent research suggests that, in China, 'organic agriculture is framed as a part of – rather than an alternative to – the large-scale, modern, industrialized agriculture and food system in media discourses' (Li 2014: 51). Curiously perhaps, the Chinese conceptual framework corresponds most closely to the current global pattern, involving longer value chains, harmonisation by transnational standards, government sponsorship, focus on international trade and export markets, and the concentration of power in large companies, either domestic or foreign (see Howard 2016: 124–143). Indeed, some authors argue that 'the displacement of the organic farming movement by the organic food industry is already taking place in most developed organic market systems' (Buck et al. 1997; see also Guthman 1998, 2014). The regulation of eco-food in China and elsewhere takes the form of legal pluralism, based on globalisation and diversity of value chains, international competition, and the intended and unintended consequences of diverse legal strategies (Snyder 2010: 90, 141).

2.5 Conclusion

Food consumption in China has changed dramatically in the previous decades (compare Campbell and Campbell 2006 with Garnett and Wilkes 2014: 71–90). Nevertheless, pockets of food insecurity remain, food safety is still a major public concern, and food quality looms as an important challenge for Chinese public health policy. This chapter presented a narrative of the current regulation of food quality in China, focusing on organic food, green food and hazard-free food. It first examined how these types of eco-food have been regulated so far. Then it argued that we can understand the co-existence of multiple regulatory schemes by referring to the theories of regulatory intermediaries, path dependence and legal pluralism.

The distinctive Chinese pattern of regulating eco-food tends to perpetuate domestic regulatory competition, tensions between different economic interests, conflicts among national food policy objectives and lack of consumer trust. However, in the short run the Chinese model may be a useful template for many other countries seeking to improve food safety and food quality, while in the long run it appears to be consistent with current global developments in the regulation of ecological food. But, what is the Chinese model? How could it be improved? From the perspective of providing safe and healthy food for ordinary people in China, the following paragraphs address these questions briefly.

First, it is important to preserve legal pluralism, except that the category Hazard-Free Food should be eliminated as soon as feasible. The current Administrative Measures on Certification of Organic Products (PRC 2015) are too much oriented toward centralisation, overly rigid, and too narrowly focused on governmental and large product/exporter interests, especially if changes in the production and marketing of eco-food in China and worldwide are taken into account. Current measures are ill adapted to the new food safety policy of social governance (Food Safety Law 2015, Article 10; Wu et al. 2018). Standards are double-edged, regulating food safety and quality while serving as barriers to market entry (Snyder 2019). Once accepted, standards, whether legally binding or not, tend to favour certain interests by harmonising rules, determining market access and even creating markets. However, the preservation of normative pluralism, using hard law and soft law, is consistent with worldwide trends concerning food quality, current conceptions of good governance and multiple modernities (Meyer and de Sales Marques 2018) and objectives of sustainable development (Capra and Mattei 2015).

Legal pluralism should also include an increase in operative participating guarantee schemes (PGS). In 2008 IFOAM defined and elaborated the basic standards for PGS, which are 'locally focused quality assurance systems', proposed as an 'affordable alternative to third-party certification' and 'particularly appropriate for small-scale farmers' (Willer and Lernoud 2019: 32–33). As of November 2017, China was reported to have no operational PSGs, with four under development including 167 producers, of which none were yet certified (Katto-Andrighetto et al. 2019: 165–166). Working initially with the EU, IFOAM has also pioneered a form of delegated group self-certification called Internal Control Systems (ICS),[2] which should be extended. Traceability and monitoring of eco-labelling, which currently is a major problem for organic and green food (Berti 2015: 54), should be improved with the help of RFID and blockchain technology (Regattieri et al. 2007; Tian 2018).

Second, all categories of certified organic food should be opened up to the Chinese domestic market. Following the example of Brazil, the current special elite-firm channels of organic production and distribution for exports and other purposes such as elite consumption (eg Tsai 2016) would be opened up to the domestic market: such firms would be encouraged to supply the regular domestic market. The same standards would apply to both export and domestic markets; higher private standards could be agreed by contract (Jensen and Zhou 2015; Zhang et al. 2015). To make this possible, established and publicly communicated deadlines should apply to all such categories, including but not limited to food exports to Hong Kong and Macau and to foreign countries. A pilot project for market convergence of elite supply chains might involve the Greater Bay Area, composed of Hong Kong, Macau and nine cities in Guangdong Province, including Shenzhen. This would be a massive project, entailing complex negotiations and calling on the combined forces of government, market actors and consumers during a period of five to 10 years. This suggestion is controversial, because it involves many well-established interests. However, in the long run it should dramatically improve food ethics and food safety and food quality for ordinary citizens.

Third, bolstered by China's dramatic economic transformation in recent decades, including the rapid growth of the domestic eco-food sector, it is likely that we shall see the continued co-existence of Green Food and organic food, and at the same time an increased role of the MOA in the field of organic food, both domestically and internationally. Green Food under the leadership of the MOA (now SAMR) is unlikely to disappear in the short term because with rising incomes its market should grow. Organic food is likely to become an increasingly diverse category due to foreign investment, improvements in domestic per capita income, increased diversity of alternative food networks and greater concerns for public health, including food safety and quality, even though as in other countries consumers of organic food are

[2] https://www.ifoam.bio/en/internal-control-systems-ics-group-certification, and https://www.ifoam.bio/en/organic-guarantee-system-ifoam-organics-international, both accessed 11 May 2019.

likely to remain a small minority who can pay the price premium. Both streams of eco-food are supported by important economic and political interests.

Government and social interests should be mobilised to sustain and improve the category of Green Food, often lauded correctly as an extremely successful Chinese innovation (Bekele et al. 2017: 66). In the future Green Food could be an achievable and sustainable model for food quality for ordinary consumers. It should be expanded to include the promotion of a healthy diet, with reduced sugar, fat and salt and little or no meat, and more attention to the environment and the ethics of farming and food safety (Freyer and Bingen 2015; Snyder 2006: 85–88) and to the improvement of public health.

References

Abbott KW, Levi-faur D, Snidal D (2017a) Theorizing regulatory intermediaries: the RIT model. Ann Am Acad Polit Soc Sci 670(1):14–35

Abbott KW, Levi-Faur D, Snidal D (2017b) Enriching the RIT Framework. Ann Am Acad Polit Soc Sci 670(1):280–228

AQSIQ (2001) 《国家质量监督检验检疫总局办公厅关于印发<无公害农产品标志管理规定>的通知》(2001)

AQSIQ (2013) Administration of Quality Supervision, Inspection and Quarantine Administrative Measures on Organic Product Certification (AQSIQ Decree No. 155) Decree No. 155, "Administrative measures on organic product certification" adopted by the Bureau executive meeting of Administration of Quality Supervision, Inspection and Quarantine (AQSIQ) on April 23 2013, is hereby promulgated and shall be put into force from April 1 2014. 《有机食品认证管理办法》Administrative Measures on Organic Product Certification (AQSIQ Decree No. 155). http://www.aqsiq.gov.cn/xxgk_13386/jlgg_12538/zjl/2013/201311/t20131120_387865.htm; https://wenku.baidu.com/view/6c75b9d5de80d4d8d15a4ff9.html (English version), 10 October 2017

AQSIQ's Notice on Issuing Pollution-Free Agricultural Product Certification Management (2001). http://www.pkulaw.cn/fulltext_form.aspx?Db=chl&Gid=70f7e912b06b5e02bdfb&keyword=无公害农产品&EncodingName=&Search_Mode=accurate&Search_IsTitle=0 (Chinese version)

Arcuri A (2015) Global food safety standards: the evolving regulatory epistemology at the intersection of the SPS agreement and the codex Alimentarius commission. In: Delimatsis P (ed) The law, economics and politics of international standardisation. Cambridge University Press, Cambridge, pp 79–103

Arthur WB (1989) Competing technologies, increasing returns and lock-in by historical events. Econ J 99:116–131

Arthur WB (1994) Increasing returns and path dependence in the economy. University of Michigan Press, Ann Arbor

Arthur WB, Ermoliev YM, Kaniovski YM (1987) Path-dependence processes and the emergence of macro-structure. Eur J Oper Res 30:294–303

Aschemann J, Hamm U, Naspetti S, Zanoli R (2007) The organic market. In: Lockeretz W (ed) Organic farming: an international history. CABI International, Boston, pp 123–151

Augustin-Jean L (2014) An approach to food quality in China: an agenda for future research. In: Augustin-Jean L, Bjørn Alperman B (eds) The political economy of agro-food markets in China: the social construction of markets in an era of globalization. Palgrave Macmillan, Basingstoke

Bavinck M (2018) Legal pluralism, governance, and the dynamics of sea food supply chains – explorations from South Asia. Marit Stud 17:275–284
Bekele GE, Zhou D, Kidane AA, Haimanot AB (2017) Analysis of organic and green food production and consumption trends in China. Am J Theor Appl Bus 3(4):64–70
Berman PS (2012) Global legal pluralism. Cambridge University Press, Cambridge
Berti R (2015) Organic food in China: the law behind Lüse Shipin and Youji Shipin. GeoProgress J 2(1):45–54
Bourdieu P (1987) The force of law: toward a sociology of the juridical field. Trans. Terdiman R. Hastings Law J 38:814–853
Braithwaite J, Drahos P (2000) Global business regulation. Cambridge University Press, Cambridge
Buck D, Getz C, Guthman J (1997) From farm to table: the organic vegetable commodity chain of northern California. Sociologia Ruralis 37(1):3–20
Büthe T, Mattli W (2011) The new global rulers: the privatization of regulation in the world economy. Princeton University Press, Princeton
CABI (2019) Integrated crop management. https://www.cabi.org/about-cabi/cabi-centres/switzerland/integrated-crop-management/. Accessed 29 Apr 2019
Campbell TC, Campbell TM II (2006) The China study. Benbella Books, Dallas
Capra F, Mattei U (2015) The ecology of law: toward a legal system in tune with nature and community. Berrett-Koehler Publishers, Inc., Oakland
CCIC (1996) China Council for International Cooperation on Environment and Development Working Group on Trade and Environment, 'China Greenfood Development and Environmental Protection'. Research report, September. https://www.iisd.org/pdf/greenfood.pdf, 13 April 2019
Chen A, Scott S, Si ZZ (2018) Top-down initiatives: state support for ecological and organic agriculture in China. In: Scott S, Si ZZ, Schmilas T, Chen A (eds) Organic food and farming in China: top-down and bottom-up ecological initiatives. Routledge, London/New York, pp 38–59
Clapp J, Fuchs D (2009) Agrifood corporations, global governance, and sustainability. In: Clapp J, Fuchs D (eds) Corporate power in global agrifood governance. The MIT Press, Cambridge, MA, pp 1–25
CNAS (2019) China National Accreditation Service for Conformity Assessment, 'CNAS Introduction'. https://www.cnas.org.cn/english/introduction/12/718683.shtml
Cody S (2019) Exemplary agriculture: independent organic farming in contemporary China. Palgrave Macmillan, Springer, Singapore
COFCC (2019) China Organic Food Certification Centre. http://www.ofcc.org.cn/en/index.php?optionid=962. Accessed 6 May 2019
Collier RB, Collier D (1991) Critical junctures and historical legacies. https://papers.ssrn.com/sol3/papers.cfm?abstract_id=1750509. Published as Chapter 1 in Collier RB, Collier D. Shaping the political arena: critical junctures, the labor movement, and regime dynamics in Latin America. Princeton University Press, Princeton
Conrad B (2017) Environmental policy: curtailing urban air pollution. In: Heilmann S (ed) China's political system. Rowman & Littlefield, London, pp 356–361
David PA (2001) Path dependence, its critics and the search for "historical economics". In: Garrouste P, Ioannides S (eds) Evolution and path dependence in economic ideas: past and present. Edward Elgar Publishing Company, Cheltenham
Delimatsis P (2015a) Introduction: continuity and change in international standardisation. In: Delimatsis P (ed) The law, economics and politics of international standardisation. Cambridge University Press, Cambridge, pp 1–16
Delimatsis P (ed) (2015b) The law, economics and politics of international standardisation. Cambridge University Press, Cambridge
Demick B (2011) In China, what you eat tells who you are. Los Angeles Times, September 16. http://articles.latimes.com/2011/sep/16/world/la-fg-china-elite-farm-20110917. Accessed 18 Sept 2011
Denmark (2015) Ministry of Foreign Affairs of Denmark, The Trade Council. Report: Organic Food and Agriculture in China, Food, Agriculture and Fisheries Team, Case No.160058, 12.02.2015. https://kina.um.dk/documents/reportfinal-januar2015.pdf

Dolan SP, Heirbaut D (eds) (2015) The law's many bodies. Duncker & Humbolt, Berlin
Ehrlich E (1936) Fundamental principles of the sociology of law (trans: Moll WL). Reprinted by Transaction Publishers, New Brunswick, 2002
Eisenman J (2018) Red China's green revolution. Columbia University Press, New York
Falk Moore S (1973) Law and social change: the semi-autonomous social field as an appropriate subject of study. Law Soc Rev 7(4):719–746. Reprinted in Falk Moore S, Law as process: an anthropological approach. Routledge & Kegan Paul, London, 1978
Fligstein N (2013) Understanding change and stability in fields. Res Org Behav 33:39–52
Food Safety Law of the People's Republic of China 2015
Foodtank (2015) 19 Organic Food Organizations and Businesses Working to Protect Consumers and the Global Environment, September 2015. https://foodtank.com/news/2015/09/nineteen-organic-food-organizations-and-businesses-working-to-protect-cons/. Accessed 16 May 2019
Foster KW (2015) Embedded within state agencies: business associations in Yantai. In: Unger J (ed) Associations and the Chinese state: contested spaces. Routledge, New York, pp 117–148. First published M.E. Sharpe, 2008
Freyer B, Bingen J (eds) (2015) Re-thinking organic food and farming in a changing world. Springer, Dordrecht
Freyer B, Bingen J, Klimek M, Paxton R (2015) Feeding the world – the contribution of IFOAM principles. In: Freyer B, Bingen J (eds) Re-thinking organic food and farming in a changing world. Springer, Dordrecht, pp 81–102
Garnett T, Wilkes A (2014) Appetite for change: social, economic and environmental transformations in China's food system. Food Climate Research Network, Oxford
Geier B (2007) IFOAM and the history of the international organic movement. In: Lockeretz W (ed) Organic farming: an international history. CABI International, Wallingford
Granovetter M (1985) Economic action and social structure: the problem of embeddedness. Am J Sociol 91:481–510
Granovetter M (1992) Economic institutions as social constructions: a framework for analysis. Acta Sociologica 35(1):3–11
Griffiths J (1986) What is legal pluralism. J Leg Pluralism Unofficial Law 18(24):1–55
Grindley P (2002) Standards strategy and policy: cases and stories. Oxford University Press, Oxford
Guthman J (1998) Regulating meaning, appropriating nature: the codification of California organic agriculture. Antipode 30(2):135–154
Guthman J (2014) Agrarian dreams: the paradox of organic farming in California, 2nd edn. University of California Press, Berkeley
Halliday TC, Shaffer G (eds) (2015) Transnational legal orders. Cambridge University Press, Cambridge
Heilmann S, Shih L (2017) The central government. In: Heilmann S (ed) China's political system. Rowman & Littlefield, London, pp 76–82
Ho N (2019) A look into traditional Chinese administrative law and bureaucracy: feeding the Emperor in Tang Dynasty China. Univ Pa Asian Law Rev 15:2. accepted for publication
Ho P, Edmonds RL (eds) (2008) China's embedded activism: opportunities and constraints of a social movement. Routledge, New York
Howard PH (2016) Concentration and power in the food system: who controls what we eat? Bloomsbury Academic, London
Hughner RS, Mcdonagh P, Prothero A, Shultz CJ, Stanton J (2007) Who are organic food consumers? A compilation and review of why people purchase organic food. J Consum Behav 6(2-3):94–110. https://doi.org/10.1002/cb.210
IFOAM (2019a) Organics international. 'Definition of organic agriculture'. https://www.ifoam.bio/en/organic-landmarks/definition-organic-agriculture. Accessed 15 Apr 2019
IFOAM (2019b) IFOAM organics international, directory of affiliates. China Organic Food Certification Center. https://directory.ifoam.bio/affiliates/759-china-organic-food-certification-center. Accessed 12 May 2019

Ilcan S, Phillips L (2006) Circulations of insecurity: globalizing food standards in historical perspective. In: Bingen J, Busch L (eds) Agricultural standards: the shape of the global food and fiber system. Springer, Dordrecht, pp 51–72
Jensen HH, Zhou JH (2015) Food safety regulation and private standards in China. In: Hammoudi A, Grazia C, Surry Y, Traversac J-B (eds) Food safety, market organization, trade and development. Springer, Heidelberg/New York/Dordrecht/London, pp 167–182
Karpik L (2010) Valuing the unique. Princeton University Press, Princeton
Katto-Andrighetto J, Kirchner C Moura e Castro F, Varini F (2019) Participatory guarantee schemes in 2018. In World 2019 at 161–166
Klein JA (2009) Creating ethical food consumers? Promoting organic foods in urban Southwest China. Soc Anthropol 17(11):74–89
Klein JA (2015) Eating green: ecological food consumption in urban China. In: Kim KO (ed) Re-orienting cuisine: East Asian foodways in the twenty-first century. Berghahn, Oxford/New York, pp 238–262
Klutzz DN (2019) The path of the law review: how interfield ties contribute toinstitutional emergence and buffer against change. Law Soc Rev 51:239–273
Lam WW-L (ed) (2018) Routledge handbook of the Chinese Communist Party. Routledge, London/New York
Lam TC, Perry JL (2001) Service organizations in China: reform and its limits. In: Lee PNS, Lo CWH (eds) Remaking China's public management. Quorum Books, Westport, pp 19–40
Lepeintre J, Sun JJ (eds) (2018) Food safety governance in China. Publications Office of the European Union, Luxembourg
Lewin K (1951) Field theory in social science: selected theoretical papers (ed D Cartwright). Harper and Row, New York
Li X (2014) The making of organic agriculture in China: boundaries, standards, and controversies. A dissertation submitted to Michigan State University in partial fulfillment of the requirements for the degree of Sociology –Doctor of Philosophy 2014
Li J (2018) China's new environment ministry unveiled, with huge staff boost, China Dialogue, 19 April 2018. https://www.chinadialogue.net/article/show/single/en/10599-China-s-new-environment-ministry-unveiled-with-huge-staff-boost. Accessed 12 May 2019
Lieberthal KG, Lampton DM (eds) (1992) Bureaucracy, politics and decision making in post-Mao China. University of California Press, Berkeley
Lieberthal K, Oksenberg M (1989) Policy making in China. Princeton University Press, Princeton
Liu R, Pienuak Z, Verbeke W (2013) Consumers' attitudes and behaviour towards safe food in China: a review. Food Control 33(1):93–104
Lockeretz W (2007) Organic farming: an international history. CABI International, Wallingford/Boston
Loconto A, Van der Kamp M (2015) Differentiating organics: performing multiple objects to organize singular markets for organic tea and biscuits in the UK. In: Freyer B, Bingen J (eds) Re-thinking organic food and farming in a changing world. Springer, Dordrecht, pp 61–80
Luo H, Song GD (2013) Soft law governance: towards an integrated approach (trans: Armour B, Tong HL, William S). Hein & Co., Buffalo
Mahoney J (2000) Path dependence in historical sociology. Theory Soc 29(4):507–548
Mamas S (2019) All your questions about organic food, answered, 28 August 2018 [written by FIELDS Online Supermarket]. http://shanghaimamas.org/all-your-questions-about-organic-food-answered/. Accessed 13 May 2019
MARA (2012) Ministry of agriculture and rural affairs of the People's Republic of China. White Paper on Food Quality and Safety, 26 April 2012. http://english.agri.gov.cn/hottopics/apq/201301/t20130115_9553.htm
Marchesini S, Hasimu H, Spadoni R (2010) An overview of the organic and green food markets in China. In: Haas R, Canavari M, Slee B, Tong C, Anurugsa B (eds) Looking east, looking west: organic and quality food marketing in Asia and Europe. Wageningen Academic Publishers, Wageningen, pp 155–172

Marques JC (2019) Private regulatory capture via harmonization: an analysis of global retailer regulatory intermediaries. Regul Gov 13:157–176
Mattli W, Seddon J (2015) The power of the penholder: the missing politics in global regulatory governance analysis. In: Delimatsis P (ed) The law, economics and politics of international standardisation. Cambridge University Press, Cambridge, pp 169–198e
Merry SE (1988) Legal pluralism. Law Soc Rev 22(5):869–896
Mertha AC (2005) The politics of piracy: intellectual property in contemporary China. Cornell University Press, Ithaca
Meyer T, de Sales Marques JL (eds) (2018) Multiple modernities and good governance. Routledge, London/New York
MOA (Ministry of Agriculture) (2012) Ministry of Agriculture of the People's Republic of China, Order No. 6 of the Ministry of Agriculture of the People's Republic of China, approved at the 7th executive meeting of the Ministry of Agriculture on June 13, 2012, promulgated on July 30, 2012, and implemented as of October 1, 2012, Article 2, signs. http://jiuban.MOA.gov.cn/zwllm/tzgg/bl/201208/t20120802_2814698.htm
OCIA International (2019) Website at: http://www.ocia.org/. Accessed 13 May 2019
OFDC (2019) Organic Food Development Center of MEP, China (OFDC-MEP), HJomepage: Home>About OFDC>Achievements>Introduction. Available at https://ofdc.org.cn/en/article.asp?c_id=67
Organic Farm (2019) Organic farm VIP delivery service, available in Beijing, Shanghai, Guangzhou and Shenzhen. http://vip.organicfarm.com.cn. Accessed 15 May 2019
Pacific Organic Policy Toolkit (2016) Organic standards and regulations. http://www.organicpa-sifika.com/poetcom
Pauli J (2007) China's organic revolution. J Organ Syst 2(1):1–11
Pierson P (2000) Path dependence, increasing returns, and the study of politics. Am Polit Sci Rev 94(2):251–267
PRC (1988) Standardization Law of the People's Republic of China. Adopted at the Fifth Meeting of the Standing Committee of the Seventh National People's Congress on December 29, 1988, promulgated by Order No.11 of the President of the People's Republic of China on December 29, 1988, and effective as of April 1, 1989
PRC (1990) Regulations for the Implementation Of The Standardization Law Of the People's Republic Of China. Promulgated by Decree No. 53 of the State Council of the People's Republic of China on April 6, 1990 and effective as of the date of promulgation. http://english.cnca.gov.cn/lawr/201512/t20151208_42262.shtml
PRC (2003a) 《无公害农产品认证程序》(2003) Procedures for the Certification of Pollution-Free Agricultural Product (2003). http://en.pkulaw.cn/display.aspx?cgid=45596&lib=law
PRC (2003b) Regulations of the People's Republic of China on Certification and Accreditation. Adopted at the 18th executive meeting of the State Council on August 20, 2003, promulgated by Decree No. 390 of the State Council of the People's Republic of China on September 3, 2003, and effective as of November 1, 2003. Article 9. https://www.cnas.org.cn/english/lawsandregulations/images/2012/12/14/245C2A684D2779E04D3DBF9E9A7A8FAB.pdf
PRC (2004) Decision of the State Council about further strengthening food safety, issued on 1 September 2004, effective date 1 September 2004, No. 23 [2004] of the State Council
PRC (2015) Administrative Measures on Certification of Organic Products (2015 Revision). Revised on 25 August 2015 pursuant to the Decision of the General Administration of Quality Supervision, Inspection and Quarantine on Revision of Certain Regulations. 中华人民共和国认证认可条例》Regulations of the People's Republic of China on Certification and Accreditation (2016 Amendment). http://en.pkulaw.cn/display.aspx?cgid=269346&lib=law. 10 October 2017
Regattieri E, Gamberi M, Manzini R (2007) Traceability of food products: general framework and experimental evidence. J Food Eng 81:347–356
Sanders R (2000) Prospects for sustainable development in the Chinese countryside: the political economy of Chinese ecological agriculture. Ashgate, Aldershot, Hants

Schmalzer S (2016) Red revolution, green revolution. University of Chicago Press, Chicago
Schmid O (2007) Development of standards for organic farming. In: Lockeretz W (ed) Organic farming: an international history. CABI International, Wallingford, pp 152–174
Schumilas T (2018) Economic, ecological, and interpersonal dimensions of alternative food networks. In: Scott S, Si ZZ, Schmilas T, Chen AJ (eds) Organic food and farming in China: top-down and bottom-up ecological initiatives. Routledge, London/New York, pp 109–127
Scorzon A, Van der Meulen B, Jiao L (2014) Organics in Chinese food law. Eur Food Feed Law Rev 9(3):179–186
Scott C (2004) Regulation in the age of governance: the rise of the post-regulatory state. In: Jordana J, Levi-Faur D (eds) The politics of regulation: institutions and regulatory reforms for the age of governance. Edward Elgar Publishing, Cheltenham, pp 145–174
Scott J (2009) The WTO agreement on sanitary and phytosanitary measures: a commentary. Oxford University Press
Scott S, Schumilas T, Si ZZ, Chen AJ (2013) Contradictions in state- and civil society-driven developments in China's ecological agriculture sector. Food Policy 45. https://doi.org/10.1016/j.foodpol.2013.08.002
Scott S, Si ZZ, Schumilas T, Chen AJ (2018a) Contradictions in state- and civil society-driven developments in China's ecological agriculture sector. Food Policy 45:158–166
Scott, S, Si, ZZ, Schmilas, T, Chen, AJ. (2018b) Organic food and farming in China: top-down and bottom-up ecological initiatives, Routledge, London/New York
Shi Y, Cheng CW, Lei P, Wen TJ, Merrifield C (2011) Safe food, reen food, good food: Chinese community supported agriculture and the rising middle class. Int J Agric Sustain 9(4):551–558
Si ZZ (2018) Farmers' markets as contested spaces: case study of the Beijing organic farmers' market. In: Scott S, Si ZZ, Schmilas T, Chen AJ (eds) Organic food and farming in China: top-down and bottom-up ecological initiatives. Routledge, London/New York, pp 128–150
Si ZZ, Scott S (2016) The convergence of alternative food networks within "rural development" initiatives: the case of the New Rural Reconstruction Movement in China. Local Environ 21(9):1082–1099
Si ZZ, Schumilas T, Scott S (2018) Bottom-up initiatives: the emergence of "alternative" food networks. In: Scott S, Si ZZ, Schmilas T, Chen AJ (eds) Organic food and farming in China: top-down and bottom-up ecological initiatives. Routledge, London/New York, pp 81–108
Skal Biocontrol (2019) Website: https://www.skal.nl/home-en-gb/about-skal/. Accessed 29 May 2019
Smythe E (2009) In whose interests? Transparency and accountability in the global governance of food: agribusiness, the codex alimentarius, and the World Trade Organization. In: Clapp J, Fuchs D (eds) Corporate power in global agrifood governance. MIT Press, Cambridge, MA, pp 92–123
Snyder F (1993) The effectiveness of European community law: institutions, processes, rules and techniques. Mod Law Rev 56(1):19–56
Snyder F (1994) Soft law and institutional practice in the European community. In: Martin S (ed) The construction of Europe: essays in Honour of Emile Noel. Kluwer, Dordrecht, pp 197–225
Snyder F (1999) Governing economic globalisation: European Union law and global economic networks. Eur Law J 5(4):334–374
Snyder F (2001) The origins of the 'Nonmarket economy': ideas, pluralism and power in EC anti-dumping law about China. Eur Law J 7(4):369–424
Snyder F (2006) Toward an international law for adequate food. In: Mahiou A, Snyder F (eds) La sécurité alimentaire/Food Security and Food Safety. E.J. Brill for The Hague Academy of International Law, Leiden, pp 103–199
Snyder F (2015) No country is an island in regulating food safety: how the WTO monitors China's food safety laws through the trade policy review mechanism. J Integr Agric 14:11, Special Issue: Food Safety in China, 2142–2156
Snyder, F. (2010). The EU, the WTO and China. Hart Publishing, Oxford

Snyder F (2016) Food safety law in China: making transnatonal law. E.J. Brill for the Xiamen Academy of International Law, Leiden

Snyder F (2019) The EU, China and product standards. In: Bersick S, Brown K, Cottey A, Gottwald J-C, Shen W (eds) Routledge handbook on EU-China relations. Routledge, London

Snyder F, Kim YS (2018) China's 2015 food safety law: crossing the river, but feeling the stones and avoiding low-hanging branches? Chin J Comp Law 6(1):1–49. https://doi.org/10.1093/cjcl/cxy004

Snyder F, Ni L (2017a) A tale of eight pesticides: risk regulation and public health in China. Eur J Risk Regul 8(3):469–505. https://doi.org/10.1017/err.2017.38

Snyder F, Ni L (2017b) Chinese apples and the emerging world food trade order: food safety, international trade and regulatory collaboration between China and the European Union. Chin J Comp Law 5(2):253–307. https://doi.org/10.1093/cjcl/cxx014

Snyder F, Ni L (2019) Three faces of China-EU cooperation: from the Beijing Olympics to one belt, one road. In: Canelas de Castro P (ed) The European Union at 60: what future for the EU? University of Macau Press, Macau

Snyder F, Ni L, Hu ZK (2019) Transnational law in the Pacific century: mapping pesticide regulation in China. In: Zumbansen P (ed) The many lives of transnational law: critical engagements with Jessup's Bold proposal. Cambridge University Press, Cambridge

Sousa Santos B (1995) Toward a new common sense. Routledge, London

Sternfeld E (2009) Organic food "Made in China". EU-China Civil Society Forum, Hintergrundfinformatienen 10/2009 – 11. August 2009. www.eu-china.net

Sun J (2018) Review of the "Law of the People's Republic of China on quality and safety of agricultural products". J Resour Ecol 9(1):106–113

Teubner G (ed) (1997) Global law without a state. Dartmouth, Aldershot

Thiers P (2005) Using global organic markets to pay for ecologically based agricultural development in China. Agric Hum Values 22:3–15

Thiers P (2006) China and global organic food standards: sovereignty bargains and domestic politics. In: Bingen J, Busch L (eds) Agricultural standards: the shape of the global food and fibre system. Springer, New York, pp 193–217

Thiers P (2009) Stretching away from the state: NGO emergence and dual identity in a Chinese government institution. In: Hasmuth R, Hsu J (eds) China in an era of transition: understanding contemporary state and society actors. Palgrave Macmillan, London, pp 145–163

Tian F (2018) An information system for food safety monitoring in supply chains based on HACCP, blockchain and internet of things. Doctoral thesis, WU Vienna University of Economics and Business

Trubek D, Dezlay Y, Buchanan R, Davis JR (1994) Global restructuring and the law: studies of the internationalization of legal fields and the creation of transnational arenas (with Yves Dezalay, Ruth Buchanan, and John R. Davis). Case West Law Rev 44:407–498

Tsai W-H (2016) Delicacies for a privileged class in a risk society: the Chinese Communist Party's special supplies food system. Issues Stud 52:2. https://doi.org/10.1142/S1013251116500053

Unger J (ed) (2008) Associations and the Chinese state: contested spaces. M.E. Sharpe, London

van der Heijden J (2017) Darker sides of intermediation: target-oriented and self-interested intermediaries in the regulatory governance of buildings. Ann Am Acad Polit Soc Sci 670(1):207–224

Vogt G (2007) The origins of organic farming. In: Lockeretz W (ed) Organic farming: an international history. CABI International, Wallingford, pp 9–29

von Benda-Beckmann F (1988) Comment on Merry. Law Soc Rev 22(5):897–902

von Benda-Beckmann F, von Benda-Beckmann K (2008) Dynamics of plural legal orders. Lit Verlag, Münster

Wang R, Si Z, Ng C, Scott S (2015) The transformation of trust in China's alternative food networks: disruption, reconstruction, and development. Ecol Soc 20(2):1–19

Wen T, Lau K, Cheng C, He H, Qiu J (2012) Ecological civilisation, indigenous culture, and rural reconstruction inChina. Mon Rev 63(9):29–44

WHO and FAO (2007) World Health Organization and UN Food and Agricultural Organization, Organically Produced Foods, Joint FAO/WHO Food Standards Commission, Codex Alimentarius Commission, Rome, 3rd edn
Wikipedia (2019) Organic Crop Improvement Association. https://en.wikipedia.org/wiki/Organic_Crop_Improvement_Association. Accessed 13
Willer H, Lernoud J (eds) (2019) The world of organic agriculture: statistics and emerging trends 2019. Research Institute of Organic Agriculture FiBL, Frick, and IFOAM – Organics International, Bonn
Wu Q (2013) Competition laws, globalization and legal pluralism: China's experience. Hart, Oxford
Wu L, Liu PP, Lv YX, Cheng XJ, Tsai F-S (2018) Social co-governance for food safety risks. Sustainability 10(4246):1–14. https://doi.org/10.3390/su10114246
Yan YX (2012) Food safety and social risk in China. J Asian Stud 71(3):705–729
Yang DL (2005) Remaking the Chinese leviathan. Stanford University Press, Stanford
Yasuda JK (2018) On feeding the masses: an anatomy of regulatory failure in China. Cambridge University Press, Cambridge
Zhang M, Qiao H, Wang X, Pu M-Z, Yu Z-J, Zheng FT (2015) The third-party regulation on food safety in China: a review. J Integr Agric 14(11):2176–2188. https://doi.org/10.1016/S2095-3119(15)61114-5
Zhou G (2017) The regulatory regime of food safety in China: governance and segmentation. Palgrave Macmillan, Springer, Cham
Zhu S (2007) Political parties in China's judiciary. Duke J Comp Int Law 17:533–560
Zhu Y (2008) An integrated approach to the translation of special terms with special reference to the Chinese term lüse shipin (green food). Transl J 12(1)
Zumbansen P (ed) (2019) The many lives of transnational law: critical engagements with Jessup's Bold proposal. Cambridge University Press, Cambridge

Francis Snyder is C.V. Starr Professor of Law, EU Jean Monnet Chair Ad Personam and Director, Centre for Research on Transnational Law, Peking University School of Transnational Law, Peking University Shenzhen Graduate School, China; Adjunct Professor, University of Macau; Emeritus Professor, CERIC, Faculty of Law and Political Science, Aix-Marseille University; and Visiting Professor, College of Europe, Bruges. Previously, he was Fellow in the Wissenschaftskolleg (Institute of Advanced Study), Berlin; Centennial Professor of Law, London School of Economics; Professor of EU Law, European University Institute (Florence); Professor of Public Law, Aix-Marseille University; and Professor of European Law, University College London. His publications include more than 40 books and 200 articles, including *Food Safety Law in China: Making Transnational Law* (E.J. Brill, Leiden, 2016) and *The EU, The WTO and China: Legal Pluralism and International Trade Regulation* (Hart, 2010). He was invited by the Chinese Central Government in 2013 to serve as lead foreign expert on reform of food safety system in China, leading to 2015 Chinese Food Safety Law. He was educated at Yale University, Sciences Po in Paris, Harvard Law School and Université de Paris I (Panthéon-Sorbonne). In 2018, he was awarded the People's Republic of China Friendship Award and also the Peking University Friendship Award.

Chapter 3
Organic Labelling Influencing Consumerism in China and Thailand: A Case for Collaborating with Mature Organic Economies

GOH Bee Chen

Abstract Consumers are interested in buying organic food products from countries where there is a reputation for organic quality and standard through the certification process. Countries like Australia, New Zealand and Canada have been well-known generally for high-quality organic food production and distribution through maintaining stringent standards in organic accreditation and certification from paddock-to-plate, or farm-to-table. It appears that an unconscious factor is also at play in influencing consumer motivation and behaviour in buying organic – trust in certification labelling. Here, consumers have to trust the organic foods that they buy are genuinely organic. A large contributing factor lies in organic food labelling including the country-of-origin labels (COOL) in organics. This Chapter will argue the case for (1) creating consumer trust in locally produced organic food in Thailand and China through implementing a regulatory framework for organics that can induce confidence in consumer behaviour; and (2) encouraging mature organic economies like Australia, New Zealand and Canada to capitalize on the country-of-origin labels in organic food confidence in order to increase their organic export trade to China and Thailand.

Keywords Asian consumers · Australia · China · Consumer trust in organics · Consumer confidence · Country-of-origin labels · Mature organic economies · New Zealand · Organic foods · Organic labelling · Thailand

---

GOH Bee Chen (✉)
School of Law and Justice, Southern Cross University, Gold Coast, QLD, Australia

Faculty of Law, Chiang Mai University, Chiang Mai, Thailand
e-mail: beechen.goh@scu.edu.au

GOH Bee Chen, R. Price (eds.), *Regulatory Issues in Organic Food Safety in the Asia Pacific*, https://doi.org/10.1007/978-981-15-3580-2_3

3.1 Introduction

> Organics is not just "chemical-free". It is a whole systems or holistic means of growing and handling food. The whole system is linked – Soil. Plants. Animals. Food. People. Environment.[1]

Why buy and consume organic foodstuffs? In recent time, we have seen a phenomenal rise in the consumption of organic foods. This trend is unsurprising, given an increased consciousness in healthy eating, personal well-being, environmental sustainability and an ethic of compassionate animal welfare.[2] Rightly or wrongly, it is often perceived that organic foods are more nutritious, totally free of or less susceptible to chemicals or toxins, and consequently can be more conducive to health and well-being.[3] 'Organic' has also been said to refer to process rather than product.[4] A rise in environmental consciousness including the enhancement of animal welfare has also led to this movement in the consumption of organic products. Climate change has increasingly become a global concern, including nowadays being championed as a climate emergency. Our treatment of the earth based on anthropocentric principles has come under scrutiny more than ever before. A sense of a holistically responsible approach to managing the earth's natural resources represents an urgent agenda. We are living in the Age of the Anthropocene.

Apart from the above factors, globally speaking, the rise of a middle class with better education and higher disposable income has contributed significantly to the demand for organic foods. In recent times, there has undoubtedly been a rapid rise of middle-class earners resulting in household wealth and consumption in organics. Studies have also shown that consumers are also more willing to pay a premium price for organic foods as opposed to conventional foods, based on the perceived health benefits or nutritional value of organic foods.[5] In Thailand, for example, organic foods can be priced at 50% more than

---

[1] Definition is derived from Australian Certified Organic: https://aco.net.au/Pages/Operators/ACOStandards.aspx

[2] Katesuda Sitthisuntikul, Pradtana Yassuck and Budsara Limnirankul, 'How does organic agriculture contribute to food security of small land holders?: A case study in the North of Thailand' in (2018) 4:1429698 *Cogent Food and Agriculture* 1, 2; Pittawat Ueasangkomsate and Salinee Santiteerakul, 'A study of consumers' attitudes and intention to buy organic foods for sustainability' in (2016) 34 *Procedia Environmental Sciences* 423, 425.

[3] Warispas Jiumpanyarach, 'The Impact of Social Trends: Teenagers' Attitudes for Organic Food Market in Thailand' in (2018) 45 (4) *International Journal of Social Economics* 682; Warunpun Kongsom and Chaiwat Kongsom, 'Consumer Behaviour and Knowledge on Organic Products in Thailand' in (2016) 10 (8) *World Academy of Science, Engineering and Technology International Journal of Economics and Management Engineering* 2612, at 2614.

[4] Song Bee Lian, 'The Effectiveness of Organic Certification Logos in influencing Consumer's Attitudes to Purchase Organic Food' in (2017) 12 (2) *Journal of Engineering and Applied Sciences* 301–306, at 301.

[5] Yaowarat Sriwaranun, Christopher Gan, Minsoo Lee and David A. Cohen, 'Consumers' Willingness to Pay for Organic Products in Thailand' in (2015) *International Journal of Social Economics* 480–510; Lijia Wang, Jianhua Wang and Xuexi Huo, 'Consumer's Willingness to Pay

conventional foods.[6] Viewed in a parallel fashion, trade in organic foods is staggering. As an example, in Australia, the Australian Organic Report of 2018 states that the conservatively estimated market value for Australian organic market is $2.4 billion dollars.[7] This is a quantum leap of almost 88% from figures in 2012.[8] That is a dramatic rise in a short span of six years. The said Australian Report further states that fruit, vegetables, nuts and meat constitute the bulk of the processing and production value, with the domestic retail value deriving from fruit, vegetables and nuts, and dairy with meat being the main export item.[9]

As a necessary corollary, conventional agriculture is seen to give way gradually to an increase in the practice of organic agriculture. Some may be inclined to remark that, in the practice of agricultural cultivation, for instance, in India, that it was always organic agriculture which later came to be dominated by conventional agriculture with the dawn of industrialization.[10] In this vein, seen in an ironic twist, we are virtually witnessing a return to the past in terms of agricultural practice.

3.2 A Preliminary Note on the Return to Organic Agriculture

As foreshadowed, one might safely assert that the current trend and hype surrounding organic food is not actually something of a new phenomenon. Our forefathers simply ate what was available in the wild before modern-day organised agriculture set in. And, prior to the advent of industrialization, traditional agriculture, globally speaking, was largely and naturally 'organic', that is to say, by and large devoid of the application of pesticides or herbicides or any other chemical used as fertilizers. In the old days, farming was generally on a small scale, and was usually seen as a family-owned occupation or communal activity in villages. Modern agriculture as we have come to know, also referred to as conventional agriculture, has been largely mechanized and undertaken on a large scale, and seemingly practised with a wanton use of chemicals and pesticides.[11]

---

a Premium for Organic Fruits in China: A Double-Hurdle Analysis' in (2019) 16 *International Journal of Environmental Research and Public Health* 126–140.

[6] Warunpun Kongsom and Chaiwat Kongsom, note 3, at 2616.

[7] Rhiannon Christie (ed), *Australian Organic Market Report 2018*, Australian Organic Ltd, 2018, 3.

[8] Ibid.

[9] Ibid.

[10] Tanmoy Mathur (ed), *The Indian Organic Market: A New Paradigm in Agriculture*, Ernt & Young LLP, 2018, 10.

[11] This has led to an environmental movement from the 1960s to our current state of a climate emergency. See the impactful work by Rachel Carson, *Silent Spring*, Houghton Mifflin, Boston, 1962.

Therefore, organic agriculture may be thus observed and said to be a return to the past, before the advent of industrialization.[12] Conventional agriculture has meant mass production, but mass production has produced disenchantment on the part of discerning consumers in the way of environmental degradation, health decline through increased consumption of chemicalized food, and animal cruelty. As alluded to above, a rising ecological consciousness on a worldwide scale has resulted in consumers favouring buying organic foods, particularly if affordability is not an issue in households with higher disposable incomes. As a natural consequence, one sees a demand-driven rise in the practice of organic agriculture.

Organic agriculture is commonly understood to mean a dispensation from using pesticides and herbicides in the farms. Standards imposed on organic agricultural practices can vary according to jurisdictions, but at the very least, the refrain from chemical application to food crops lies at the very foundation of organics. Organic agriculture in emerging organic economies like China and Thailand provide clear examples pointing to a global consciousness on the concept of health and environmental sustainability.

China, since ancient times, has largely remained an agrarian society. Farming practices in the bygone era were, as aforementioned, typically traditional and small-scale. In every true sense, organic.[13] As has been observed by Qiao, "All forms of sustainable agriculture in China are based on 4000 years of traditional practice crop rotation, compost application with organic matter recycling as well as some traditional ecological systems like mulberry trees combined with fish ponds, which help to maintain soil fertility and ecosystems."[14] In modern times, the introduction of machinery and the use of chemicals in agriculture on a wide scale has paved the way for conventional farming and has also meant that mass food production has become the norm. However, in China (as elsewhere around the world), in recent times, grave concerns surrounding food safety and environmental degradation have propelled, arguably, a return to organic agriculture. There has been a rapid rise in organic agriculture in order to meet domestic and international consumer demand for organic food products. In a telling way, in a study undertaken more than a decade ago by the European Union, it was found then that China had ranked third in the world after Australia and Argentina in having land certified for organic farming.[15] It has been observed that, in the case of China, the emergence, or as posited here – re-emergence – of the organic agriculture movement in China occurred in the 1930s.[16]

---

[12] See, for example, Thai media article: https://www.adb.org/results/helping-farmers-go-organic-thailand

[13] See, generally, F.H. King, *Farmers of Forty Centuries: Organic Farming in China, Korea and Japan*, Dover Publications, (Dover Edn) 2004.

[14] Yuhui Qiao, "Organic Agriculture Development in China" in H Willer and L Kilcher (eds), *The World of Organic Agriculture: Statistics and Emerging Trends 2011*, IFOAM Bonn and FiBL Frick, 2011, 132–136, at page 132.

[15] EU-China Trade Project, Report on "Organic Agriculture in China: Current Situation and Challenges" May 2008, at page 4.

[16] Huichen Gao, Hong Park and Akihiko Sakashita, 'Conventionalization of Organic Agriculture in China: A Case Study of Haobao Organic Agricultural Company in Yunnan Province' in (2017) 19 *Japanese Journal of Agricultural Economics* 37–42, at page 37.

That movement has since rapidly escalated and organic production and certification has now been brought under a series of laws and regulations.[17] Organic agriculture has also been hailed as a means to help alleviate poverty in China.[18]

And, in the case of the rise of organic agriculture in Thailand, for instance, it is clear that organic agriculture in the traditional sense has always been in place before being replaced by conventional agriculture, seen to be more profitable and expedient. In the concurring words of Chinvarasopak, "Organic agriculture is not a totally new practice; it is a revival of a style of farming that was once the standard for local farming in Thailand".[19] The movement in Thailand promoting organic agriculture gained prominence about four decades ago as a result of wide-scale campaigning by non-governmental organisations in aid of environmental sustainability and healthy eating.[20] As has been remarked, "Organic farming in Thailand was initially implemented by local NGOs in the early 1980s. These groups established the Alternative Agriculture Network (AAN)... The Green Net Cooperative and Earth Net Foundation were established in 1993 by the AAN."[21] Many of the campaigns by the environmental activists occurred particularly in the northern province of Chiang Mai, with climate and available land leaning in favour of organic agriculture.[22] Various Thai governmental initiatives as well as Thai Royal Projects supporting sustainable agriculture soon followed suit. Significantly, in 2005, Thailand launched its National Agenda on Organic Agriculture.[23] It was implemented over a 5-year period (2006–2010), with the Thai Government injecting a 16-million baht budget for this scheme.[24] Nowadays, in Thailand, the promotion of organic agriculture is undertaken by the Thai Government as well as by non-governmental organisations and the private sector.

---

[17]Yumei Xie, Hailei Zhao, Karolina Pawlak and Yun Gao, "The Development of Organic Agriculture in China and the Factors Affecting Organic Farming" in (2015) 2(36) *Journal of Agribusiness and Rural Development* 353–361, at page 353–354.

[18]http://p.china.org.cn/2017-08/24/content_50003182.htm

[19]Pasupha Chinvarasopak, 'Key Factors Affecting the Success of Organic Agriculture in Thai Communities: Three Case Studies in Ubon Ratchathani and Srisaket Provinces' in (2015) 13(2) *Thai Journal of Public Administration* 105–130, at page 108.

[20]Yaowarat Sriwaranun et al., note 5, at pages 480–481.

[21]Ibid.

[22]See Arpaphan Pattanapant and Ganesh P. Shivakoti, "Opportunities and Constraints of Organic Agriculture in Chiang Mai Province, Thailand' in (2009) 16 *Asia-Pacific Development Journal* 115–147. In this connection, the author wishes to express her gratitude to the Faculty of Law, Chiang Mai University, for organising beneficial visits to organic farms and organic markets in November 2018 on field trips to Mae Tha Sub-District in Chiang Mai Province, Thailand. In particular, grateful thanks are extended to Dr Nuthamon Kongcharoen at the Law Faculty who accompanied the author on these field trips.

[23]Pasupha Chinvarasopak, note 19, at page 107.

[24]Yaowarat Sriwaranun et al, note 5, at page 481.

In this connection, it may further be observed that, increasingly in Asia, organic farming, and as argued here, a return to organic farming, is seen as a lifestyle choice, particularly by the younger generation of professionals.[25]

3.3 Organic Food Labelling: The Australian Example

As the foregoing demonstrates, consumers are willing to pay more for organic foods based on the perceived inherent health value of organic foods. Organic foods constitute credence products.[26] That they are credence claims is due to the fact that the organic quality is inherent in the food production and cannot easily be externally detected in the final products. Consumers have to rely on the organic labelling of the organic foods and trust that these credence claims are genuinely satisfied. With regard to such credence claims, it must be noted that the information pertaining to what is organic is considered asymmetrical information. This is because consumers do not bear equal access to the organic production information as the organic farmers or producers.[27]

An initial caveat here is necessary. The scope of this Chapter does not permit a comparison of the entire process of organic food regulatory regimes in countries such as Australia and New Zealand – taken here to represent mature organic economies - with China and Thailand – taken here to represent emerging organic economies - in order to yield a comprehensive understanding of the regulatory framework on organics. Rather, the focus here is only on the labelling of organic, or certified organic, foods which represents an important regulatory process in the market distribution of organic foods. Further, what will be analysed in this section on organic labelling will be by way of using only the Australian example. In particular, it will be shown that labelling derived from organic certification inspires confidence in the purchase of organic foods, leading to consumer willingness to pay a premium price.[28]

---

[25] See, for example, media posts on Beijing in China 'Return to Organic Farming: Next Generation of Chinese Farmers': http://www.xinhuanet.com/english/2019-02/07/c_137805072.htm; 'Meet the woman leading China's new organic farming army': https://www.aljazeera.com/indepth/features/2015/11/woman-leading-china-organic-farming-army-beijing-151123140338900.html; Johor Bahru in Malaysia about FOLO and Medini Green Parks: https://vulcanpost.com/624248/malaysia-social-enterprise-folo-farms/;https://theaseanpost.com/article/organic-farming-southeast-asia; Thailand and ASEAN generally: https://www.adb.org/results/helping-farmers-go-organic-thailand

[26] Ibid; M Canavari, N Cantore, E Pignatti and R Spadoni, 'Role of Certification Bodies in the Organic Production System' in Rainer Hass, Maurizio Canavari, Bill Slee, Tong Chen and Bundit Anugugsa (eds), *Looking East, Looking West: Organic and Quality Food Marketing in Asia and Europe*, Wageningen Academic Publichers, Wageningen, 2010, at 86.

[27] Ibid, at 85; Yaowarat Sriwaranun et al, note 5, at 481.

[28] Lijia Wang, Jianhua Wang and Xuexi Huo, note 5, at 136.

In Australia, it has been judicially remarked that the legal framework for organic food labelling represents a 'murky environment'.[29] Australian consumers interested in organic foods are continually confused by the various use of 'organic' and 'certified organic' in organic food labels, unaware of the significant difference in such usages. The variance is due to the fact that the standards which apply depend on the categories of organic foods referred to. At present, there are three categories of organic foods attracting different applicable standards in organic food labelling: (a) organic foods for domestic consumption; (b) organic foods for exports; and (c) organic foods imported into Australia. They will be dealt with in turn.

As a starting point, it may be observed that food labelling falls within the purview of the Food Standards Australia and New Zealand ('FSANZ).[30] Labelling requirements are covered under the Food Standards Australia and New Zealand Code, as developed by FSANZ.[31] The Code carries legislative effect and has been respectively adopted in State and Territories legislation like the Food Act 2001 (Australian Capital Territory), the Food Act 2003 (New South Wales), the Food Act 2006 (Queensland), and the Food Act 2008 (Western Australia), to name some State jurisdictions. Food labelling is, in turn, influenced by the requirement for food certification.

There are two applicable standards in terms of organic food labelling requirements in Australia: the National Standard for Organic and Bio-Dynamic Produce ('the National Standard') which applies to organic food exports, and the Australian Standard 6000–2009 ('AS 6000') which applies to domestic organic foods and imported organic foods. It is noteworthy that the AS 6000 has been modelled on the National Standard and as such, the two standards, though different in their application, are rather similar in their content.[32] Do has succinctly made the following observation with regard to the organic labelling regime in Australia:[33]

> The Australian Commonwealth Parliament adopted a co-regulatory approach with respect to the two organic standards. The two organic standards are 'complementary and additional to other health, agricultural or food standards or regulatory requirements recognised by or enacted by the Commonwealth, States and Territories'. Whilst the two organic standards have no legal effect on their own, they are legally enforced in conjunction with existing legislative instruments. Organic produce that is intended for exportation is regulated by the National Standard in conjunction with the *Export Control (Organic Produce Certification) Orders* (Cth) and *Export Control Act 1982* (Cth). Organic produce that is intended for the domestic market is regulated by the AS 6000–2009 in conjunction with the *Competition and Consumer Act 2001* (Cth). Finally, organic produce imported into Australia is also regulated by the AS 6000–2009 in conjunction with the *Competition and Consumer Act 2001* (Cth) and the *Imported Food Control Act 1992* (Cth).

---

[29] Martin J in *Marsh v Baxter* (2014) 46 WAR 377. For an insightful academic exposition, see Christina Do, 'Organic Food Labelling in Australia: A 'Murky Environment' in Need of Reform' in (2015) 34(1) *University of Queensland Law Journal* 123.

[30] See Food Standards Australia and New Zealand Act 1991 (Commonwealth).

[31] Australia New Zealand Food Standards Code, Standard 1.1.1.

[32] See Standards Australia, Australian Standards 6000–2009 Organic and Bio-Dynamic Products.

[33] Do, note 29, at 126.

Domestic consumers of organic foods in Australia may be surprised to learn that organic food certification is a voluntary process.[34] As earlier stated, there is a difference in the organic food labelling as to whether or not such food is labelled 'organic' or 'certified organic'. If a particular organic food is labelled 'certified organic', then there must be evidence to support such a credence claim. In other words, some third party certification for the organic food has been complied with and obtained.

Currently, in Australia, there exist five organic certification bodies, as below:[35]

1. Australian Certified Organic
2. National Association for Sustainable Agriculture Australia (NASAA)
3. AUS-QUAL Pty Ltd
4. Bio-Dynamic Research Institute
5. Organic Food Chain

Since organic certification for domestic consumption is a voluntary exercise, a product claiming to be certified organic must have undergone the organic certification process by any one of these recognised bodies. Each of the above five certifying bodies carries its own identification logos. An appropriate logo may then be attached for the 'certified organic' label. However, if an organic food product carries the 'certified organic' label without having obtained the appropriate certification in the first instance, the marketing of such organic foods will appear to breach the Australian Consumer Law for misleading or deceiving the consumer pursuant to Section 18 of the Australian Consumer Law.[36]

Australian organic foods for the export market are required to undergo a stringent mandatory organic certification process. Similarly, imported organic foods into Australia must comply with mandatory organic standards. The Australian Department of Agriculture is charged with the regulatory responsibility to ensure that exporters comply with applicable rules and regulations.[37]

3.4 Country-Of-Origin Labels (COOL)

As discussed in the foregoing section on organic food labelling, generally speaking, labelling is an essential part of the marketing and distribution of foods generally, and in particular, organic food labelling for satisfying credence claim. As has been observed, 'organic' is a credence claim and, where mandatory, must comply with

---

[34] https://www.agriculture.gov.au/ag-farm-food/food/organic-biodynamic

[35] Andrew Lawson, Amy Cosby, Derek Baker, Shawn Leu, Ed Lefley, Amarjit Sahota and Nick Bez, 'Australia' in Helga Willer and Julia Lernoud (eds), *The World of Organic Agriculture: Statistics and Emerging Trends 2019*, Research Institute of Organic Agriculture (FiBL), Frick, and IFOAM Organics International, Bonn, 2019, at page 298.

[36] The Australian Consumer Law is contained in Schedule 2 of the Competition and Consumer Act 2010.

[37] https://www.agriculture.gov.au/export/controlled-goods/organic-bio-dynamic

prevailing certification standards. Research has further found that labelling plays a role in influencing consumer behaviour in food safety.[38] In the case of organic foods, labelling plays a more important role as a case of differentiation between what consumers would be prepared to pay, and be willing to pay a premium price, for perceived higher quality foods that organic foods might provide.

Indeed, both the requirements for organic certification and labelling can lead to increased costs and time. The applicable regulatory framework can come across as rather complex, despite the fact that various countries are attempting to re-investigate the regulatory framework in organics in order to try to reduce the costs and bureaucracy involved.

In Australia, for instance, food labelling indicating the country of origin has shifted from a voluntary to mandatory regime for retail food sale. The mandatory requirement for country-of-origin food labelling contained in the Australia New Zealand Food Standards *Code to the Country of Origin Food Labelling Information Standard 2016* came into force on 1 July 2018.[39] This was preceded by a two-year transitional period from 1 July 2016. To assist with and facilitate implementation, the Australian Competition and Consumer Commission has published a series of guides and information sheets for the food industry.[40]

And, in both China and Thailand, there exist requirements for country-of-origin labelling. In China, the Food Labelling Standards of China (GB7718-2011) apply to imported foods stating the country of origin. For Thailand, the Food Act of 1979 is the principal legislative instrument governing food matters. Specifically, Thai Ministry of Public Health Notification Numbers 367 and 383 provide for labelling to state the country of origin.[41] By way of further example, other Asian jurisdictions such as Japan, Malaysia and Vietnam similarly have requirements for country-of-origin food labelling.[42]

[38] Pongthai Thairotin, Kiyokazu Ujiie and Hisato Shuto, 'An Evaluation of Consumers' Preference on Food Safety Certificates and Product Origins: A Choice Experiment Approach for Fresh Oranges in Metropolitan Bangkok, Thailand' in (2015) 24 (2) *Agricultural Information Research* 74–80.

[39] The 2016 Standard was made under Section 134 of the Australian Consumer Law, contained in Schedule 2 of the Competition and Consumer Act 2010 (Commonwealth).

[40] See Australian Competition and Consumer Commission, 'Country of Origin Claims and the Australian Consumer Law', March 2019; Australian Competition and Consumer Commission, 'Country of Origin Food Labelling Guide', March 2019.

[41] See Wimalin Rimpeekool, Sam-ang Seubsman, Cathy Banwell, Martyn Kirk, Vasoontara Yiengprugsawan and Adrian Sleigh, "Food and nutrition labelling in Thailand: a long march from subsistence producers to international traders" (2015) 56 *Food Policy* 59–66.

[42] Japan: Food Labelling Act and the Labelling Standards Ordinance; Malaysia: Food Act 1983 and Food Regulations 1985; Vietnam: National Standard on Organic Agriculture issued by the Ministry of Science and Technology in December 2017 (No. TCVN11041:2017) with labelling requirements covered under Part 1.

3.5 Consumer Trust

A central argument of this chapter relates to consumer trust in buying organic food products at premium prices, even though it has been found that rising consumer affluence is an influential factor in consumers' willingness to pay more for organic food products.[43] From the point of view of psychology, one may assert that trust operates in the subconscious sphere which may go to influence consumer spending behaviour. On the one hand, we may be well aware that the consumption of organic food, as it were, carries positive connotations such as healthy eating, and caring for the environment and animal welfare.[44] However, when it comes to consumer spending in organics, trust inevitably plays a critical role.[45] Questions such as 'am I *really* buying organic which is what the label says?' or 'how organic?' in terms of the credence claim prey on the minds of the consumers when intending to purchase organic food products.

As has been observed by Smed et al, "Concerning trust it is important that the consumers view the label as a guarantee of the compliance with the organic production rules and that the consumer trusts the organic producers."[46] Labelling inspiring consumer confidence in this context entails both credence labelling (that is, foods are organic or are certified organic) and country-of-origin labelling, especially with regard to the country of origin being one from a mature organic economy.[47]

As alluded to earlier, country-of-origin labelling is seen as instrumental in keeping the faith of organic consumption, particularly if such organic products originate from mature organic economies like Australia, New Zealand, Canada or the European Union, to name some. This is because there is a perceptual bias that these organic products have undergone compliance with stringent organic certification standards in these countries seen to be mature organic economies.[48] As we have seen earlier, in the case of Australia, even though organic certification is voluntary for the domestic consumption of organic foods, stringent and mandatory organic standards

---

[43] Yaowarat Sriwaranun et al, note 5, at page 504.

[44] Sinne Smed, Laura M Andersen, Niels Kaergard and Carsten Daugbjerg, 'A Matter of Trust: How Trust Influence Organic Consumption' in (2013) 5 (7) *Journal of Agricultural Science* 91–106, at 93.

[45] P Pomsanam, K Napompech and S Suwanmannepong, 'Factors Driving Thai Consumers' Intention to Purchase Organic Foods' in (2014) 7 (4) *Asian Journal of Scientific Research* 434, 442.

[46] Sinne Smed et al, note 44, at page 101.

[47] On a relevant study in relation to Thailand and China, see Susanne Pedersen, Jessica Aschemann-Witzel and John Thorgesen, 'Consumer Evaluation of Imported Organic Food Products in Emerging Economies in Asia', International Food Marketing Research Symposium, Dubrovnik, Croatia, 2017. See also media article on Australian organic food products: https://www.intheblack.com/articles/2017/09/01/organic-food-demand-australia

[48] Ibid; See also John Thorgesen, Susanne Pedersen, Maria Paternoga, Eva Schwendel and Jessica Aschemann-Witzel, 'How Important is Country-of-Origin for Organic Food Consumers? A Review of the Literature and Suggestions for Future Research' in (2017) 119 *British Food Journal* 542–557, at page 550.

apply for organic food exports to other countries. Consumer trust is thus engendered readily in such authentic credence claims for organic foods.[49]

Therefore, it may be further argued that countries with mature organic economies like aforementioned Australia, New Zealand, Canada or the European Union, among others, should endeavour to benefit from exporting organic products to countries such as China or Thailand.

As a necessary corollary, it can be argued that emerging organic economies like China and Thailand, whilst embarking upon the regulation of their respective organic certification standards, can explore reciprocal regulatory frameworks to increase consumer confidence in their domestic organic food products. Such reciprocity may be achieved through the entry into mutual recognition regimes for the certification of organic products, canvassed in the succeeding section. There are, definitely, advantages to be enjoyed in such reciprocal arrangements beyond cutting through complex bureaucracies involved in organic food importation and exportation.

3.6 Mutual Recognition Regime in Organics

When there is jurisdictional co-operation in the matter of organics, especially a mutual recognition regime in organic food certification, the flow of international trade as such can be vastly enhanced.[50] There will be smoother facilitation in the export and import of organic food, with signatory countries removing the necessary bureaucratic inspections and complex regulatory compliance.[51]

From the point of view of emerging organic economies like China and Thailand, such mutual recognition achieves desirable outcomes in both domestic and international markets for organic foods. This is because, on the one hand, a mutual recognition regime seeks to inculcate trust by domestic consumers in locally produced organic foods based on such mutuality in organic certification with a mature organic economy. And, on the other hand, in the export of organic foods, the emerging organic economy can enjoy the side-stepping or waiver of stringent and bureaucratic compliance regulations by the mature organic economy which reduces costs and saves time.

There are already instances of such co-operative instruments, for example, between Australia and New Zealand;[52] the European Union with Japan, the United

[49] See Shijiu Yin, Fei Han, Yiqin Wang, Wuyang Hu and Shanshan Lv, 'Ethnocentrism, Trust and the Willingness to Pay of Chinese Consumers for Organic Labels from Different Countries and Certifiers' in (2019) *Journal of Food Quality* 1–13, at pages 10–11.

[50] Such mutual recognition may be variously known as 'mutual equivalency' in Canada or 'mutual authentication' in Japan.

[51] See Diane Bowen (in collaboration with Ulrich Hoffmann), "Plurilateral Regulatory Cooperation on Organic Agriculture and Trade", UNFSS Discussion Paper No.5, April 2015, United Nations: United Nations Forum on Sustaibility Standards, at pages 2–3.

[52] Trans-Tasman Mutual Recognition Arrangement with effect from 1998: http://www.agriculture.gov.au/import/goods/food/importing-food-from-new-zealand

States of America, Canada, and Switzerland;[53] and Canada with the United States, the European Union, Japan, Switzerland and Costa Rica.[54]

Importantly, too, in alignment with the arguments posited in this Chapter, the value of consumer trust in the consumption of organic food certified by mature organics economies should not be under-estimated. As a corollary, where a mutual recognition regime is in place, the perception that each has achieved the high organics standards will undoubtedly be present in the minds of consumers willing to pay a premium price for trusted organic foods.

In aid of a relevant discussion for this Chapter, what follows is a brief snapshot of the recent mutual recognition instrument entered into between China and New Zealand.[55]

China and New Zealand: The date of 14 November 2016 marked a momentous occasion as it witnessed the entry into a co-operative international regime with regard to organic foods in the form of the signing of the Mutual Recognition Arrangement for Certified Organic Products between China and New Zealand.[56] It is believed that, currently, actual implementation is yet to occur. However, this has been significantly hailed as the first-ever international instrument in mutual recognition in organics that China has entered into with another country. As indicated, the immediate positive consequence is the improvement in bilateral trade relations in the enhanced implementation of the Free Trade Agreement between China and New Zealand. The bilateral trade in the main organic food products from China to New Zealand have been noted to be coffee, frozen vegetables, grains, and pet foods, and from New Zealand to China the main organic products seem to be dairy products, meat, fruits and wine.[57]

It appears that there are negotiations in process between China and Thailand, Denmark and the United Kingdom,[58] as well as the European Union[59] for entry into similar mutual recognition regimes in organics. And, as has been observed, China and the European Union can learn from each other in organic food safety and

---

[53] https://ec.europa.eu/agriculture/newsroom/82_en

[54] Government of Canada - Canadian Food Inspection Agency: https://www.inspection.gc.ca/food/requirements-and-guidance/organic-products/equivalence-arrangements/eng/1311987562418/1311987760268

[55] Further reference may be made to the co-authored Chapter 14 on New Zealand by Tracey Epps and Danae Wheeler in this edited volume.

[56] See http://english.cnca.gov.cn/news/201612/t20161213_53011.shtml; See also http://www.oanz.org/news/22/24/China-and-New-Zealand-sign-organic-products-mutual-recognition-deal.html

[57] Ibid.

[58] See http://www.chinadaily.com.cn/china/2017-03/22/content_28634556.htm

[59] See https://ec.europa.eu/agriculture/newsroom/82_en; Alberto Scorzon, Bernd van der Meulen and Li Jiao, 'Organics in Chinese Food Law' in (2014) 3 *European Food and Feed Law Review* 179.

bilateral trade in organics.[60] Such international co-operation can go a long way to enhance trade relations.

3.7 Conclusion

Organics heralds a significant era in consumption when discerning markets rise globally in response to a wake in food safety, health and well-being, in most instances, attributable to a rise in people having a better education and more disposable income. As this Chapter has remarked, growing concerns for the protection of the natural environment and animal welfare further represent impactful contributors in the consumption of organics. Consumer attitudes have been changing to embrace food production in terms of how food is produced, not just food consumption in the sense of what to eat. Undoubtedly, with better education and a rise in affluence, it is no wonder that where possible, organic food is in high demand.

However, the picture may not always be rosy in the consumption of organics. Consumers need to be wary of issues pertaining to organic certification and organic labelling due to the fact that 'organic' is a credence claim. As such, the quality of organics is inherent in the product with the total reliance by consumers on such credence claims. Information is said to be asymmetrical about credence claims on the part of the producers *vis-à-vis* the consumers. Rigorous processes need to be in place to ensure that there is strict compliance with organic certification from farm to table. As a safeguard, like in Australia, consumer legislation has been enacted against falsifying claims on products alleged to be certified organic when these products are, in fact, non-organic. In these instances, rogue producers or distributors will be held accountable for misleading or deceiving consumers.

This Chapter has attempted to argue that there is an immense economic opportunity for mature organic economies like Australia, New Zealand or the European Union, among other jurisdictions, to export organic foods into emerging organic economies like Thailand and China. As has been observed, there has been a rapid rise of organic consumption in the case of Thailand and China representing emerging organic economies. Thai and Chinese consumers who have a preference for organic food consumption tend to trust and display a bias for foreign-imported organic foods, based on the belief that organic food certification and the consequential organic food labelling of such organic products in these mature organic economies are more stringent. This further leads to the belief that the credence claims are genuine. Such consumer trust is fundamental in their willingness to pay a premium price for organic as opposed to conventional foods.

[60] Snyder, Francis and Ni, Lili, Three Faces of China – EU Cooperation, from the Beijing Olympics to One Belt, One Road (April 24, 2018). European Union Academic Programme – Macau Bi-annual Conference, '60 Years after the Treaties of Rome: What is the Future for the European Union?', Faculty of Law, University of Macau, 27–28 November 2017; Peking University School of Transnational Law Research Paper No. 18-8. Available at SSRN: https://ssrn.com/abstract=3167981, at 14–25.

This Chapter similarly urges emerging Asian organic economies like Thailand and China to conclude mutual recognition instruments in organics. An important example is the case of China and New Zealand, as instanced by the signing of the Mutual Recognition Arrangement for Certified Organic Products in November 2016. It is hoped that such mutual recognition will smoothen the path for domestic organic consumption in building or enhancing consumer trust in domestically produced organic foodstuffs, facilitated by the mutual recognition regimes. The fact that mutual recognition reduces bureaucratic complexities in import and export compliance in organic foods is itself highly desirable. Another possible positive outcome may lie in reducing costs as bureaucracy is side-stepped. Overall, it represents a win-win solution. An enhancement in international trade, particularly in organic food trade, for trusted quality and better pricing can only serve as a welcome outcome. The greater good, ultimately, is a cleaner planet, with people and animals the primary beneficiaries.

References

Aljazeera. https://www.aljazeera.com/indepth/features/2015/11/woman-leading-china-organic-farming-army-beijing-151123140338900.html

Asian Development Bank. https://www.adb.org/results/helping-farmers-go-organic-thailand

Australian Certified Organic. https://aco.net.au/Pages/Operators/ACOStandards.aspx

Australian Competition and Consumer Commission. Country of origin claims and the Australian consumer law. March 2019

Australian Competition and Consumer Commission. Guide to country of origin food labelling. March 2019

Australian Government Department of Agriculture. https://www.agriculture.gov.au/ag-farm-food/food/organic-biodynamic

Australian Government Department of Agriculture. https://www.agriculture.gov.au/export/controlled-goods/organic-bio-dynamic

Australian Government Department of Agriculture. http://www.agriculture.gov.au/import/goods/food/importing-food-from-new-zealand

Bowen D (in collaboration with Hoffmann, U) (2015) Plurilateral regulatory cooperation on organic agriculture and trade. UNFSS Discussion Paper No.5, United Nations: United Nations Forum on Sustaibility Standards, April 2015

Canavari M, Cantore N, Pignatti E, Spadoni R (2010) Role of certification bodies in the organic production system. In: Hass R, Canavari M, Slee B, Chen T, Anurugsa B (eds) Looking east, looking west: organic and quality food marketing in Asia and Europe. Wageningen Academic Publichers, Wageningen

Carson R (1962) Silent spring. Houghton Mifflin, Boston

China Daily. http://www.chinadaily.com.cn/china/2017-03/22/content_28634556.htm

China's Poverty Reduction Online. http://p.china.org.cn/2017-08/24/content_50003182.htm

Chinvarasopak P (2015) Key factors affecting the success of organic agriculture in Thai communities: three case studies in Ubon Ratchathani and Srisaket Provinces. Thai J Public Admin 13(2):105–130

Christie R (ed) (2018) Australia organic market report 2018. Australian Organic Ltd.

Competition and Consumer Act 2010 (Commonwealth of Australia)

Do C (2015) Organic food labelling in Australia: a 'Murky environment' in need of reform. Univ QLD Law J 34(1):123

EU-China Trade Project (2018) Report on "Organic agriculture in China: current situation and challenges", May 2018

European Commission. https://ec.europa.eu/agriculture/newsroom/82_en

Gao H, Park H, Sakashita A (2017) Conventionalization of organic agriculture in China: a case study of Haobao Organic Agricultural Company in Yunnan Province. Jpn J Agric Econ 19:37–42

Government of Canada Food Inspection Agency. https://www.inspection.gc.ca/food/requirements-and-guidance/organic-products/equivalence-arrangements/eng/1311987562418/1311987760268

In the Black. https://www.intheblack.com/articles/2017/09/01/organic-food-demand-australia

Jiumpanyarach W (2018) The impact of social trends: teenagers' attitudes for organic food market in Thailand. Int J Soc Econ 45(4):682–699

King FH (2004) Farmers of forty centuries: organic farming in China, Korea and Japan, Dover edn. Dover Publications

Kongsom W, Kongsom C (2016) Consumer behaviour and knowledge on organic products in Thailand. World Academy of Science, Engineering and Technology. Int J Econ Manag Eng 10(8):2612–2616

Lawson, A, Cosby, A, Baker, D, Leu, S, Lefley, E, Sahota, A and Bez, N, 'Australia' Willer, H, and Lernoud, J, The world of organic agriculture: statistics and emerging trends 2019, Research Institute of Organic Agriculture (FiBL), Frick, and IFOAM Organics International, Bonn, 2019

Mathur T (ed) (2018) The Indian organic market: a new paradigm in agriculture. Ernst & Young LLP

Organics Aotearoa New Zealand. http://www.oanz.org/news/22/24/China-and-New-Zealand-sign-organic-products-mutual-recognition-deal.html

Pattanapant A, Shivakoti GP (2009) Opportunities and constraints of organic agriculture in Chiang Mai Province, Thailand. Asia-Pac Dev J 16:115–147

Pedersen S, Aschemann-Witzel J, Thorgesen J (2017) Consumer evaluation of imported organic food products in emerging economies in Asia. International Food Marketing Research Symposium, Dubrovnik, Croatia

People's Republic of China Certification and Accreditation Administration. http://english.cnca.gov.cn/news/201612/t20161213_53011.shtml

Pomsanam P, Napompech K, Suwanmaneepong S (2014) Factors driving Thai consumers' intention to purchase organic foods. Asian J Sci Res 7(4):434–446

Qiao Y (2011) Organic agriculture development in China. In: Willer H, Kilcher L (eds) The world of organic agriculture: statistics and emerging trends 2011. IFOAM/FiBL, Bonn/Frick, pp 132–136

Rimpeekool W, Seubsman S, Banwell C, Kirk M, Yiengprugsawan V, Sleigh A (2015) Food and nutrition labelling in Thailand: a long march from subsistence producers to international traders. Food Policy 56:59–66

Scorzon A, van der Meulen B, Jiao L (2014) Organics in Chinese food law. Eur Food Feed Law Rev 3:179

Sitthisuntikul K, Yassuck P, Limnirankul B (2018) How does organic agriculture contribute to food security of small land holders?: a case study in the North of Thailand. Cogent Food Agric 4:1–12

Smed S, Andersen LM, Kaergard N, Daugbjerg C (2013) A matter of trust: how trust influence organic consumption. J Agric Sci 5(7):91–106

Snyder F, Ni L (2018) Three faces of China – EU Cooperation, from the Beijing Olympics to One Belt, One Road' (April 24, 2018). European Union Academic Programme – Macau Bi-annual Conference, '60 years after the treaties of Rome: what is the future for the European Union?', Faculty of Law, University of Macau, 27–28 November 2017; Peking University School of Transnational Law research paper no.18-8

Song BL (2017) The effectiveness of organic certification logos in influencing consumer's attitudes to purchase organic food. J Eng Appl Sci 12(2):301–306

Sriwaranun Y, Gan C, Lee M, Cohen DA (2015) Consumers' willingness to pay for organic products in Thailand. Int J Soc Econ 42:480–510

Thaiyotin P, Ujiie K, Shuto H (2015) An evaluation of consumers' preference on food safety certificates and product origins: a choice experiment approach for fresh oranges in Metropolitan Bangkok, Thailand. Agric Inf Res 24(2):74–80

The Asean Post. https://theaseanpost.com/article/organic-farming-southeast-asia

Thorgesen J, Pedersen S, Paternoga M, Schwendel E, Aschemann-Witzel J (2017) How important is country-of-origin for organic food consumers? A review of the literature and suggestions for future research. Br Food J 119:542–557

Ueasangkomsate P, Santiteerakul S (2016) A study of consumers' attitudes and intention to buy organic foods for sustainability. Proc Environ Sci 34:423–430

Vulcanpost. https://vulcanpost.com/624248/malaysia-social-enterprise-folo-farms/

Wang L, Wang J, Huo X (2019) Consumers' willingness to pay a premium for organic fruits in china: a double-hurdle analysis. Int J Environ Res Public Health 16:126

Xie Y, Zhao H, Pawlak K, Gao Y (2015) The development of organic agriculture in China and the factors affecting organic farming. J Agribus Rural Dev 2(36):353–361

Xinhua Net. http://www.xinhuanet.com/english/2019-02/07/c_137805072.htm

Yin S, Han F, Wang Y, Hu W, Lv S (2019) Ethnocentrism, trust and the willingness to pay of Chinese consumers for organic labels from different countries and certifiers. J Food Qual 19:1–13

GOH Bee Chen A Former Malaysian Rhodes Scholar, Professor GOH Bee Chen is Professor of Law and Director of the Judge-in-Residence Programme, School of Law and Justice, Southern Cross University, Australia. She has recently been a Visiting Professor in the Faculty of Law, Chiang Mai University, Thailand. She is a Director and Fellow of the Australian Academy of Law and Fellow of Cambridge Commonwealth Society and of the Society for Advanced Legal Studies in London. Besides organic regulatory framework, her current collaborative research project deals with Law and Theatre. Her scholarly interests include mediation and ADR, especially on cross-cultural (Sino-Western) dispute resolution, and international law of peace. Her publications include *Negotiating with the Chinese* (Dartmouth/Routledge, 1996), *Law Without Lawyers, Justice Without Courts: On Traditional Chinese Mediation* (Ashgate/Routledge, 2002), *Activating Human Rights and Peace: Theories, Practices and Contexts* (Ashgate/Routledge, 2012, with Offord and Garbutt); and *Scholarship, Practice and Education in Comparative Law: A Festschrift in Honour of Mary Hiscock* (Springer, 2019, with Farrar and Lo).

Chapter 4
Chinese Organic Food Law and Its Impacts on Climate Change

Ying Shen

Abstract China is among the world's largest producers in organics and its organic agriculture has been boosting since the turn of the Millennium with the rapid development of both domestic and international production and trade in organic food. To instill transparency and trust among stakeholders, the Chinese government has issued a series of new regulations relating to food safety and organic certification. Organic agriculture covered by these regulations is not only affected by climate change but also contributes to it, both adapting to changes and offering options for mitigation (i.e. reducing greenhouse gas emissions and storing carbon). This chapter aims to provide an overview of Chinese organic food law and investigate its impacts on climate change. Through investigating stakeholders involved in and regulations relating to organics, challenges to the enforcement of food safety law as well as organic certification issues are identified and analyzed. Particularly, the impacts of these regulations on climate change mitigation and adaptation are discussed in order to examine the development of organic agriculture in China.

Keywords Organic agriculture · Food safety law · Organic standards · Organic certification · Chinese · Climate change adaptation · Climate change mitigation

Glossary of Acronyms

CO_2	Carbon dioxide
CNCA	Certification and Accreditation Administration of the People's Republic of China
CNOPS	Chinese National Organic Product Standard
COFCC	Chinese Organic Food Certification Center
CPC	Communist Party of China
FHL	Food Hygiene Law of the People's Republic of China
FSL	Food Safety Law of the People's Republic of China

Y. Shen (✉)
Postdoc Fellow of Käte Hamburger Kolleg/Centre for Global Cooperation Research, University of Duisburg-Essen, Duisburg, Germany

GOH Bee Chen, R. Price (eds.), *Regulatory Issues in Organic Food Safety in the Asia Pacific*, https://doi.org/10.1007/978-981-15-3580-2_4

AQSIQ	General Administration of Quality Supervision, Inspection and Quarantine
GMF	Genetically modified food
GMOs	Genetically modified organisms
GHG	Greenhouse gas
IPCC	Intergovernmental Panel on Climate Change
IFOAM	International Federation Organic Agriculture Movement
LAPQS	Law on Agricultural Product Quality and Safety of the People's Republic of China
MOA	Ministry of Agriculture
MOFCOM	Ministry of Commerce
MEE	Ministry of Ecological Environment
MOH	Ministry of Health
N_2O	Nitrous oxide
OFDC	Organic Food Development and Certification Center of China
RCA	Regulations of the People's Republic of China on Certification and Accreditation
SAIC	State Administration for Industry and Commerce
SEPA	State Environmental Protection Agency
SFDA	State Food and Drug Administration
SFSA	State Food and Safety Administration
US	United States
USDA	United States Department of Agriculture

4.1 Introduction

China[1] has quite a long history of sustainable agriculture, with over 4000 years of traditional agricultural practices like crop rotation, compost application with organic matter recycling, and some traditional ecological systems such as mulberry trees combined with fish ponds, which are helpful to maintain soil fertility and ecosystems.[2] The rural component of the Chinese always played an important role in the history of China, which had been emphasized by the Communist Party of China (CPC) during the Chinese Communist Revolution and later proved to be one of the keys to the success of the Communist Revolution. However, little attention was paid to Chinese agriculture after the CPC took helm of the country. It was not until 1978

[1] Unless otherwise stated, the legislation and policies on China discussed in this chapter refers to those on the mainland of the People's Republic of China, regardless of the legislation and policies on the Special Administrative Regions of Hong Kong and Macao.

[2] In his book 'Farmers of Forty Centuries', the American agronomist F. H. King describes many successful cases of traditional Chinese agricultural practices. See F. H. King, *Farmers of Forty Centuries: Organic Farming in China, Korea, and Japan* (1911, reprinted in 2004 by Dover Publications).

that the agriculture regained its place as one of the four top priorities for the Chinese government.[3] In the context of economic reform starting in the late 1970s, rural China changed dramatically under decollectivization and the new family responsibility system. Since then, Chinese agriculture has received considerable development, with unprecedented increase in the diversity and variety of food available. The most significant step for Chinese agriculture was the introduction of organic agriculture around 30 years ago. According to the Organic Food Development and Certification Center of China (OFDC), organic farming is defined as '[a] kind of plant and animal production system that rejects synthetic chemical substances such as pesticides, fertilizers, growth regulators, fodder additives, etc. and genetically modified organisms (GMOs) and products thereof, and instead, by following the law of nature and the principle of ecology, harmonizes crop cultivation with animal rearing and by adopting a series of sustainable-development-oriented agricultural techniques maintains a stable sustainable agriculture'.[4] Though the concepts of organic agriculture or organic farming were taken by Chinese agronomists from the United States (US), these concepts from the US themselves were inspired to a great extent by farming techniques found in Asia as described in King's book *Farmers of Forty Centuries* a century ago.[5] China's reconsideration on organic agriculture in the 1990s arose from people's concern on food safety and environmental issues. The first period from 1989 to 2005 can be called the organic agriculture with Chinese characteristics,[6] since international standards for organic food had not been adopted by the Chinese government during this period. In 1989, the so-called 'Green Food Program' – i.e., a certified voluntary quality system similar to organics and other low input production systems – was introduced by the Ministry of Agriculture (MOA) as a countermeasure to food safety and environmental issues.[7] Though the 'Green Food' was initially considered to be a replication of the western organic food standards, it actually applies the product standards rather than the process standards.[8] Compared to organic products which do not incorporate any chemicals, green food products are allowed to incorporate limited amounts of chemicals and fertilizers. 'Green Food' is thus deemed a 'half-way house' between conventional

[3] The top priority in China after 1978 is the modernization of agriculture, industry, national defence, and science and technology, which were set forth by Deng Xiaoping in the slogan of 'four modernizations'.

[4] Organic Food Development and Certification Center of China (OFDC), *OFDC Organic Certification Standards* (took effect on 1 March 2015) art.3.2.

[5] John Paull, 'The making of an agricultural classic: Farmers of Forty Centuries or Permanent Agriculture in China, Korea and Japan, 1911–2011' (2011) 2(3) *Agricultural Sciences* 175, 175–180.

[6] Riccardo Berti, 'Organic Food in China: The Law behind Lüse Shipin and Youji Shipin' (2015) 2(1) *GeoProgress Journal* 45, 46.

[7] USDA Foreign Agricultural Service, 'Organic Products and Agriculture in China' (GAIN Report No.CH6405, 21 June 2006). https://www.flexnews.com/files/china220606.pdf (last accessed on 29 June 2019).

[8] Ibid.

and organic agriculture.[9] In 2005, to seek more opportunities to export Chinese organic food abroad, the first national standards dealing with organics – the Chinese National Standards for Organic Produce – was developed by the Certification and Accreditation Administration of China. Since then, organic food and green food have been coexisting in the field of sustainable agriculture in China.

Currently, China is among the world's largest producers in organics. According to the 2019 International Federation Organic Agriculture Movement (IFOAM) report on the world's organic agriculture, three million hectares are dedicated to organic agriculture in China, making China the third largest country in the world in terms of areas of organic agricultural land.[10] With the rapid development of both domestic and international production and trade in organic food, China's organic agriculture has been booming since the turn of the Millennium. To instill transparency and trust among stakeholders, the Chinese government has issued a series of new regulations relating to food safety and organic certification. Organic agriculture ensured by these regulations is not only affected by climate change but also contributes to it, both adapting to changes and offering options for mitigation (i.e. reducing greenhouse gas emissions and storing carbon). This chapter aims to provide an overview of Chinese organic food law and investigate its impacts on climate change. Through investigating stakeholders involved in and regulations relating to organics, challenges to the enforcement of food safety law as well as organic certification issues are identified and analyzed. Particularly, the impacts of these regulations on climate change mitigation and adaptation are discussed as an outlook on the development of organic agriculture in China.

4.2 Overview of Chinese Organic Food Law

Up to now, there is no specific organic food law in China and the food regulatory regime in China remains fragmented. The introduction of organic agriculture in China resulted from people's increasing concern on food safety. The crisis of confidence on food safety also promoted the development of organic standards and certification in China. In this context, the starting point of the discussion on Chinese organic food law here is thus a brief introduction to China's food safety regulations as well as organic standards and certification.

---

[9] John Paull, 'The Greening of China's Food – Green Food, Organic Food, and Eco-labelling' (Sustainable Consumption and Alternative Agri-Food Systems Conference, Liege University, Arlon, Belgium, 27–30 May 2008) 11.

[10] In 2017, 69.8 million hectares of organic agricultural land, including in-conversion areas, were recorded. The countries with the largest areas of organic agricultural land are Australia (35.6 million hectares), Argentina (3.4 million hectares), and China (3 million hectares). See Helga Willer and Julia Lernoud (eds), *The World of Organic Agriculture: Statistics & Emerging Trends 2019* (FiBL & IFOAM – Organics International, 2019) 25.

When discussing the regulatory system of organic food in China, it is also necessary to simultaneously keep an eye on the relevant policies, since China is a country where policies achieve better implementation than the legal one. As the barometer of China's policy priorities, the 'No. 1 Central Document'[11] has focused on rural matters since 2003.[12] According to the No. 1 Central Document issued in January 2014, food security and rural environment was placed in the top policy priorities for 2014 in China.[13] Given that the political discourse often exerts influence on the Chinese legal system, the law might not be enforced in the same way as it is written in the legal text. Differences between what the law stipulates and how it works in practice are taken into account when discussing challenges to the enforcement of food safety law and organic certification issues.

4.2.1 *Regulations Relating to Food Safety and Organic Certification*

4.2.1.1 Evolution of Food Safety Regulations

The Chinese government has long been adopting the command-and-control method to regulate food safety through a series of national food safety standards. For example, after the infamous food safety scandal, the 'melamine milk' incident occurred in September 2008, the Ministry of Health (MOH) issued the *Good Manufacturing Practice for Powdered Formulae for Infants and Young Children, National Food Safety Standards, GB 23790-2010* to set the minimum safety and process standards in the field of powdered formulae for infants and young children.[14] In 2011, the MOH issued the *General Provisions on Pre-packaged Food Labeling, National Food Safety Standards, GB 7718-2011*, which is a mandatory information provision in the field of pre-packaged food labeling.[15] The *Guidelines for the Use of Food Additives, National Food Safety Standards, GB 2760-2014* issued by the MOH in

[11] The 'No. 1 Central Document' is the first formal document issued by the Central Committee of the CPC every January.

[12] Alberto Scorzon, Bernd van der Meulen and Li Jiao, 'Organics in Chinese Food Law' (2014) 9(3) *European Food and Feed Law Review* 179, 180.

[13] Reuters, 'Food security, rural environment top China's policy agenda in 2014' (19 January 2014) *Reuters*. https://www.reuters.com/article/us-china-agriculture/food-security-rural-environment-top-chinas-policy-agenda-in-2014-idUSBREA0I04920140119 (last accessed on 29 June 2019).

[14] MOH, *Good Manufacturing Practice for Powdered Formulae for Infants and Young Children, National Food Safety Standards, GB 23790-2010* [*GB 23790-2010* 食品安全国家标准 粉状婴幼儿配方食品良好生产规范] (issued on 26 March 2010, took effect on 1 December 2010) (in Chinese).

[15] MOH, *General Provisions on Pre-packaged Food Labelling, National Food Safety Standards, GB 7718-2011* [*GB 7718-2011* 食品安全国家标准 预包装食品标签通则] (issued on 20 April 2011, took effect on 20 April 2012) (in Chinese).

2014 provides conditions of use requirements for food additives.[16] These national food safety standards are convenient for the Chinese government to control some specific food safety issues, but are not comprehensive enough to cover all issues involved.

In addition to those national food safety standards, specific laws on food safety have been successively promulgated by the Chinese government since 1982.[17] These laws though took longer time to be drafted and formulated before being passed and implemented, are more comprehensive and effective than the fragmented national standards. The first law dedicated to food safety in China is the *Experimental Food Hygiene Law of the People's Republic of China* (hereinafter referred to as Experimental Law) promulgated by the Standing Committee of the National People's Congress on 19 November 1982 and took effect on 1 July 1983.[18] After 12 years' implementation, this Experimental Law was replaced by the *Food Hygiene Law of the People's Republic of China* (hereinafter referred to as FHL) in 1995.[19] Though the FHL was based on the amendments and updates of the Experimental Law, it remained a general and poor regulation, without for example unified food standards, stringent punishment and the standardization of food inspection agencies. One weakness of the FHL was its narrow scope which covered mainly the industrial production of food, regardless of the quality and safety management of agricultural product.[20] This gap was filled in 2006 when the Standing Committee of the National People's Congress passed the *Law on Agricultural Product Quality and Safety of the People's Republic of China* (hereinafter referred to as LAPQS).[21]

---

[16] MOH, *Guidelines for the Use of Food Additives, National Food Safety Standards, GB 2760-2014* [*GB 2760-2014* 食品安全国家标准 食品添加剂使用标准] (issued on 24 December 2014, took effect on 24 May 2015) (in Chinese).

[17] Before the promulgation of the *Experimental Food Hygiene Law of the People's Republic of China* in 1982, the Chinese government had only issued several administrative regulations on food safety, such as the *Regulations on Food Hygiene Administration for Trial Implementation of the People's Republic of China* promulgated by the State Council of the People's Republic of China on 17 August 1965 and the *Regulations for the Administration of Food Hygiene of the People's Republic of China* promulgated by the State Council of the People's Republic of China on 27 August 1979. See, *Regulations on Food Hygiene Administration for Trial Implementation of the People's Republic of China* [中华人民共和国食品卫生管理试行条例] (promulgated on 17 August 1965, abolished on 27 August 1979) (in Chinese); *Regulations for the Administration of Food Hygiene of the People's Republic of China* [中华人民共和国食品卫生管理条例] (promulgated on 27 August 1979, abolished on 1 July 1983) (in Chinese).

[18] *Experimental Food Hygiene Law of the People's Republic of China* [中华人民共和国食品卫生法(试行)] (promulgated on 19 November 1982, took effect on 1 July 1983, abolished on 30 October 1995) (in Chinese).

[19] *Food Hygiene Law of the People's Republic of China* [中华人民共和国食品卫生法] (promulgated on 30 October 1995, took effect on 30 October 1995, abolished on 1 June 2009) (in Chinese).

[20] Shan-shan Chung and Chris K C Wong, 'Regulatory and Policy Control on Food Safety in China' (2013) 67(6) *Journal of Epidemiology and Community Health* 476, 476.

[21] *Law on Agricultural Product Quality and Safety of the People's Republic of China* [中华人民共和国农产品质量安全法] (promulgated on 29 April 2006, took effect on 1 November 2006, amended on 26 October 2018) (in Chinese).

The LAPQS was formulated to ensure the quality and safety of agricultural products as well as public health (Article 1 LAPQS).[22] It covers issues regarding agricultural production management and related safety criteria such as the license to use pesticides, veterinary drugs, feed and feed additives, fertilizer, and veterinary instruments (Article 21 LAPQS),[23] the mandatory record keeping of inputs and processes in agricultural production (Article 24 LAPQS),[24] packaging and labeling in agricultural products (Chapter V LAPQS).[25] In terms of responsibilities and legal liabilities, the LAPQS sets provisions for different stakeholders involved such as the agricultural product testing institute (Article 44 LAPQS),[26] producers (Articles 46, 47 LAPQS)[27] and sellers (Articles 48, 49, 50, 54 LAPQS),[28] respectively.

The above mentioned laws, though have touched upon some issues relating to food safety, do not comprehensively cover all food safety issues. For example, food safety systems of risk assessment, unified food standards and food labeling management are beyond the scope of the FHL. Given the increasing food safety incidents in China, a basic law in the field of food safety is increasingly in demand. This demand peaked in the year 2008 when the 'melamine milk' incident came to light and attracted attentions both at home and abroad.[29] As a country among the top food exporting countries in the international market,[30] food safety issues in China are no longer domestic but also international issues. In response to food safety concerns while expressing its strong interest in accessing the international market, the Chinese government has continuously improved its food safety regulation through a series of legislative reforms and harmonization of standards for primary production and food safety issues.[31] Around 1 year after the exposure of the 'melamine milk' incident, a new food safety law, namely the *Food Safety Law of the People's Republic of China* (hereafter referred to as FSL) took effect on 1 June 2009, replacing the FHL.[32] The FSL covers a wide range of issues relating to food safety, including food production,

---

[22] LAPQS, art 1.

[23] LAPQS, art 21.

[24] LAPQS, art 24.

[25] LAPQS, arts 28–32.

[26] LAPQS, art 44.

[27] LAPQS, arts 46, 47.

[28] LAPQS, arts 48, 49, 50, 54.

[29] For example, the United Nations system launched a paper to facilitate China to improve its food safety within around 2 months after the exposure of the 'melamine milk' incident. See World Health Organization, 'UN Issues Paper on Food Safety in China' (2008). http://www.wpro.who.int/china/mediacentre/releases/2008/20081022/en/. (accessed on 30 June 2019).

[30] See Knoema, 'World Exports and Imports of Agricultural Products' (19 December 2018) *Knoema*. https://knoema.com/cduhihd/world-exports-and-imports-of-agricultural-products (last accessed on 30 June 2019).

[31] Scorzon et al., above n 12, 180–181.

[32] *Food Safety Law of the People's Republic of China* [中华人民共和国食品安全法] (promulgated on 28 February 2009, took effect on 1 June 2009, first amended on 24 April 2015, second amended on 29 December 2018) (in Chinese).

process and catering services, food additives, packaging materials, food storage and transportation as well as related safety management (Article 2 FSL).[33] In terms of safety issues regarding edible agricultural products, the quality and safety management of these products still applies the LAPQS, whereas the marketing of edible agricultural products, the standard-setting of quality safety, the disclosure of safety information as well as regulations on agricultural inputs are subject to the FSL (Article 2 FSL).[34] Realizing that the low level of governmental control and a set of poorly coordinated food safety standards were the main reasons of previous food safety violations in China,[35] the FSL created a new governmental department – the Food Safety Committee – exclusively dedicated to food safety oversight and enforcement of applicable standards and regulations (Article 5 FSL).[36] One of the breakthroughs of the FSL is that it has established an accountability system to which local government leaders shall be subjected in case a food safety scandal occurring in their prefecture (Chapter 7 FSL).[37] It also set the mandatory national food safety standards (Chapter 3 FSL)[38] and a surveillance and assessment system for food safety risks (Chapter 2 FSL).[39] Under such a food safety early-warning system, consumers could avoid human health damage before the occurrence of unintended 'human testing'.[40]

Since its implementation in 2009, the FSL has been amended and revised twice, with its most extensive revision in 2015. Compared to the 2009 FSL, the 2015 FSL greatly expanded its length and content, with the number of its provisions increased from 104 to 154 and the number of its words doubled from 15,000 to almost 30,000. The highlights of this amendment are manifested in the following six aspects. First, the supervision system provided in the 2015 FSL are clearer than that provided in the 2009 FSL. The 2009 FSL was silent on how the duty to monitor food production, circulation and servicing should be divided among the General Administration of Quality Supervision, Inspection and Quarantine (AQSIQ), the State Administration for Industry and Commerce (SAIC), and the State Food and Drug Administration (SFDA). To address the issue caused by the ambiguous division, the 2015 SFL clearly sets the division and cooperation among different authorities in Article 5.[41] Second, the 2015 FSL stresses the principle of prevention primarily while improving the food risk monitoring and assessment system. In terms of food risk monitoring

[33] FSL, art 2.

[34] Ibid.

[35] Lu Xiaojing, 'The Cause and Effect Analysis of the Melamine Incident in China' (2011) 5(3) *Asian Journal of Agricultural Research* 176, 185.

[36] FSL, art 5. The creation of the Food Safety Committee was initially set in Article 4 of the 2009 FSL and was amended to be Article 5 of the 2015 FSL.

[37] FSL, chapter 7.

[38] FSL, chapter 3.

[39] FSL, chapter 2.

[40] Hong-Gang Ni and Hui Zeng, 'Law enforcement is key to China's food safety' (2009) 157 *Environmental Pollution* 1990, 1990.

[41] FSL, art 5.

and assessment, the 2015 FSL clearly provides the assessment factors, situations requiring risk assessment, the impacts of risk assessment results as well as the circulation of risk assessment information between stakeholders involved (such as related administrative departments, local people's governments, food producers and traders, inspection and certification agencies, industry associations, consumer association, and news media). Third, the 2015 FSL expands the scope of food safety supervision, strengthening its supervision on food additives, food related products and online food trading. Particularly, it requires providers of the third-party online food trading platforms to register the legal names of food traders admitted to the platform and check the licenses of those that are required to obtain licenses (Article 62 FSL).[42] If the provider of a third-party online food trading platform fails to register the legal names of food traders admitted to the platform or examine their licenses, it will bear the joint and several liability with the food trader and compensate the consumer whose lawful rights and interests are harmed (Article 131 FSL).[43] Fourth, the 2015 FSL establishes the whole-process supervision system. To implement the farm-to-fork all-round and full-chain supervision, the whole-process food safety tracing system established by the 2015 FSL requires food producers and traders to establish a food safety tracing system to ensure the traceability of food (Article 42 FSL),[44] encourages food producers and traders to gather and retain production and trade information by digital means to establish a food safety tracing system (Article 42 FSL),[45] encourages the mass production of food and the chain food operation or distribution (Article 43 FSL),[46] encourages food producers and traders to participate in food safety liability insurance (Article 43 FSL).[47] Fifth, the 2015 FSL strengthens its supervision on special foods such as health foods (Articles 75–79 FSL),[48] infant formula foods (Articles 81–83 FSL)[49] and genetically modified food (GMF) (Article 69 FSL).[50] Sixth, the 2015 FSL strengthens its provisions on criminal, administrative and civil legal liabilities regarding food safety. For example, the maximum penalty has increased significantly, with 30 times of the goods value as the highest penalty (Article 123 FSL).[51] If a food producer or trader has cumulatively received three punishments other than being ordered to cease production or trade or revocation of its license in a year, the food and drug administrative department shall order it to cease production or trade, or even revoke its license

[42] FSL, art 62.

[43] FSL, art 131.

[44] FSL, art 42.

[45] Ibid.

[46] FSL, art 43.

[47] Ibid.

[48] FSL, arts 75–79.

[49] FSL, arts 81–83.

[50] FSL, art 69.

[51] FSL, art 123.

(Article 134 FSL).[52] For anyone who knowingly provides production or trade premises or other conditions for the illegal acts, the administrative penalty will be imposed by the food and drug administrative department of the people's government (Article 122 FSL).[53] In terms of the civil legal liability, the 2015 FSL establishes the so-called 'pay damages in advance' responsibility system for consumer compensation – that is, consumers who have suffered any loss from food failing to meet the food safety standards may claim damages against the trader or producer. The producer or trader, whichever is the first to receive such a claim, shall pay damages in advance, and shall not shuffle off the responsibility onto the other (Article 148 FSL).[54] The punitive compensation is also applied in the 2015 FSL, where a consumer may require a producer of food failing to meet the food safety standards or a trader knowingly dealing in such food to pay an indemnity of ten times the price paid or three times the loss (Article 148 FSL).[55] With these stringent provisions, the 2015 FSL is thus called 'the harshest-ever' food safety law in China.[56]

In 2018, the FSL was revised again during the Seventh Session of the Standing Committee of the 13th National People's Congress. Compared to last amendment, little revision was made this time. All revisions in the 2018 FSL are in accordance with the institutional reform of the State Council. For example, the former 'food and drug administrative department' has been changed to the 'food and safety administrative department', the former 'quality supervision department' has been deleted and changed to the 'food and safety administrative department', and the former 'environmental protection department' has been changed to the 'department of ecological environment'. The substance remains the same as the 2015 FSL.

4.2.1.2 Development of Organic Standards and Certification

The development of certified organic agriculture in China was built entirely on the concept, standards, organization, accreditation, monitoring, and trade developed in the Western countries,[57] since most certified organic products are export-oriented. The launch of organic production in China dates back to 1990 when organic tea from Lin'an County of the Zhejiang Province was exported for the first time with SKAL certification of the Netherlands.[58] Since then, more kinds of organic agricul-

---

[52] FSL, art 134.

[53] FSL, art 122.

[54] FSL, art 148.

[55] Ibid.

[56] Xinhuanet, 'Where is the "harsh" in "the harshest-ever" food safety law?' ["史上最严"食品安全法"严"在哪儿?] (25 April 2015) *Xinhuanet*. http://www.xinhuanet.com/politics/2015-04/25/c_127730790.htm (last accessed on 30 June 2019) (in Chinese).

[57] Yuhui Qiao, 'Organic Agriculture Development in China' in Helga Willer and Lukas Kilcher (eds) *The World of Organic Agriculture: Statistics and Emerging Trends 2011* (FiBL and IFOAM, 2011) 132, 132.

[58] International Fund for Agricultural Development (IFAD), *Organic Agriculture and Poverty Reduction in Asia: China and India Focus Thematic Evaluation* (IFAD Report No. 1664, 2005) 13.

tural products have been certified in China, and Chinese organic agriculture has been booming with the rapid development of international production and trade in organic foods.

Certified organic products have to be distinguished from agricultural products with the green food label, since they are certified under different standards. Standards of certified green food products (Green Food Standards) are specified by the MOA, which are different from the IFOAM Standard.[59] The original Green Food Standard, also known as 'Standard A', allows the use of a certain amount of synthetic agricultural chemicals. Green food products certified under such a standard shall not be deemed organic products but a transition between chemical and organic food. To meet the export demand of green food, a more stringent standard – the so-called 'Standard AA' – was introduced into the Green Food Standard in 1995. Green food products certified under the 'Standard AA' are substantially equivalent to the standards of organic food, since the 'Standard AA' excludes the use of synthetic pesticides and fertilizers and is harmonized with organic standards. Both 'Standard A' and 'Standard AA' are product standards rather than process standards, which focus on the end product rather than the process. Instead of monitoring the actual use of chemicals during the whole process of production, the Green Food Standard only tests the products themselves for chemical residues. Despite the difference between the Green Food Standard and the international organic standards, the bifurcation of Green Food Standard laid the foundation for the rapid articulation from Green Food certification to organic certification in China.

Though certified organic products have occurred since 1990, it was not until 2000 that food safety and eco-labeled products became the major themes in delegates' proposals at every session of the People's Congress and Chinese People's Political Consultative Conference.[60] Since then, a series of rules and regulations were introduced by the Chinese government to regulate the organic sector. In 2001, the former State Environmental Protection Agency (SEPA) successively issued the *Organic Food Certification and Management Measures*[61]and the standard *Technical Norm on Organic Food (HJ/T 80-2001)*,[62] and approved the OFDC which is among the first organic certification bodies in China. On 1 November 2003, the *Regulations of the People's Republic of China on Certification and Accreditation* (hereinafter referred to as the RCA) came into force.[63] The RCA laid down principal rules on certification and accreditation, including unified supervision and management sys-

[59] The IFOAM Standard is an internationally applicable organic standard that can be used directly for certification. For more information on the IFOAM Standard, see https://www.ifoam.bio/en/ifoam-standard (last accessed on 30 June 2019).

[60] Qiao, above n 57, 134.

[61] SEPA, *Organic Food Certification and Management Measures* [有机食品认证管理办法] (promulgated on 19 June 2001, took effect on 19 June 2001) (in Chinese).

[62] SEPA, *Technical Norm on Organic Food* [有机食品技术规范] (HJ/T 80-2001) (in Chinese).

[63] State Council, *Regulations of the People's Republic of China on Certification and Accreditation* [中华人民共和国认证认可条例] (promulgated on 3 September 2003, took effect on 1 November 2003, amended on 6 February 2016) (in Chinese).

tems on certification and accreditation, unified accreditation system, approval on establishment of certification bodies, voluntary and mandatory certification schemes, verification for qualifications of laboratories and inspection bodies, as well as supervision on certification bodies and training, consultant bodies. According to the rule of unified supervision and management systems on certification and accreditation (Article 4 RCA),[64] the management of the organic product certification and accreditation was handed over from the former SEPA to the Certification and Accreditation Administration of the People's Republic of China (CNCA)[65] in 2004. Local quality and technical supervision departments and entry-exit inspection and quarantine departments become the local organic product certification and supervision departments. To eliminate the inconsistency of different certification standards and requirements, the *Administrative Measures for Organic Product Certification* (AQSIQ Decree No. 67),[66] the *Chinese National Organic Product Standard* (CNOPS) (GB/T 19630–2005)[67] and the *Rules for Implementing the Certification of Organic Products* (CNCA-OG-001)[68] were successively issued in 2004 and 2005 by the AQSIQ and the CNCA, respectively. By 2005, mandatory organic standards and supervision systems had been introduced for organic certification bodies operating in China, and all organic products, including imported organic products, must thus comply with these national rules and standards. The CNOPS (GB/T 19630-2005) was set on the basis of learning from international and developed countries' organic product standards, taking into full consideration the principle of keeping in line with the international standards. Moreover, it was even stricter than the standards of major developed countries in many aspects. For example, the scope of the input allowed for use in the organic production in the CNOPS (GB/T 19630-2005) was even narrower than that of the EU, the US and Japan, which was extremely strict. The CNOPS (GB/T 19630–2005) also provided that the amount of the residue of pesticides in organic products may not be more than 5% of the residue of pesticides limits provided in GB2763.[69] However, 5% of residue of pesticides limits in most cases is nearing the detection limit of the instrument. Moreover, it was the first time that the management system was listed as one part of

---

[64] RCA, art 4.

[65] The CNCA was established by the State Council of the PRC in August 2001. Its primary mandate is to unify and streamline management of standards and certifications, which were previously managed by various departments before the implementation of the RCA.

[66] AQSIQ, *Administrative Measures for Organic Product Certification* (AQSIQ Decree No. 67) [《有机产品认证管理办法》(国家质量监督检验检疫总局令第67号)] (promulgated on 5 November 2004, took effect on 1 April 2005, abolished on 1 April 2014) (in Chinese).

[67] AQSIQ and China National Standardization Management Committee, *Chinese National Organic Product Standard* (CNOPS) (GB/T 19630-2005) [《有机产品》国家标准 GB/T 19630-2005] (promulgated on 1 April 2005, took effect on 1 April 2005, abolished on 1 March 2012) (in Chinese).

[68] CNCA, *Rules for Implementing the Certification of Organic Products* (CNCA-OG-001) [《有机产品认证实施规则》(CNCA-OG-001)] (promulgated on 1 June 2005, took effect on 1 June 2005, abolished on 1 March 2012) (in Chinese).

[69] CNOPS (GB/T 19630-2005).

Fig. 4.1 Certification mark of China organic product

the national standards all over the world, showing that the Chinese government attached great importance to standardize the production and processing system of organic product. In addition to using the Certification Mark of China Organic Product (Fig. 4.1), the product marked with 'Organic' in China might use the Certification Mark of China Organic Conversion Product (Fig. 4.2) as well.[70]

After 5 years' implementation, some rules needed to be modified, and in spring 2010, the CNCA commissioned experts to revise the standards.[71] The new standard on organic food – CNOPS (GB/T 19630-2011) – came into force on 1 March 2012,[72] which is stricter than the CNOPS (GB/T 19630-2005). The new standard adopts a zero-tolerance policy towards certain residues, and introduces stricter standards to reduce the cropping pressure, especially on rice. A numeric code on the label is also introduced to ensure the traceability of organic products. The *Rules on Implementation of Organic Products Certification* was revised at the same time.[73] On 1 April 2014, the *Administrative Measures for Organic Product Certification* (AQSIQ Decree No.

[70] CNOPS (GB/T 19630-2005), art 7.

[71] Qiao, above n 57, 132.

[72] AQSIQ and China National Standardization Management Committee, *Chinese National Organic Product Standard* (CNOPS) (GB/T 19630-2011) [《有机产品》国家标准 GB/T 19630–2011] (promulgated on 5 December 2011, took effect on 1 March 2012) (in Chinese).

[73] CNCA, *Rules for Implementing the Certification of Organic Products* (CNCA-N-009) [《有机产品认证实施规则》(CNCA-N-009)] (promulgated on 1 December 2011, took effect on 1 March 2012) (in Chinese).

Fig. 4.2 Certification mark of China organic conversion product

155) took effect,[74] replacing the former AQSIA Decree No. 67. Given the misuse of the 'Organic Conversion Product' label (Fig. 4.2) by some producers that could confuse the public as well as the already crowded scenario of the eco-labeling in China, the AQSIA Decree No. 155 abolished the Certification Mark of China Organic Conversion Product (Fig. 4.2). Since then, there is only one organic label – the Certification Mark of China Organic Product (Fig. 4.1) – for all categories of organic products in China.

4.2.2 Challenges to the Enforcement of Food Safety Law

With the promulgation and revision of the FSL as well as other related laws and regulations during recent years, China has established a comprehensive legal regime that provides a sound legal foundation and a favorable regulatory environment for ensuring food safety, enhancing food quality and controlling food imports and exports. The main reason leading to the still daunting food safety problems in China is not the lack of new 'basic food laws' but weaknesses involved in the law enforcement.

One of the challenges to the enforcement of food safety law in China relates to the multistage food safety management system. Vertically, the regulatory control of

[74]AQSIQ, *Administrative Measures for Organic Product Certification* (AQSIQ Decree No. 155) [《有机产品认证管理办法》(国家质量监督检验检疫总局令第155号)] (promulgated on 15 November 2013, took effect on 1 April 2014) (in Chinese).

food safety is a responsibility shared among national, provincial and local government authorities. Horizontally, China's food safety supervision system is a multi-department inspection framework, involving various government departments such as the State Food and Safety Administration (SFSA), the MOA, the MOH, the AQSIQ, the Ministry of Commerce (MOFCOM), the SAIC, the Ministry of Ecological Environment (MEE), and others. These departments are each responsible for different parts of food safety supervision according to relevant laws and regulations. Though the 2015 FSL has improved the ambiguous division on food safety supervision in 2009 FSL, it still has inherent flaws involved in the multistage management system. For one thing, the multistage food safety management system always creates 'blind spots' in food safety supervision as a result of the ambiguities of responsibility of various government departments.[75] The chains of accountability for food safety between different levels of government may also be not clear. For another, it is not surprising that different government departments often pass the buck due to the blurred lines of responsibility.[76] The blind spots during the food safety supervision and the buck-passing among different government departments caused by the multistage food safety management system have occurred in many food safety scandals such as the 'melamine milk' incident. The 2015 FSL has made its effort to clarify the chains of accountability for food safety between different levels of government and harmonize relevant government departments under the food safety supervision system. However, it is still impossible to avoid conflicts among these departments. Since various government departments participate in the supervision and monitoring of food safety, the target of an efficient coordination is still far.

Another challenge lies in the high cost of monitoring. Due to the large number of small production points involved in the food safety monitoring which are also more likely to produce substandard products, it is particularly expensive to thoroughly monitor the food product safety and quality.[77] One possible solution to this issue is increasing the available funds of food safety monitoring agents. However, the poor financial capacity of public agencies, particularly in China's western region, has often been considered the reason for regulatory and government failure.[78] Given the tight budget of many local food safety authorities in China, the types and quantities of food samples purchased for routine monitoring and testing are very likely to be insufficient. In particular, higher-priced food items tend to be exempted from the routine monitoring and testing. To reduce the cost of monitoring, China launched an Inspection Exemption Certification policy in 2001, according to which a food producer could be exempted from inspection if it enjoyed a large market share, continuously passed quality inspections three times, and implemented food safety

[75] Ni and Zeng, above n 40, 1991.

[76] Ibid.

[77] Chung and Wong, above n 20, 476.

[78] United Nations in China, 'Advancing Food Safety in China' (Occasional Paper, March 2008).

standards above national ones.[79] The purpose of this policy was to encourage food producers to voluntarily enhance production standards and perform well in formal inspections so that they would enjoy a good reputation with the 'inspection-free' label. Ironically, the first brand of the infamous 'melamine milk' – Sanlu milk power – was itself such an 'inspection-free' brand, showing the failure of the Inspection Exemption Certification policy. Within 1 week after the exposure of the 'melamine milk' incident, the AQSIQ abolished the Inspection Exemption Certification policy,[80] and no product can be exempted from inspection since then. The failure of the Inspection Exemption Certification policy shows that regular and routine inspections for all products are significant for effective law enforcement, which cannot be sacrificed to reduce the cost of monitoring. Another possible solution to address the high cost issue is establishing a power-sharing public supervision system, under which the general public, especially some social interest groups, is empowered to co-supervise food safety with the Chinese government.[81] Through disseminating food quality and safety information among the public, information asymmetries between food producers and consumers could be eliminated, and an early-warning system in food safety supervision is more likely to be established as well. To reduce the cost involved in food safety monitoring, it is also suggested to apply a risk-based approach in food safety legislation, where industry is responsible for proving compliance rather than regulators having to demonstrate non-compliance.[82] This approach, however, may ratchet up the regulatory burden on food producers, creating additional compliance costs. The government thus needs to pay attention to the trade-off between the traditional command-and-control approach and the risk-based approach.

4.2.3 Organic Certification Issues

Similar to the food safety management in China, the regulatory control of organic agriculture in China is a responsibility shared among different departments including the MOA, the MEE and the AQSIQ. The competent department in charge of the implementation of the organic product standard is the CNCA, which is also responsible for the authorization and supervision of Chinese organic certification bodies. In the absence of a basic organic food law which could specifically and clearly

[79] AQSIQ, *Administrative Methods for Product Inspection Exemption* (AQSIQ Decree No. 9) [《产品免于质量监督检查管理办法》(国家质量监督检验检疫总局令第9号)] (promulgated on 4 December 2001, took effect on 4 December 2001, abolished on 18 September 2008) (in Chinese).

[80] AQSIQ, *On the Decision to Abolish the 'Administrative Methods for Product Inspection Exemption'* (AQSIQ Decree No. 109) [《关于废止<产品免于质量监督检查管理办法>的决定》(国家质量监督检验检疫总局令第109号)] (promulgated on 18 September 2008, took effect on 18 September 2008) (in Chinese).

[81] Ni and Zeng, above n 40, 1991.

[82] United Nations in China, above n 78.

define the division of work among various departments responsible for organic food-related duties, responsibilities for organic agriculture and the implementation of the organic production standard are defined separately in each decree of the departments involved. As a result, it may encounter similar problems indicated in the food safety management. Departments involved may either compete against each other or pass the buck to other departments due to the overlapping or blurred lines of responsibilities.

In terms of monitoring organic certification, the quality of eco-labeling in China may be questioned by countries importing Chinese organic foods. Though an increasing number of domestic customers are purchasing organic foods nowadays, Chinese organic products have long been catering for the needs of overseas markets and are thus export-oriented. Food safety scandals involving organic food during the past years have eroded the trust of countries importing Chinese organic products.[83] For example, in July 2007, the US inspectors discovered that the ginger labeled as organic food sold in the Whole Foods Market (which is a supermarket chain that specializes in organic food grown in China) contained a dangerous pesticide.[84] This incident raised concerns about the reliability of Chinese organic products. Since foreigners importing Chinese organic products are not allowed to inspect Chinese farms themselves, they have to rely on Chinese certifiers. However, the Chinese Organic Food Certification Center (COFCC) under the MOA, the agency responsible for certifying all organics in China, inspects only 30% of organic products, leaving the rest being certified by third-party vendors accredited by the CNCA.[85] Due to the lack of the United States Department of Agriculture (USDA)-accredited certifiers, the USDA has to outsource to Chinese third parties who may neglect to comply with all regulations. Moreover, China does not recognize organic standards from most foreign countries except New Zealand.[86] The lack of equivalency in organic standards and certification between the US and China further leads to discrepancies on the understanding of chemicals and agricultural practices allowed in organics between China and the US. The weak enforcement of organic standards and corruption involved in rent-seeking would also contribute to abuses and illegal activities occurring in organic certification. The frequent occurrence of fraudulent

[83] Irene Luo, '5 Reasons You Shouldn't Trust 'Organic' From China' (26 July 2015) *The Epoch Times*. https://www.theepochtimes.com/5-reasons-you-shouldnt-trust-organic-from-china_1612468.html (last accessed on 30 June 2019).

[84] Nicholas Zamiska and David Kesmodel, 'Tainted Ginger's Long Trip From China to US Stores' (19 November 2007) *The Wall Street Journal*. https://www.wsj.com/articles/SB119543404007297464 (last accessed on 30 June 2019).

[85] Luo, see above n 83.

[86] The Mutual Recognition Arrangement for Certified Organic Products between China National Certification and Accreditation Administration and New Zealand Ministry of Primary Industry was signed on 14 November 2016, which is China's first international mutual recognition of organic products certification signed with foreign countries. See CNCA, 'China and New Zealand Ink an Arrangement on Mutual Recognition of Organic Product Certification' (14 November 2016) *Certification and Accreditation Administration of the People's Republic of China*. http://english.cnca.gov.cn/news/201612/t20161213_53011.shtml (last accessed on 8 August 2019).

Chinese organic certificates found in the US has destroyed the US government's trust in Chinese organic products.[87] Fraudulent labeling has become one of the main reasons for the distrust of Chinese organic products. If the Chinese government could not effectively address this issue, its export of organic products will be negatively influenced. It is suggested that a mutual recognition regime on organic standards can be widely established between China and more foreign countries. This mutual recognition regime could help promote not only export but also import of organic products in China.

4.3 Impacts of Chinese Organic Food Law on Climate Change

The impacts of climate change on food production and food security are often discussed, which is predicted by the Intergovernmental Panel on Climate Change (IPCC) that climate change will affect food security by 2050, with the largest numbers of food-insecure people located in South Asia.[88] However, the relationship between climate change and food production and food security is often a two-way street. Though food production and food security are significantly influenced by climate change, their developments in turn affect the trajectory of climate change. This section discusses the reversed effects of China's organic food law on climate change – i.e., how the regulations of Chinese organic food influence climate change adaptation and mitigation.

4.3.1 Impacts on Climate Change Adaptation

It is well-known that China is among the world's largest producers and consumers of many crops nowadays. Its agriculture has global impact on both global market and global environment. China's changes in its food production and consumption necessarily bring environmental effects that cause both domestic and international concerns.

The key staple crop in China is rice. According to an Asia-wide study, climate change scenarios would reduce rice yield over a large portion of the Asian continent, with eastern China as one of the most vulnerable regions.[89] One possible adap-

[87] As shown by the USDA, 9 out of 23 incidents of fraudulent organic certificates found in the US between February 2011 and June 2013 involved Chinese companies. See Luo, see above n 83.

[88] IPCC, *Climate Change 2014: Impacts, Adaptation, and Vulnerability* (Working Group II Contribution to the Fifth Assessment Report of the Intergovernmental Panel on Climate Change, Cambridge University Press, 2014) 490.

[89] Yuji Masutomi, Kiyoshi Takahashi, Hideo Harasawa and Yuzuru Matsuoka, 'Impact assessment of climate change on rice production in Asia in comprehensive consideration of process/parameter

tive response to climate change which is likely to have negative impacts on rice yield is making efforts to increase the production capability of rice. The increase of production capability, however, is always challenged by climate change where farmers have to face challenges from more unpredictable rainfall, soil degradation, and new or different pests and diseases. In addition to challenges from climate change, increasing deterioration of agricultural land quality resulted from the excessive use of pesticides and fertilizers in conventional agriculture would threaten the crop production capability as well. To help farmers adapt to climate change, they are encouraged to promote organic agriculture. With sustainable production systems, high soil organic matter content and soil cover could help prevent nutrient and water loss. Especially in the case that increasingly frequent extreme climate events are posing an increasing threat to food security in China, enhancing agricultural land quality through organic farming can help promote China's disaster prevention and mitigation capabilities, making soils more resilient to floods, droughts, and land degradation processes. Moreover, the impacts of crops irrigation on water resource are essential to climate change adaptation. Unlike conventional agriculture relying on pesticides and fertilizers that are harmful to water resource, organic agriculture could effectively prevent water pollution. In addition to maintaining soil and water quality, organic farming also attempts to preserve seed and crop diversity, which increases crop resistance to pests and disease. Maintaining seed and crop diversity also helps farmers develop new cropping systems to adapt to climate changes. Overall, organic agriculture enables farmers to minimize the risk and vulnerability of conventional agricultural production, leading to stable agro-ecosystems and yields as well as lower production costs.

As defined by the IPCC, climate change adaptation means 'adjustment in natural or human systems in response to actual or expected climate stimuli or their effects, which moderates harm or exploits beneficial opportunities'.[90] In short, adaptation for the human system is seeking benefits and avoiding harm, whereas adaptation for the natural system is moving towards the desired direction through human intervention. All interventions through actively developing organic agriculture to maintain soil and water quality as well as seed and crop diversity are effective adaptation strategies against climate change. The problem is whether China's current legal system on organic food could promote the development of organic agriculture in China. Challenges to the enforcement of food safety law and organic certification issues identified and discussed above may bring about negative impacts on people's trust and preference on organic food. If the public distrusts or dislikes organic food, the market share of organic food will decline, thereby negatively affecting the development of organic agriculture. This result may accelerate the deterioration and

uncertainty in general circulation models' (2009) 131(3–4) *Agriculture, Ecosystems & Environment* 281, 281–291.

[90] IPCC, *Climate Change 2001: Synthesis Report* (A Contribution of Working Groups I, II and III to the Third Assessment Report of the Intergovernmental Panel on Climate Change, Cambridge University Press, 2001) 398.

contamination of agricultural land and water resource qualities, ultimately being disadvantaged to climate change adaptation.

4.3.2 *Impacts on Climate Change Mitigation*

Climate change mitigation, as defined by the IPCC, refers to 'an anthropogenic intervention to reduce the sources or enhance the sinks of greenhouse gases'.[91] When discussing topics of 'climate change' and 'China' simultaneously, attention paid on agriculture is less than that on energy and manufacturing. Notwithstanding, agriculture's contribution to greenhouse gas (GHG) emissions is not negligible. Though agriculture contributes little to carbon dioxide (CO_2) emissions, it is the major source of non-CO_2 GHG (i.e., methane and nitrous oxide (N_2O)) emissions. According to a 2017 report provided by the PBL Netherlands Environmental Assessment Agency, agriculture accounts for 38% of China's methane emissions, with rice cultivation as the largest agricultural source of agricultural methane emissions.[92] Further, the primary sources of N_2O emissions are synthetic and natural fertilizers which are commonly used in agricultural soil management.[93] Excessive use of fertilizer not only significantly contributes to N_2O emissions but also increases the risk of water pollution. These negative impacts, however, could be mitigated through the promotion of organic agriculture.

Firstly, organic agriculture can contribute to combating climate change by mitigating GHG emissions, especially N_2O emissions. This is particularly important since N_2O is an important GHG, whose impact on warming the atmosphere is 298 times higher than that of CO_2.[94] There is a direct relation between N_2O emissions and the amount of nitrogen fertilizers applied to agricultural land.[95] Given that organic farming prohibits the use of synthetic nitrogen fertilizers, focusing instead on the establishment of closed nutrient cycles, minimizing nutrient losses via run-off, volatilization and emissions, nitrogen levels on organic farms tend to be lower per hectare than on conventional farms which can help mitigate climate change while delivering enough food.[96]

---

[91] Ibid.

[92] J.G.J. Olivier, K.M. Schure and J.A.H.W. Peters, 'Trends in Global CO_2 and Total Greenhouse Gas Emissions: 2017 Report' (PBL Netherlands Environmental Assessment Agency, December 2017) 24.

[93] Ibid.

[94] Diana Signor, Carlos Eduardo and Pellegrino Cerri, 'Nitrous Oxide Emissions in Agricultural Soils: A Review' (2013) 43(3) *Pesquisa Agropecuária Tropical* 322, 334.

[95] Ibid.

[96] Organic Without Boundaries, 'How Organic Agriculture Helps Mitigate Climate Change' (12 September 2018) *Organic Without Boundaries*. https://www.organicwithoutboundaries.bio/2018/09/12/climate-change-mitigation (last accessed on 14 July 2019).

Secondly, organic agriculture can help mitigate climate change through storing carbon in the soil. Many soil management practices used by organic agriculture – such as minimum tillage, returning crop residues to the soil, the use of cover crops and rotations, and the greater integration of nitrogen fixing legumes – can help store carbon in the soil while increasing productivity.[97] Through returning more carbon to the soil in the process of organic agricultural production, less carbon is emitted into the atmosphere.

Third, compared to conventional agriculture that requires significant amounts of energy to manufacture synthetic fertilizers and pesticides, organic agriculture can minimize energy consumption by 30–70% per unit of land by eliminating the energy required to manufacture synthetic fertilizers and pesticides, and by using internal farm inputs, thus reducing fuel used for transportation.[98] As a result, GHG emissions resulted from agriculture-related energy consumption can be significantly reduced.

Given the fact that organic agriculture is an effective strategy to mitigate GHG emissions and sequester carbon, regulations promoting organic food can help Chinese government achieve its emissions reduction target and fulfill its commitment in the Paris Agreement. Weaknesses involved in the enforcement of these regulations may, however, undermine China's effort on climate change mitigation.

4.4 Conclusion

Since China's reconsideration on organic agriculture in 1990s, there has been a gradual establishment of an organics regulatory framework by the Chinese Government. Though there is no specific law on organic food, regulations relating to food safety and organic certification have been issued and implemented to promote organic agriculture during the past three decades. The key problems related to food safety issues before 2009 were the low level of governmental control and a set of poorly coordinated food safety standards, both leading to the frequent occurrence of food safety incidents in China. Realizing the seriousness of these problems, the Chinese government issued the FSL in 2009 and made significant amendments on this law in 2015. Meanwhile, a series of national standards on organic products have been issued and modified since China launched its organic production in 1990. Despite China's efforts on regulating organic food during the past 30 years, challenges still remain in the enforcement of these regulations due to the multistage management system and high costs of monitoring. The lack of universal mutual recognition regime on organic standards has further hindered international trades on organic products between China and most countries. Given that organic agriculture

[97] Ibid.

[98] IFOAM, 'Organic Agriculture Countering Climate Change' (IFOAM Factsheet, 2018). https://www.ifoam.bio/sites/default/files/ifoam_climate_change_eng_web.pdf (last accessed on 14 July 2019).

can help farmers to adapt to changing climate conditions as well as contribute to climate change mitigation, weaknesses identified in the enforcement of organic food-related regulations may have negative impacts on climate change. Improvements on these regulations are therefore of great importance to the promotion of organic agriculture as well as climate change adaptation and mitigation.

References

AQSIQ (2008a) Administrative Methods for Product Inspection Exemption (AQSIQ Decree No. 9) [《产品免于质量监督检查管理办法》(国家质量监督检验检疫总局令第9号)] (promulgated on 4 December 2001, took effect on 4 December 2001, abolished on 18 September 2008) (in Chinese)

AQSIQ (2008b) On the decision to abolish the 'Administrative methods for product inspection exemption' (AQSIQ Decree No. 109) [《关于废止<产品免于质量监督检查管理办法>的决定》(国家质量监督检验检疫总局令第109号)] (promulgated on 18 September 2008, took effect on 18 September 2008) (in Chinese)

AQSIQ (2014a) Administrative Measures for Organic Product Certification (AQSIQ Decree No. 67) [《有机产品认证管理办法》(国家质量监督检验检疫总局令第67号)] (promulgated on 5 November 2004, took effect on 1 April 2005, abolished on 1 April 2014) (in Chinese)

AQSIQ (2014b) Administrative Measures for Organic Product Certification (AQSIQ Decree No. 155) [《有机产品认证管理办法》(国家质量监督检验检疫总局令第155号)] (promulgated on 15 November 2013, took effect on 1 April 2014) (in Chinese)

AQSIQ and China National Standardization Management Committee (2012a) Chinese National Organic Product Standard (CNOPS) (GB/T 19630-2005) [《有机产品》国家标准 GB/T 19630-2005] (promulgated on 1 April 2005, took effect on 1 April 2005, abolished on 1 March 2012) (in Chinese)

AQSIQ and China National Standardization Management Committee (2012b) Chinese National Organic Product Standard (CNOPS) (GB/T 19630–2011) [《有机产品》国家标准 GB/T 19630-2011] (promulgated on 5 December 2011, took effect on 1 March 2012) (in Chinese)

Berti R (2015) Organic food in China: the law behind Lüse Shipin and Youji Shipin. GeoProgress J 2(1):45–57

Chung S-s, Wong CKC (2013) Regulatory and policy control on food safety in China. J Epidemiol Community Health 67(6):476–477

CNCA (2012a) Rules for implementing the certification of organic products (CNCA-OG-001) [《有机产品认证实施规则》(CNCA-OG-001)] (promulgated on 1 June 2005, took effect on 1 June 2005, abolished on 1 March 2012) (in Chinese)

CNCA (2012b) Rules for implementing the certification of organic products (CNCA-N-009) [《有机产品认证实施规则》(CNCA-N-009)] (promulgated on 1 December 2011, took effect on 1 March 2012) (in Chinese)

CNCA (2016) China and New Zealand Ink an arrangement on mutual recognition of organic product certification (14 November 2016). Certification and Accreditation Administration of the People's Republic of China. http://english.cnca.gov.cn/news/201612/t20161213_53011.shtml

Experimental Food Hygiene Law of the People's Republic of China [中华人民共和国食品卫生法(试行)] (promulgated on 19 November 1982, took effect on 1 July 1983, abolished on 30 October 1995) (in Chinese)

Food Hygiene Law of the People's Republic of China [中华人民共和国食品卫生法] (promulgated on 30 October 1995, took effect on 30 October 1995, abolished on 1 June 2009) (in Chinese)

Food Safety Law of the People's Republic of China [中华人民共和国食品安全法] (promulgated on 28 February 2009, took effect on 1 June 2009, first amended on 24 April 2015, second amended on 29 December 2018) (in Chinese)
IFAD (2005) Organic agriculture and poverty reduction in Asia: China and India focus thematic evaluation (IFAD Report No. 1664, 2005)
IFOAM (2018) Organic agriculture countering climate change (IFOAM Factsheet, 2018). https://www.ifoam.bio/sites/default/files/ifoam_climate_change_eng_web.pdf
IFOAM Standard. https://www.ifoam.bio/en/ifoam-standard
IPCC (2001) Climate change 2001: synthesis report, A contribution of working groups I, II and III to the third assessment report of the intergovernmental panel on climate change. Cambridge University Press, Cambridge
IPCC (2014) Climate change 2014: impacts, adaptation, and vulnerability, Working group II contribution to the fifth assessment report of the intergovernmental panel on climate change. Cambridge University Press, Cambridge
King FH (2004) Farmers of forty centuries: organic farming in China, Korea, and Japan (1911, reprinted in 2004 by Dover Publications)
Knoema (2018) World exports and imports of agricultural products (19 December 2018). Knoema. https://knoema.com/cduhihd/world-exports-and-imports-of-agricultural-products
Law on Agricultural Product Quality and Safety of the People's Republic of China [中华人民共和国农产品质量安全法] (promulgated on 29 April 2006, took effect on 1 November 2006, amended on 26 October 2018) (in Chinese)
Luo I (2015) 5 reasons you shouldn't trust 'Organic' from China (26 July 2015). The Epoch Times. https://www.theepochtimes.com/5-reasons-you-shouldnt-trust-organic-from-china_1612468.html
Masutomi Y, Takahashi K, Harasawa H, Matsuoka Y (2009) Impact assessment of climate change on rice production in Asia in comprehensive consideration of process/parameter uncertainty in general circulation models. Agric Ecosyst Environ 131(3–4):281–291
MOH (2010) Good manufacturing practice for powdered formulae for infants and young children, National Food Safety Standards, GB 23790-2010 [GB 23790-2010 食品安全国家标准 粉状婴幼儿配方食品良好生产规范] (issued on 26 March 2010, took effect on 1 December 2010) (in Chinese)
MOH (2012) General provisions on pre-packaged food labelling, National Food Safety Standards, GB 7718-2011 [GB 7718-2011 食品安全国家标准 预包装食品标签通则] (issued on 20 April 2011, took effect on 20 April 2012) (in Chinese)
MOH (2015) Guidelines for the use of food additives, National Food Safety Standards, GB 2760-2014 [GB 2760-2014 食品安全国家标准 食品添加剂使用标准] (issued on 24 December 2014, took effect on 24 May 2015) (in Chinese)
Ni H-G, Zeng H (2009) Law enforcement is key to China's food safety. Environ Pollut 157:1990–1992
OFDC (2015) OFDC organic certification standards (took effect on 1 March 2015)
Olivier JGJ, Schure KM, Peters JAHW (2017) Trends in global CO2 and total greenhouse gas emissions: 2017 report (PBL Netherlands Environmental Assessment Agency, December 2017)
Organic Without Boundaries (2018) How organic agriculture helps mitigate climate change (12 September 2018). Organic Without Boundaries. https://www.organicwithoutboundaries.bio/2018/09/12/climate-change-mitigation
Paull J (2008) The greening of China's food – green food, organic food, and Eco-labelling. (Sustainable consumption and alternative agri-food systems conference, Liege University, Arlon, Belgium, 27–30 May 2008)
Paull J (2011) The making of an agricultural classic: farmers of forty centuries or permanent agriculture in China, Korea and Japan, 1911-2011. Agric Sci 2(3):175–180
Qiao Y (2011) Organic agriculture development in China' in Helga Willer and Lukas Kilcher. In: The world of organic agriculture: statistics and emerging trends 2011. FiBL and IFOAM, Frick, pp 132–136

Regulations for the Administration of Food Hygiene of the People's Republic of China [中华人民共和国食品卫生管理条例] (promulgated on 27 August 1979, abolished on 1 July 1983) (in Chinese)
Regulations on Food Hygiene Administration for Trial Implementation of the People's Republic of China [中华人民共和国食品卫生管理试行条例] (promulgated on 17 August 1965, abolished on 27 August 1979) (in Chinese)
Reuters (2014) Food security, rural environment top China's policy agenda in 2014 (19 January 2014). Reuters. https://www.reuters.com/article/us-china-agriculture/food-security-rural-environment-top-chinas-policy-agenda-in-2014-idUSBREA0I04920140119
Scorzon A, van der Meulen B, Jiao L (2014) Organics in Chinese food law. Eur Food Feed Law Rev 9(3):179–186
SEPA (2001a) Organic food certification and management measures [有机食品认证管理办法] (promulgated on 19 June 2001, took effect on 19 June 2001) (in Chinese)
SEPA (2001b) Technical norm on organic food [有机食品技术规范] (HJ/T 80-2001) (in Chinese)
Signor D, Eduardo C, Cerri P (2013) Nitrous oxide emissions in agricultural soils: a review. Pesqui Agropecu Trop 43(3):322–338
State Council (2016) Regulations of the People's Republic of China on certification and accreditation. [中华人民共和国认证认可条例] (promulgated on 3 September 2003, took effect on 1 November 2003, amended on 6 February 2016) (in Chinese)
United Nations in China (2008) Advancing food safety in China (Occasional paper, March 2008)
USDA Foreign Agricultural Service (2006) Organic products and agriculture in China (GAIN Report No.CH6405, 21 June 2006). https://www.flexnews.com/files/china220606.pdf
Willer H, Lernoud J (eds) (2019) The world of organic agriculture: statistics & emerging trends 2019. FiBL & IFOAM – Organics International, Frick
World Health Organization (2008) UN issues paper on food safety in China. http://www.wpro.who.int/china/mediacentre/releases/2008/20081022/en
Xiaojing L (2011) The cause and effect analysis of the melamine incident in China. Asian J Agric Res 5(3):176–185
Xinhuanet (2015) Where is the "harsh" in "the harshest-ever" food safety law? ["史上最严"食品安全法"严"在哪儿?] (25 April 2015) Xinhuanet. http://www.xinhuanet.com/politics/2015-04/25/c_127730790.htm (in Chinese)
Zamiska N, Kesmodel D (2007) Tainted Ginger's long trip from China to US stores (19 November 2007). The Wall Street Journal. https://www.wsj.com/articles/SB119543404007297464

Ying Shen is an early career researcher with multijurisdictional educational qualifications, having earned her LLB, LLM and PhD from China, Germany and Australia, respectively. Her expertise lies in the field of environmental law, with a special focus on climate change issues. She was an Endeavour Research Fellow and the first Australia-APEC Women in Research Fellow representing China in 2015 and has since held different research positions in several universities and institutions in both Australia and Germany. She was awarded the German Chancellor Fellowship by the Alexander von Humboldt Foundation in 2019. She is the Author of *China's Way to Carbon Emissions Reduction: The Choice of Regulatory Instruments and Its Legal Challenges*, a monograph published by Wolters Kluwer in 2015.

Part II
Histories and Influences on Standards

Chapter 5
Evolution of the Organic Japanese Agricultural Standard System: A 20-Year History

Satoshi Kodera

Abstract This chapter aims to evaluate the development of the organic Japanese Agricultural Standard (JAS) system and to examine the influence of the system on organic food safety in Japan. It takes an overview of the 20-year history of organic JAS system in Japan and examines the content of the current system. It also discusses the several challenges faced by the organic JAS system in Japan.

Keywords Organic JAS · JAS laws · Food safely · Food labelling

5.1 Introduction

This chapter aims to evaluate the development of the organic Japanese Agricultural Standard (JAS) system and to examine the influence of the system on organic food safety in Japan. The organic JAS system is based on the JAS Law and allows operators approved by registered certifying bodies to attach the organic JAS logo to products that were produced or manufactured in accordance with relevant organic JAS standards. The system was introduced in 1999 when the Codex Alimentarius Commission adopted the Guidelines for the Production, Processing, Labelling and Marketing of Organically Produced Foods.

In 1999, Japan revised the Law Concerning Standardization and Proper Labeling of Agricultural and Forestry Products and established organic JAS standards. Under the new system, only operators approved by registered certifying bodies are permitted to attach the organic JAS logo to products that were produced or manufactured in accordance with relevant organic JAS standards. It has been approximately

I am grateful to N. Yukimaru for her support and encouragement. This research was supported by JSPS KAKENHI Grant Number JP19K21680.

S. Kodera (✉)
Seinan Gakuin University, Fukuoka, Japan
e-mail: kodera@seinan-gu.ac.jp

GOH Bee Chen, R. Price (eds.), *Regulatory Issues in Organic Food Safety in the Asia Pacific*, https://doi.org/10.1007/978-981-15-3580-2_5

20 years since the introduction of the organic JAS system. The aim of this chapter is to examine how the system has evolved and to clarify the role that it has played in the development of organic food safety in Japan and the challenges it faces today.

The chapter begins by looking at the history of the organic JAS system in Japan. It explains the background of how and why this system was introduced. It then takes an overview of the path of the organic JAS system. The second section of the chapter examines the content of the current organic JAS system. The system sets the standards for four kinds of products: organic plants, organic processed foods, organic feeds, and organic livestock. The chapter takes an overview of the organic JAS standards, the certification methods, and the mutual recognition among countries related to the import and export of these four kinds of products. The third section examines the challenges that the current organic JAS system faces. This section specifically aims at clarifying the criticism directed at the organic JAS system.

5.1.1 History of the Organic JAS System in Japan

5.1.1.1 Organic Farming in Japan

Some farming practices that did not use chemical materials could be found in Japan before World War II (Nishio 2019: 61–65). However, as Honjoh (2004: 40–48) summarizes, organic farming in Japan largely started as part of the relationship between organic farmers and consumers. Around the 1960s, people became skeptical about food safety due to various pollution-related diseases resulting from the high growth of the Japanese economy. Then, in the 1970s, consumers began to buy organic agricultural products directly from organic farmers in the form of group purchasing. Organic farming thus started gradually to spread throughout Japan through this "partnership (*teikei*)" between farmers and consumers (Honjoh 2004: 40–48).

At that time, there were no significant trends toward setting up standards for organic farming products, because the distribution of organic farming products was limited to consumers and their partner farmers. Customers purchased organic farming products through a face-to-face relationship with organic farmers and hence there was no need specifically to establish any standards for organic produce.

However, after the 1980s, an interest in organic farming products grew all around Japan. Organic farming products started to be circulated through general distribution. Especially after the 1990s, organic farming products became "commercialized" because the restaurant industry began focusing on these products. The demand for organic farming products rapidly increased and an "organic bubble" occurred in Japan (Saito 2015: 201–203).

5.1.1.2 Standardization of Organic Produce

As a result of the "organic bubble," organic produce started to be represented in a misleading way, which became a serious problem. Once organic produce was disseminated through general distribution, the definition of organic produce and the distinction between organic and nonorganic produce became a significant issue that needed to be addressed. In September 1988, the Japan Fair Trade Commission accused producers of agricultural products of providing a misleading representation of their products by designating them as "chemical free" and produced through "fully organic cultivation" (Honjoh 2004: 74). As a result of this charge, the public started to pay more attention to the setting up of standards for organic products. The "organic bubble" that occurred in the 1990s led to a demand for organic products. Now it was necessary to establish standards for these products.

The movement to create such standards was seen both within and outside of Japan. In the United States, the Organic Foods Production Act came into effect in 1990. The European Union (EU) also set up standards based on the law, such as instituting a rule regarding organic produce in 1991 (Compagnoni 2010). The Codex Alimentarius Commission, too, started its review of guidelines regarding organic produce in 1991.

In Japan, the Ministry of Agriculture, Forestry and Fisheries (MAFF) started to devise guidelines on organic farm products. In the beginning, the guidelines had no force of law. The MAFF enacted the Guidelines for Organic Agricultural Products and Specially Cultivated Agricultural Products in October 1992. These guidelines underwent several revisions and led to revision of the JAS Law in July 1999.

The JAS Law was originally enacted in 1950. The purpose of the JAS Law is "to improve quality, increase productivity, improve the fairness and simplicity of transactions in, and rationalize the use or consumption of agricultural and forestry products other than foods and drinks by enacting and disseminating appropriate and rational standards for products, and to help consumers choose products by enforcing proper quality labeling of them, in conjunction with measures by the Food Labeling Act (Act No. 70 of 2013), thereby contributing to their smooth production and distribution, promotion of agricultural production based on consumer demand, and protection of the interests of consumers" (Article 1 of the JAS Law). Under the JAS Law, standards related to the quality and the labeling of agricultural and forestry products were established.[1]

Originally, the JAS Law did not include standards for specific processes and production methods (PPMs). Therefore, in 1993, the MAFF revised the JAS Law to establish a new system covering PPMs and product labeling. In 1999, the MAFF

---

[1] In April 2015, the Food Labeling Act went into force (Act No. 70 of June 28, 2013). This act integrated the labeling system in Japan. As a result, provisions regarding food labeling in the JAS Law were moved into the Food Labeling Act. Accordingly, the name of the JAS Law was changed from the Law Concerning Standardization and Proper Labeling of Agricultural and Forestry Products to the Law Concerning Standardization, and so forth of Agricultural and Forestry Products.

submitted a bill to the Diet to revise the JAS Law in order to include organic agricultural products and so forth, and it came into effect in July of that year. In January 2000, the JAS for Organic Plants (Notification No. 59 of the Ministry of Agriculture, Forestry and Fisheries of January 20, 2000) and the JAS for Organic Processed Foods (Notification No. 60 of the Ministry of Agriculture, Forestry and Fisheries of January 20, 2000) were established, based on the revised JAS Law. After the Codex Alimentarius Commission added provisions related to organic livestock in 2001, two new standards were enacted in 2005: the JAS for Organic Feeds (Notification No. 1607 of the Ministry of Agriculture, Forestry and Fisheries of October 27, 2005) and the JAS for Organic Livestock Products (Notification No. 1608 of the Ministry of Agriculture, Forestry and Fisheries of October 27, 2005) (Honjoh 2004: 74–86).

5.1.2 Current Organic JAS System in Japan

5.1.2.1 Overview of the Organic JAS System

The current organic JAS system allows certified operators to attach "organic JAS logos" to their products after a third-party organization approves them based on an inspection verifying that the production conforms to JAS standards. As mentioned above, there are organic JAS standards for organic plants, organic processed foods, organic feeds, and organic livestock products.

In order for farms and manufacturers to be able to use the organic JAS logo on their products (see Fig. 5.1), they need to be approved as certified operators by a registered certifying body that has been accredited by the MAFF. As of February 28, 2020, there were 55 registered certifying bodies in the country and 15 outside the country (MAFF 2020). Regarding operators, as of March 31, 2019, there were 7780 certified operators (MAFF 2019).

5.1.2.2 Certification Process

Operators must submit an application to the registered certifying body to be certified, which then conducts assessments based on technical criteria for certifying and provides certification. For certification, document reviews and on-site assessments are conducted to confirm that.

Fig. 5.1 Organic JAS logo

(a) fields or processing plants meet organic production criteria, that is, organic JAS standards; and
(b) production management and documentation of production management are conducted appropriately so that production will conform with said standards (MAFF 2015).

The registered certifying body investigates certified production farms or manufacturers to determine whether their products are produced according to organic JAS standards. This investigation is conducted at least once a year (see Fig. 5.2).

5.1.2.3 Organic JAS Standards[2]

5.1.2.3.1 Organic Plants

The JAS for Organic Plants provide principles of organic production, criteria for production methods, and a system of labeling. First, regarding principles of organic production, Article 2 of the JAS for Organic Plants stipulates the following: "To sustain and enhance natural recycling in agriculture, the productivity of the

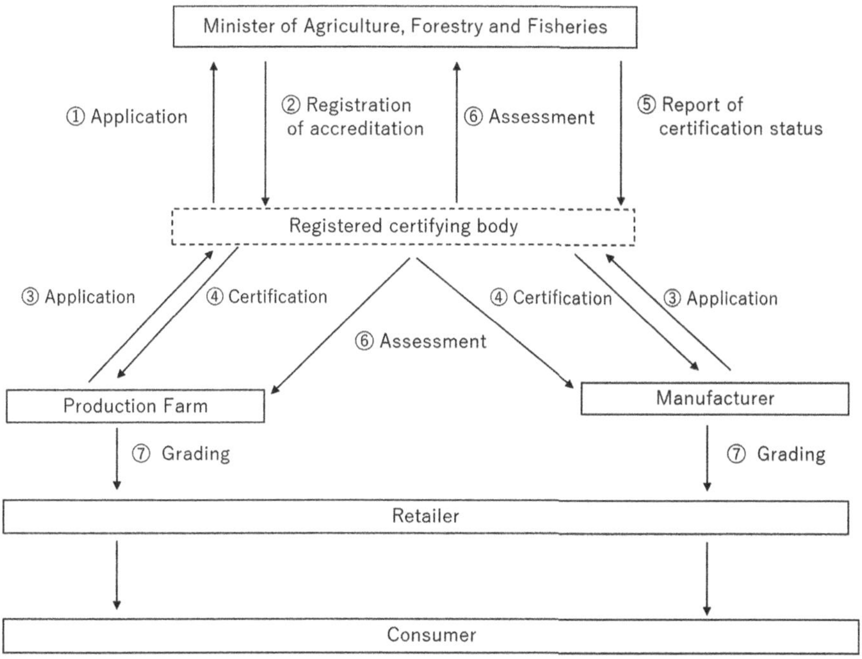

Fig. 5.2 Organization of inspection certification system. (Reproduced from MAFF 2015)

[2] For a provisional translation of the organic JAS standards explained in this chapter, see http://www.maff.go.jp/e/jas/specific/criteria_o.html

farmland derived from the soil properties shall be generated by avoiding the usage of chemical synthetic fertilizer and agricultural chemicals, and organic agricultural products shall be produced in fields adopting such cultivation management methods as reducing the load derived from the agricultural production on the environment as much as possible."

Second, Article 4 specifies criteria for production methods with regard to:

1. fields, collection area;
2. seeds or seedlings to be used in fields;
3. manuring practices in fields;
4. cultivation sites, fungus spawn, and cultivation management in cultivation sites;
5. control of noxious animals and plants in fields or cultivation sites;
6. general management, and management for raising seedlings; and
7. management concerning harvest, transportation, selection, processing, cleaning, storage, packaging, and other postharvest processes.

The system requires that farmers avoid the direct usage of "prohibited substances" such as chemical synthetic fertilizers and agricultural chemicals as well as take the necessary action to prevent such "prohibited substances" from being transported into their fields. Moreover, "soil fertility shall be maintained and enhanced by compost, etc. and over a period of at least 2 years before the sowing or planting (at least 3 years before the first harvesting of perennial plants," "synthetic fertilizers and agricultural chemicals shall not be used, in principle," and "seeds or seedlings shall not be produced by recombinant DNA technology" (Article 3 of the JAS for Organic Plants).

Article 5 states that the names of organic plants shall be labeled according to the following methods if they are in accordance with food labeling standards (Cabinet Office Ordinance No. 20 in 2015). Texts with their English translation are shown in parentheses, and the general name of the organic plant shall be filled out in the "○○":

1. "有機農産物" (organic plant)
2. "有機栽培農産物" (organically grown plant)
3. "有機農産物○○" or "○○(有機農産物)" (organic plant ○○ or ○○ (organic plant).)
4. "有機栽培農産物○○" or "○○(有機栽培農産物)" (organically grown plant ○○ or ○○ (organically grown plant))
5. "有機栽培○○" or "○○(有機栽培)" (organic farming ○○ or ○○ (organic farming)).
6. "有機○○" or "○○(有機)" (organic ○○ or ○○(organic))
7. "オーガニック○○" or "○○(オーガニック)" (organic ○○ or ○○(organic))

5.1.2.3.2 Organic Processed Foods

Regarding principles for organic processed foods, Article 2 of the JAS for Organic Processed Foods stipulates the following: "organic processed foods [shall be produced] by processing methods utilizing physical and biological functions, avoiding the use of chemically synthesized additives and chemical agents, and by preserving the characteristics of organic plants and organic livestock products as ingredients in the manufacturing and processing process".

Specifically, according to Article 4, the weight of nonorganic products and food additives shall be no more than 5% of the weight of all ingredients of organic processed foods, except for salt, water, and processing aids. Also, recombinant DNA technology shall not be used. Organic processed foods shall be produced or processed in facilities (MAFF 2015: 4).

Article 5 states that the names of organic livestock shall be labeled according to the following methods, if they are in accordance with food labeling standards. Texts with their English translation are shown in parentheses, and the general name of the organic processed foods shall be filled out in the "○○":

1. (1)"有機○○" or "○○(有機)" (organic ○○ or ○○(organic))
2. "オーガニック○○" or "○○(オーガニック)" (organic ○○ or ○○ (organic))

5.1.2.3.3 Organic Livestock Products

Concerning principles for organic livestock products, Article 2 of the JAS for Organic Livestock Products stipulates the following: "organic livestock products are, in principle, manufactured from organic livestock raised in deference to their physiological and behavioral needs, by providing feeds produced through reducing pressures on the environment as much as possible, and avoiding the use of veterinary drugs, to sustain and enhance the natural cyclical function of agriculture".

Article 4 establishes criteria for production methods for organic livestock products. According to these criteria, organic feeds shall be mainly provided. Also, livestock shall be raised by minimizing stress through pasturing in the open air and so on. Moreover, antibiotics and so on shall not be used, in order to prevent disease. Finally, recombinant DNA technology shall not be used (MAFF 2015: 4).

Article 5 states that the names of organic livestock products shall be labeled through the following methods if they are in accordance with food labeling standards. Texts with their English translation are shown in parentheses, and the general name of the livestock products shall be filled out in the"○○":

1. "有機畜産物" (organic livestock product);
2. "有機畜産物○○" or "○○ (有機畜産物)" (organic livestock product ○○ or ○○ (organic livestock product));

3. "有機畜産○○" or "○○ (有機畜産)" (organic livestock ○○ or ○○ (organic livestock));
4. "有機○○" or "○○ (有機)" (organic ○○ or ○○ (organic)); or
5. "オーガニック○○" or "○○ (オーガニック)" (organic ○○ or ○○ (organic))

5.1.2.3.4 Organic Feeds

Regarding the principles for organic feeds, Article 2 of the JAS for Organic Feeds stipulates the following: "organic feeds shall be produced to ensure that the characteristics of organic agricultural products, organic processed foods, organic livestock products, and physical and biological functions are applied as processing methods without the use of feed additives and substances synthesized chemically".

Specifically, according to Article 4, the weight of nonorganic ingredients shall be no more than 5% of all ingredients. The use of feed additives shall be within the minimum amount necessary. Recombinant DNA technology shall not be used. Organic agricultural products, organic processed foods, organic milk, and organic feeds used for ingredients shall be controlled and not mixed with other agricultural and livestock products or processed foods.

Article 5 states that the names of organic feeds shall be labeled according to the following methods, if they are in accordance with food labeling standards. Texts with their English translation are shown in parentheses, and the general name of the livestock and so forth shall be filled out in the "○○":

1. "有機飼料" or "オーガニック飼料" (organic feed);
2. "有機飼料○○" or "○○(有機飼料)" (organic feed ○○ or ○○ (organic feed)); or
3. "オーガニック飼料○○" or "○○(オーガニック飼料)" (organic feed ○○ or ○○ (organic feed))

5.1.2.4 Mutual Authentication of the Organic Certification System among Countries

Article 12 of the JAS Law provides for a mutual authentication of the organic certification system. This mutual authentication regards the systems of Japan and other countries as equivalent and treats the organic certified products of other countries like those of Japan.

Under this system, the following requirements are necessary in order for overseas organic products to be distributed with organic JAS logos. First, Japanese importers need to be approved by a registered certifying body. Second, the products must be produced in a country verified by the Japanese government as having an equivalent grading system. Third, the products must be certified as organic in the equivalent countries and bear a certificate issued by governmental organizations or quasi-governmental organizations of the equivalent countries (MAFF 2016: 22–23).

As of the time of writing (August 2019), the countries whose organic certification system is regarded as equivalent to the organic JAS system are the United States, Argentina, Australia, Canada, Switzerland, New Zealand, and the member states of the EU (MAFF 2015: 5).

5.1.3 Challenges of the Organic JAS System in Japan

5.1.3.1 Divergence from the Codex Guidelines

The JAS basically follows the international standards of the Codex Alimentarius Commission. However, the JAS diverges from the Codex Guidelines in certain ways. As Honjoh (2004) criticizes, this divergence is clear with regard to "factory" farming and genetically engineered/modified organisms (GEO/GMO).

First, Annex 2 of the Codex Guidelines enumerates "Permitted Substances for the Production of Organic Foods" and stipulates that so-called factory farming sources are not permitted in farmyard and poultry manure. According to footnote 20 of Annex 2, "factory" farming refers to industrial management systems that are heavily reliant on veterinary and feed inputs not permitted in organic agriculture. However, under the current JAS system, there is no regulation on "factory" farming for farmyard and poultry manure. The Japanese government justifies this divergence from the Codex Guidelines based on the Precautions that have been added at the beginning of Annex 2. Paragraph 4 of the Precautions states that "the following lists do not attempt to be all inclusive or exclusive, or a finite regulatory tool but rather provide advice to governments on internationally agreed inputs. A system of review criteria as detailed in Section 5 of these Guidelines for products to be considered by national governments should be the primary determinant for acceptability or rejection of substances."

Second, as Honjoh (2004) notes, there is divergence regarding genetically modified products. The JAS prohibits the use of seeds and seedlings produced by using GMO technology. However, the current organic JAS system does not have any provisions to strictly prohibit the use of composts derived from genetically modified crops (MAFF 2016: 54). In contrast, the Codex Guidelines state, "all materials and/or the products from genetically engineered/modified organisms (GEO/GMO) are not compatible with the principles of organic production (either the growing, manufacturing, or processing) and therefore are not accepted under these guidelines"(Article 1.5). The MAFF has tried to justify this gap based on the Precautions discussed above. However, as Honjoh (2004: 91) points out, such an interpretation is unreasonable because the Codex Guidelines directly prohibit the use of genetically modified crops.

5.1.3.2 Other Criticism of the Organic JAS System

Various criticisms have been directed toward the organic JAS system in addition to those discussed above. First, the burden on organic farmers, such as the cost for certification, has been seen as problematic. Honjoh (2004: 102–104) observes that the burden of cost for certification and of operating procedures such as keeping records are reasons that organic JAS has not been popular. According to the MOA Nature Farming and Culture Foundation (2011, cited in Honjoh 2017: 16), among 11,856 organic farmers, only 3994 are certified as organic JAS farmers, and 7865 are not certified. In addition, a questionnaire survey for noncertified organic JAS farmers revealed that the high cost and the complexity of the application form are the reasons why farmers do not apply for certification (MOA Nature Farming and Culture Foundation 2011, cited in Honjoh 2017: 16).

Second, a lack of understanding regarding organic farming on the part of the consumer needs to be taken into consideration. Saito (2015) points out that an incorrect image of organic farming exists throughout society. In other words, although the organic JAS system and the Codex Guidelines both see the main aim of organic farming as environmental conservation, this aim is not understood well enough by the general public. According to the Japan Organic Agriculture Association, consumers' answers to the question "What do you recall from the term 'organic farming'?" were "Safety/Security" (63.7%), "Good for health" (48.9%), "Environmentally friendly" (40.6%), and "Tasty" (33.3%) (Japan Organic Agriculture Association (2012), cited in Saito (2015) pp. 18–19). However, as Nishio (2019) shows, the debate still continues throughout the world concerning whether organic produce is safer or better in terms of nutrition. It cannot be said that people understand sufficiently the original purpose of the organic JAS system, natural recycling in agriculture, and the reduction of the load from agricultural production on the environment.

5.2 Conclusion

Since the organic JAS system was introduced in 1999, labeling regulations for organic farming products have become stricter. However, the organic JAS system only regulates labeling and does not cover the whole of organic farming. As this chapter posits, this "bias toward labeling regulation" has been criticized (Honjoh 2004).

Nevertheless, the Act on Promotion of Organic Agriculture (Act No. 112 of 2006) was enacted in December 2006 and went into force in April 2007. Article 4 of this act stipulates the responsibility of the national and local government to comprehensively formulate and implement policies for the promotion of organic agriculture. Also, Japan has implemented various policies related to the promotion of organic farming based on this Act, which is considered "the basic law of agriculture" (Honjoh 2017: 26).

However, in spite of the organic JAS system and the Act on Promotion of Organic Agriculture, organic farming is far from spreading all over Japan. Nishio (2019) points out the following three reasons for this: (1) the demand for organic products has not increased due to economic stagnation and a reduced average income; (2) there is a lack of effort to provide consumers with a correct understanding of organic farming and organic products, and organic farming has not been incorporated into domestic agricultural policies, and (3) there has been not enough backing for producing and supporting autonomous organic farmers.

As Honjoh (2004: 275–276) discusses, organic farming is something that seeks to rectify the "fragmentation" between producers and consumers as well as that between human life and nature. In other words, organic farming has the potential to reconstruct the relationship between producers and consumers, and between human life and nature. The organic JAS system is not more than to regulate the labeling of organic agricultural products and so forth. The system itself is important, but it is not sufficient to reconstruct these two kinds of relationships. What are now required are policies to systematically push forward organic farming under the Act on Promotion of Organic Agriculture.

References

Compagnoni A (2010) Organic food labels: history and latest trends. In: Albert J (ed) Innovations in food labeling. Woodhead Publishing Limited, Cambridge, pp 75–93

Honjoh N (2004) Nihon no yuki nogyo: Seisaku to hoseido no kadai (Organic farming in Japan: challenges in policy and legal system). Noson Gyoson Bunka Kyokai, Tokyo

Honjoh N (2017) Yuki nogyo suishin ho no seika to kadai (I)(II) (Enactment and enforcement of the organic agriculture promotion act, their achievement and remaining issues (I) (II): reviewing the past ten years' operation of the act). Yuki Nogyo Kenkyu 9(2):6–28

Japan Organic Agriculture Association (2012) Yuki nogyo eno shohisha no rikai zoshin chosa hokoku: Shohisha ishiki anketo to seisansha・shohisha no koryu jirei (Research report on the promotion of understanding of consumers of organic agriculture: Attitude survey of consumers and cases of exchange between farmers and consumers). http://www.joaa.net/research/syohisya-h24.pdf

Ministry of Agriculture, Forestry and Fisheries (2019) Yuki Toroku Ninsho Kikan Betsu Ninsho Jigyosha Su (Heisei 31 Nen 3 Gatsu 31 Nichi Genzai) (The number of certificated operators according with the registered certifying body (As of March 31, 2019). https://www.maff.go.jp/j/jas/jas_kikaku/attach/pdf/yuuki_old_jigyosya_jisseki_hojyo-56.pdf. Accessed February 28 2020

Ministry of Agriculture, Forestry and Fisheries (2020) Yuki Toroku Ninsho Kikan Ichiran (List of the organic registered certifying bodies). https://www.maff.go.jp/j/jas/jas_kikaku/yuuki_kikan.html. Accessed February 28 2020

Ministry of Agriculture, Forestry and Fisheries, Food Safety and Consumer Affairs Bureau (2015) The inspection certification system for organic products. http://www.maff.go.jp/e/jas/specific/pdf/organic_products_system_1501.pdf

Ministry of Agriculture, Forestry and Fisheries, Food Safety and Consumer Affairs Bureau, the Labeling and Standards Division (2016) Questions and answers on the Japanese agricultural standards for organic plants and organic processed foods (Preliminary translation). http://www.maff.go.jp/e/policies/standard/jas/specific/.../organic-1.pdf

MOA Nature Farming and Culture Foundation (2011) Yuki nogyo kiso deta sakusei jigyo hokokusho (Business report on the formation of fundamental dataset of organic farming). https://moaagri.or.jp/manage/wp-content/themes/moaagri/pdf/hojojigyo/H22_yukikiso_houkokusho.pdf
Nishio M (2019) Kensho Yuki Nogyo: Global kijun de yomi toku rinen to kadai (Examining organic farming: principles and problems from the perspective of global standards). Noson Gyoson Bunka Kyokai, Tokyo
Saito S (2015) Yuki yasai wa uso o tsuku (Organic vegetables tell lies). SB Creative, Tokyo

Satoshi Kodera is a Professor in the Department of Law, Seinan Gakuin University, Fukuoka, Japan. His work focuses specifically on public international law, international economic law and international law of development. He is a Coauthor of *The Introduction to WTO and FTA Law* (Horitsu Bunkasha, 2016, in Japanese).

Chapter 6
Science and Food Production: Reviewing the Red Revolution of China Through a Green Lens (1950–1979)

Rohan Price

Abstract The rise of hundreds of millions of people into China's middle class has created unprecedented demand for food that is clean, green and produced sustainably. This trend is at odds with China's history since 1949. Marked by a struggle to produce a very different kind of 'green revolution', China slowly built scientific and technological investment in agriculture to maximise production levels to give the country a balance of trade requisite to industrialisation. Although China will never realise its dream of agricultural production sovereignty, or achieve widespread organic production domestically, the standards it applies to food imported on behalf of its new middle class will take its inspiration from the Western idea of the Organic Guarantee.

Keywords Agricultural policy · Agricultural history · China

6.1 Introduction

To this day, the ruling party Chinese Communist Party (CCP) holds dear to an isolationist ideal for food production. Chairman Xi Jinping observed in a farm visit to prosperous Heilongjiang in 2018: 'the Chinese people should hold the rice bowl in their own hands, and the bowl should contain the food produced by ourselves'.[1] Reminiscent of Mao, this comment's primitivism long circulated in communist China's twentieth century agricultural history as two national sovereignty issues:

1. the cost of its ever-growing food import dependence; and
2. access to internationally collaborative food crop science.

[1] Ma Chi and Cao Zinan, 2018. President Xi's quotes on food security. China Daily. http://www.chinadaily.com.cn/a/201810/16/WS5bc545baa310eff303282938.html Accessed 10 June 2019.

R. Price (✉)
School of Law and Justice, Southern Cross University, Gold Coast, QLD, Australia
e-mail: rohan.price@scu.edu.au

GOH Bee Chen, R. Price (eds.), *Regulatory Issues in Organic Food Safety in the Asia Pacific*, https://doi.org/10.1007/978-981-15-3580-2_6

China's ambition of agrarian self-reliance has been conditioned by the consumption needs of hundreds of millions of people and the spectre of famine. These factors have tended to create a regulatory narrative favouring quantity and availability of food, not its quality. These days, China's young, urban middle class, standing at perhaps half a billion people, has no direct experience of the starving days and its quantity-driven and go-it-alone mentality. The bright young things follow the mantra of the West, stressing as it does, clean and green agricultural practices producing high quality food. As China's wealth and trade influence continue to increase, the dominant issue will not be how, when or if it will make a transition to domestic organic food production. It will be how the West deals with China expressing its historical ambition for food production sovereignty. China will realise this via the quality standards it places on food grown elsewhere.

In a special report in June 1970, the CIA's Directorate of Intelligence notes that China's "current food supplies appear adequate whether measured by annual production of grain or the size of the average diet" but that the country "is still far from achieving a breakthrough similar to the 'Green Revolution' that has occurred elsewhere in Asia".[2] There had been local productive advances in the production of rice and grain including use of fertilisers and pesticides, deployment of the rural masses in drainage and irrigation schemes and introduction of rural communes and land collectivisation creating greater scales of economy. They were not, together, sufficient to tackle the Malthusian problem of exponential population growth and providing for ever-increasing food demand. This is why the cost to the environment or the quality of food have not historically been prominent in China.

China's slow agricultural rise had international dimensions. According to the 1970 CIA report, the key element lacking in China's approach that had produced the green revolution in India, Pakistan and other Asian countries was the introduction of new, high-yielding varieties of seed.[3] This omits a key context: for over two decades from 1950, the CIA monitored the trade blockade by the United States on China by recommending sanctions and tariffs against Asian countries who it discovered was involved in trading with the communist regime. Nevertheless, until rice surpluses could be produced consistently from the five local rice strains and exported to Burma or Ceylon in sufficient quantity, the Chinese government had to dip into its meagre cash reserves to import grain for domestic consumption. This impeded state capital flows necessary for industrialisation. In the 1970s, China's 'green revolution' weakly anticipated by the CIA report was defined by vast, highly productive vistas of wheat and rice feeding millions of people; it was not an allusion to environmentally sensitive production or a concerted push to improve the appearance, taste and wholesomeness of food for an increasingly choosy eating public.

The recent transformation from food quantity to food quality in China's discourse tends to obscure its key food supply question from the mid twentieth century until the end of the agricultural collectives in the late 1970s: what was the objective

[2] CIA-RDP85T00875R001500020030-7: 'China Stays Even in Food/Population Race' (1970) 1.
[3] CIA-RDP85T00875R001500020030-7: 5.

of local agricultural science, such as it was? Was it to develop the productivity of the rice harvest to feed hundreds of millions of its people as efficiently as locally possible, or were its proceeds a national bounty to be used to improve China's balance of trade and international export prestige? This became a vexed question because the policy dominance of the balance of trade approach in spurring on state investment in industrialisation had a tragic outcome in the late 1950s.

These days, China is a wealthy enough nation to be a net importer of wheat, rice and soybeans. Although this has changed the structure and outlook of world agriculture forever, the question of food security in China remains a politically touchy one because of the past. China's Great Famine (1958–1961) dominates the historical memory of the nation's agricultural achievements. A tragic confluence of regional grain harvest failures, the state's confiscatory over-commitment to repay Russian development loans with grain, contemporaneous establishment of rural communal kitchens providing throttle points of food supply, and an enervating push toward industrialisation, the famine left upwards of 36 million people dead. As immeasurably sad as it was, a tendency has emerged for the famine to eclipse other ways of interpreting the early agricultural achievements of China (1950–1957) or the eventual triumph of agricultural collectivism (1962–1978).

This chapter challenges generally held assumptions that the Great Famine was the only story of twentieth century Chinese agriculture, or that the CCP was subsequently indifferent in policy terms to the tragedy years. The 1970 CIA report contended that the CCP held an isolationist policy that China did not need the world to solve its problems: 'Peking seems only mildly aware that the current breather in the food/population race must be used to make fast progress in food raising technology. Thus, the regime still appears unwilling to provide the increased resources necessary to bring about a technological breakthrough'.[4] The international shunning in part producing Chinese self-reliance was real too. But after 1961, the state kept import reserves of cash in case grain was needed on an emergency basis. After 1970, China achieved its green revolution based on increased state investment in science and technology.

The development of scientific farming as a response in part to the Great Famine, according to Schmaltzer, 'came into its own amid the intensifying politics of 1965, flourished throughout the Cultural Revolution and remains relevant even today'.[5] Eisenman recognised that the commune system taxed household resources to finance capital accumulation in agriculture and technological innovation through the so-called 'research and extension' system.[6] He contended that although this system kept households living in 'austere' circumstances amounting to 'subsistence level conditions' its value lay in the fact that it 'bankrolled productive investment that generated the agricultural surpluses needed to kick-start a long-run cycle of

[4] CIA-RDP85T00875R001500020030-7: 3.

[5] Sigrid Schmaltzer. 2016. *Red revolution, green revolution: Scientific farming in socialist China* Chicago: Chicago University Press, 27.

[6] Joshua Eisenman 2018. *Red China's green revolution: Technological innovation, institutional change and economic development under the commune* New York: Columbia University Press 25.

sustained investment and output growth'.[7] The West's revisionism that China's commune system was abandoned as an abject productive failure, or that it crushed the individual without making the life of the country better, are beginning to yield to economic research demonstrating the contrary. It is important to credit the collective's influence alongside science.

Agricultural workers and the technocrats of the communist state reconciled to make a productive success of the 1970s paving the way to the organics discourse that we now take for granted. This chapter tells the story from the documentary record compiled on China through the satellites, listening stations, binoculars and reading glasses of the biggest overseas concentration of American spies in the world – the CIA's China Desk in the U.S. Consulate in British Hong Kong throughout the 1950s and 1960s.

This chapter is a view from the West with suitable injections in the name of balance. Schmaltzer's three elements of policy interpretation, namely, social good, environmental constraints and public utility[8] is used thread the narrative between oppositional histories.

6.2 The Early Days (1950–1957)

Some problems in Chinese agriculture are age old: how to merge China's patchwork of small holdings into larger farms without causing rural dislocation or failing to reap the benefit of a larger scales of economy.[9] The leadership of the CCP has, over the last decade, emphasised other issues of comparatively long standing, too, including urban migration's effects on food security, the protection of arable land at all costs, needs for water conservancy plans and continuous technology investment.[10] Others issues are new, including how China's modest arable land resource can meet the demand for its fast-changing dietary preferences as diverse an American-style diet and healthfulness.

Due to the emergence of a Chinese middle class and its new nutritional sensibilities, food politics in China have begun to tackle health and quality problems related to food consumption i.e. consumer expectation has joined productive efficiency as an important goal in the national psyche. The preoccupations of food governance in China in the twenty-first century would be unrecognisable to its population half a century ago. The national government is focussed on food contamination scandals. Making fraught bureaucracies accountable for maintaining regulatory distinctions between organic, green and pollution-free food has also hit the headlines. Not so

[7] Eisenman, 25.

[8] Schmaltzer, 26.

[9] Tracie Macmillan, 'How China Plans to Feed 1.4 Billion Growing Appetites' (2 February 2019) Avail at: https://www.nationalgeographic.com/magazine/2018/02/feeding-china-growing-appetite-food-industry-agriculture/ (accessed 10 June 2019).

[10] Ma Chi and Cao Zinan, 1.

long ago, however, the amount of rice and grain produced by the country in a harvest year was more or less the only fact that mattered.

According to CIA statistics, in the years 1951 and 1952, China increased its milled equivalent rice exports from 120, 000 metric tons to 193, 000 metric tons.[11] In 1953 and 1954, 272, 000 and 263,000 metric tons were exported[12] (principally to Ceylon in exchange for rubber). Favourable growing conditions in the Yangtze valley were largely responsible for an excess of rice. The adoption of agricultural collectivisation in 1955 did little to dampen yield improvements right up until 1957 when it was reported by the CCP that rice acreage had increased over 13% and total production was almost 27% higher than on 1952 levels.[13] Taken at face value, modest agricultural productivity improvements in the early 1950s should be celebrated. Coming as they did after decades of civil war, foreign occupation and ruthless purging of the countryside, they are little short of remarkable. The Chinese countryside was back in the game.

The Chinese communist government promised to liberate the peasants in 1949. By moderating private interests in land, however, the role of often involved a high degree of military compulsion and its targets could be arbitrary. Food production in this era was politics played in its crudest form. Although the collectivisation of the land only became national policy at the end of 1955, there were plenty of early examples of *ad hoc* rural property confiscation and a relocation. They formed part of a wave of CCP military drives against resistant KMT elements designed to confirm control of the south of the country (1950–1951).[14] In August 1951, local cadres in Chungshan, Guangzhou province (Zhongshan these days) closed down two rice mills and transferred their equipment to Canton (Guangdong) on the grounds that 'both mills were owned by rich landlords'.[15]

In another incident in Chungshan at around the same time, a summer grain levy was introduced, the harvest was subject to 'armed supervision', local peasants were 'forced' to sell seventy boatloads of grain which were sent to Canton to 'replenish the military food supply'.[16] Such administrative arrangements as moving privately-owned rice mills, or military requisitioning of harvests can be thought quite common in early communist China. They also made the 'social needs' model of food production raised by Schmaltzer a highly contestable matter even in the idealistic early years of the new communist regime. The policy of initially tipping out some landlord farmers, and then kicking them all out in 1955–1958, can be seen as a necessary readjustment if one hazards the view that CCP deserves credit, after making

[11] CIA-RDP64-00014A000100130029-7: 'Asian Rice Problems and the Cold War' (1956) 3.

[12] CIA-RDP64-00014A000100130029-7: 3.

[13] CIA-RDP04-01460R000100050001-2, CCP Report 'Cultivation of Chinese Paddy Rice' (1979),12.

[14] See generally, R.B.E. Price. 2019. *Resistance in colonial and communist China: Anatomy of a riot* Oxford: Routledge, 2019, 10-18.

[15] CIA-RDP82-00457R008400550011-1: 'Rice Production in Communist China' (1951), [1].

[16] CIA-RDP82-00457R008400550011-1: [2].

mistakes, for pulling the national agricultural fortunes of the country up by the bootstraps to successfully feed hundreds of millions of people.

When compared to the landlords turned out of their mills, the case of Ch'en Yung-k'ang strongly suggested that the CCP's agricultural policies in the early 1950s were as much a patchwork as China's many small farm holdings. Almost as soon as it came to power, and in its haste to assert leadership over rice growing practices, the CCP fostered scientific trial farms and publicised individual farmers who achieved exceptional yields attributable to their age old technical understandings and/or adoption of science. A South Kiangsu farmer, Ch'en was designated in 1952 as a 'Model Farmer of East China' for setting a record of 1443 catties of rice (873 kilograms) per mou (666 square metres) of land.[17] His methods included, 'double and deep ploughing, ample and seasonable applications of fertiliser, seed selection techniques, proper handling and spacing of plants, careful handling of water supply and complete elimination of weeds'.[18] He had also ploughed under clover, used 1600 catties of hog manure and 20 catties of ammonium sulphate.[19] In 1952, the neighbours of Mr Ch'en copied his techniques 'with good success'.[20] However, it would be mistaken to over-estimate the penetration of model farming practices as wide or deep.

Mr Ch'en was lauded by CCP cadres in 1952. He was described almost unremarkably in the CCP report as 'a well-to-do farmer of the middle class who owns 16 mou of land'.[21] Yet despite his achievement, he and his district could not have avoided the full effect of the policy change in 1955 toward socialist agricultural cooperatives that the CCP called 'another high tide of socialist revolution [that] swept the villages'.[22] Formalising and expanding the 'Resolution Concerning the Cooperative Organisation of Agriculture' adopted in early 1953,[23] after then end of 1955 the 'preliminary' farm cooperatives reputedly developed into 'the mature type of truly socialist agricultural cooperatives'.[24] Thus came to a close an era of tentative individualism on the land under which the CCP supported private landowners to develop harvests by fostering distribution of technical expertise on adaptive practices concerning judicious use of local strains, as well as tips on fertiliser use and managing crops in areas of high salinity/alkalinity.

When the CCP framed rice growing as 'a continuous struggle against nature'[25] it had China's notorious seasonal variability in mind, as well as how specialist growing techniques were needed to allow for saline and ammoniated water and thus urged on

[17] CIA-RDP80-00809A000700120362-5: 'East China Authorities Authenticate High Production Record' (1953) 2.

[18] CIA-RDP80-00809A000700120362-5: 2.

[19] CIA-RDP80-00809A000700120362-5: 2.

[20] CIA-RDP80-00809A000700120362-5: 3.

[21] CIA-RDP80-00809A000700120362-5, 1.

[22] CIA-RDP04-01460R000100050001-2, 11.

[23] CIA-RDP04-01460R000100050001-2, 11.

[24] CIA-RDP04-01460R000100050001-2, 11.

[25] CIA-RDP04-01460R000100050001-2, 1.

farming within pre-existing contamination constraints. The CCP promoted a theory of 'a culture system for each type of land with obvious results of higher yields'.[26] However, it arguably relied too much on the technical insights developed haphazardly over thousands of years of experience of peasant rice growing, and was not outward-looking for reasons of international political hostility and a policy of under-rating continual scientific investment.

There is a reason to believe the story of Mr Ch'en. As praiseworthy as his efforts were, the Party nevertheless had some technical advice for him to improve his yields yet further by increasing his use of fertiliser including phosphorous and potassium, using vegetable ash fertiliser instead of mixing soil and pig manure; he also needed to soak seeds in salt water before planting and attend to cross pollination issues.[27] Whatever can be said about the Party's infallibility, or its over-prescription of fertilisers and chemical compounds to control pests and weeds, there seems little doubt about its commitment to developing a process of production attuned to high yields albeit at the limits of its isolated science. What the verification and public recognition for successful farmers did, however, was to promote a sugar high of leading examples rather than a country-wide extension of investment in scientific, if nutrient and pesticide-heavy, rice and wheat growing practices.

Leading the charge for the causes of the Great Famine to be understood in less complex terms than they have been, Yang Jisheng maintained that the forcible seizure of grain by the government was the principal cause of the famine and that seizure policy was driven by 'absurdly inflated estimates of harvests following the adoption of irrational and unscientific farming methods'.[28] Lack of systematic investment modelled on what science was available after inheriting an isolated country ravaged by war and impoverishment comes closer to the mark. Underneath the growing, quality and distribution of China's food lay a political tussle over the proper objective of technical expertise and empiricism. I would emphasise poor dissemination of rice trial results for various land types and the CCP's under-estimation of the huge scientific effort required for China to kick start a 'green revolution' as legitimate complaints about its performance.

Even as a historical proposition, Schmaltzer's nomination of environmental needs in China being ignored in the stampede toward personal profit needs to be interpreted with some care. In 1951, the provinces of central and southern China including Kiangsi, Kwangsi, Kwangtung and Canton District recorded 'bumper harvests of early rice' and that 'the expected summer famine … [was] averted'.[29] In some provinces, yields increased on the previous year by 20% and this was attributed to a flood prevention program, construction of reservoirs and drainage ditches

---

[26] CIA-RDP04-01460R000100050001-2, 11.

[27] CIA-RDP80-00809A000700120362-5: 3.

[28] Jasper Becker. 2013. Review of *Tombstone: the untold story of Mao's great famine and The great famine in China, 1958–1962: a documentary history* Asian Affairs 42(2): 340.

[29] CIA-RDP80-00809A000700010235-8: 'Reports of Bumper Harvests of Early Rice in Central and Southern China Region' (1951), [1]–[4].

but that 'credit must be given to the farmers for their meticulous care of the land' as well as 'the government for loaning large quantities of fertiliser'.[30]

Let us bracket the fact that such a report appeared in the pro-CCP *Ta Kung Pao* newspaper where an adequate harvest could become a 'bumper' one at the pen point of a communist propagandist. Yet, the CCP was not in the habit of praising the unpraiseworthy either. The point raised by the bumper harvest news item was that care of the land *and* concentrated mineral fertilisation were not seen as contradictory because feeding millions of people was the objective. Moreover, the Party's concern that the quantity of arable land did not diminish was more or less constant in the three decades up to the 1970s. Ammoniated water from overuse of fertiliser became a condition for growing rather than an environmental limit.

Much of the Western narrative about land degradation under widespread scientific and industrial farming relates to the risk of its devastation for agricultural purposes. One example purported that land affected by highly saline and alkaline brackish water at the Hang-ku Mechanised State Farm could successfully grow rice up to 1334 catties per mou when proper techniques of land levelling, harrowing, flooding depth and seed drill spacing were used.[31] Although falling short of improving strain selection, or introducing new ones, it is misleading to suggest that Chinese state was not investing at all in locally reflective and empirical scientific techniques to make use of coastal land degraded by salt water.

The CIA report of 1970 referred to in my introduction pointed to the possibility of China importing rice seed from the Philippines or wheat seed from Mexico.[32] In dismissing both options as not properly adapted to Chinese conditions, the CIA masked its own culpability in making trade with such countries problematic through its China blockade. An element of Schmaltzer's trivium, the social or distributive element of food production, was reflected in the famine-free, steady growth of yields and rice exports in China from 1950 to 1957 attested by statistics that the CIA accepted. These year-on-year increases exports were little short of remarkable. They were too modest and annually undulating to be a figment of a communist statistician's imagination. The Chinese peasants, having returned from war, and given a few years of good growing climate, had nevertheless been pushed back on hard by the state to produce the surpluses for export and given little support. A visiting Pakistani trade unionist made this clear in 1954:

> During my long journeys in China, I did not witness any proof of modernisation nor did I see tractors, tubewells or other modern agricultural implements.[33]

Even if technology in agriculture investment was a low state priority, the year-to-year increase in overall yield was not unduly affected by radical land reform or establishment of rural collectives. If anything, the Chinese state, with its

[30] CIA-RDP80-00809A000700010235-8: [8].

[31] CIA-RDP80-00809A000700170395-4: 'Improvements in Rice Growing Techniques in China' (1954) 1–2.

[32] CIA-RDP85T00875R001500020030-7: 5.

[33] CIA-RDP78-02771R000300180002-5, 'A Source Paper on Agricultural Cooperatives/Collectives in Communist China' (1957) 5.

confiscatory production quotas and taxes on peasants, slipped easily into the hated role of the self-absorbed KMT landlord that it had eradicated. But the peasants produced beyond expectations, and despite the technological inadequacies, justified Mao's earthy awe.

6.3 The Famine Years (1958–1961)

In 1979, a few years after Mao had died, when the Gang of Four was on trial and Deng Xiaoping was about to take the reins, the CCP Central Committee took the opportunity to publish a report, 'Cultivation of Chinese Paddy Rice'. A nostalgic odyssey of the history and science of Chinese rice production, it stood at several hundred pages in length. On the Great Famine, the CCP conceded that the drought in the Yangtze in 1959 and 1960 'caused a natural calamity all over the country'.[34] Nevertheless, it maintained that 'the people's communes demonstrated a tremendous power of production' and that 'the total production of rice was maintained at the level as 1957 and in some areas the harvests were even better than that level'.[35] There is, dare it be said, a kernel of CCP confession in this statement although the enormous half-truth was intended to deflect criticism away from the Party's intensified grain confiscation in the late 1950s that caused the famine. If rice production did not fall, then why was there a national calamity that claimed at least 36 million lives? Yang Jisheng called that bluff. He found, despite the drought conditions in the Yangtze valley region, 'as people starved, cadres focussed on meeting procurement quotas'.[36] An astonishing total of 7.2 million tons of grain were exported in the famine years of 1959 and 1960.[37] The government put an end to the famine by importing over 5 million tons of grain and dropping exports to 1.355 million tons in 1961.[38]

In 1958, to initiate the so-called, Great Leap Forward in agriculture, Mao Tse-tung framed an eight-word constitution for the sector: 'soil, fertiliser, water, seed, density, [land] protection, management and labour'.[39] Thus, the expertise of farmers combined with the research efforts of Chinese scientists allowed 'a technical reform movement in rice culture [to be] launched'.[40] Under this program, agricultural colleges and research bodies were said to have 'seriously carried out the party's directive' of combining theory with practice and making science serve production'.[41]

[34] CIA-RDP04-01460R000100050001-2, CCP Report 'Cultivation of Chinese Paddy Rice' (1979), 13.

[35] CIA-RDP04-01460R000100050001-2, 13.

[36] Yang Jisheng. 2012. *Tombstone: the untold story of Mao's great famine*. London: Allen Lane, 33.

[37] Yang Jisheng, xx and xxi.

[38] Yang Jisheng, xxiv.

[39] CIA-RDP04-01460R000100050001-2, 14.

[40] CIA-RDP04-01460R000100050001-2, 14.

[41] CIA-RDP04-01460R000100050001-2, 14.

The historical and political problem caused by the Great Famine was that agricultural production was defined by the Party leadership as a grain target to be met and exported, not as mouthfuls of rice eaten to sustain millions of every day Chinese people.

Yang Jisheng found that the Great Famine commencing in 1958 was caused by rice harvest over-estimation and accordingly disproportionate levels of confiscation. The government's response to the alleged 'false reporting of output and rice withholding' of peasants was fuelled in part by the culture of protections and quotas, as well as over-estimation encouraged by the 'sputnik fields' produced by intensive nutrition inputs.[42] The sputnik fields gave a misleading impression of the harvest to come. They were, nevertheless, a demonstration of what wonders China's limited scientific resources could produce. They were whisked up and politically abused by the CCP. As the example of Mr. Ch'en example showed, the sputnik fields took hold in the early years after Liberation in a spirit of civic recognition, science and crop density improvement. By the late 1950s, however, the sputnik system had become a deadly cat and mouse game between the cadres and the peasants. They were a false representation of what was being produced but at once a perfect pretext for political persecution of those arbitrarily labelled 'Rightist' and a way to justify continuing grain exports.

During the Great Famine, the bounty and provision of applied science took a noticeable back seat to the government's economic balance sheet and balance of trade targets. In the case of rice agriculture in China, the CCP sought 'to reform and revolutionise farming techniques' through 'a three way cooperation between the leadership, technical staff and the farming masses'.[43] This self-image could be maintained from 1950 until 1957. Yet after 1962, until the formal end of agricultural collectivisation in 1979, narcissism and obfuscation characterised the CCP's summary of the terrible years of famine. During them, 'the rice producing regions carried out many projects of basic construction, technical improvements and mechanised farming regardless of the natural calamities'.[44] Quite aside from the fact that the Report itself recognised that mechanisation of the paddies had only begun to seriously occur in 1970s, this deflection joins the claim about increased rice production throughout the famine as an example of CCP history being a convenient story to tell rather than conveying frank ownership of the bad times along with the good, not that this has necessarily been a virtue of Western democracies.

6.4 Agricultural Communes and Collectivism (1958–1978)

Quite a bit of my discussion has concerned the frailty of state support for the scientific and technological improvement of food production in communist China despite its developmental objective of lifting grain exports to propel industrialisation. In

[42] Yang Jisheng, 28–29.

[43] CIA-RDP04-01460R000100050001-2, 13.

[44] CIA-RDP04-01460R000100050001-2, 13.

this connection, there was an odd pessimism in a CCP policy paper on rural communes and collectivisation in 1953:

> Without large scale investment by the state, agricultural producers' cooperatives cannot be organised and will not operate successfully even if they are organised. Thus at the present time China's agricultural production problem is not how to organise agricultural producers' cooperative but how to use the supply and marketing cooperatives as a means to stimulating production.[45]

This 1953 policy paper also incorrectly envisaged a cooperative system that peasants opted into on a voluntary basis. When the system was actually instituted a few years later, the voluntarism principle was abandoned 'in the head-long rush to force the farmer into the collectives and stripping from him not only the ownership of his farm, but also his implements and livestock, grain, seeds and fruit trees.'[46] Hardliner and Mao Tse-tung ally, Liu Shao-chi, proudly described the agricultural communes program in 1958 as 'organisers of the entire life of the people'.[47] By September 1958, a purported 90.4% of China's peasants had been forced into rural communes.[48] Under this system, the CCP not only confiscated private landholdings of the peasants but attempted to destroy family loyalties by centralising institutional mess halls, kindergartens, nurseries and tailoring services to free up women with children to work in labour brigades on farms or in factories.[49]

Denis Bloodworth, writing for *The Observer* wrote that he had witnessed: 'individual farmhouses demolished and great tenements with communal kitchens and canteens, designed to accommodate a hundred families, rising incongruously in the fields'.[50] Although there can be no doubt that the rural communes and the so-called 'collective life' were a shock to the body politic of peasant China, and an unambiguous attempt to replace affections for family with obedience to the state, they were also intended to free up women to drive agricultural productivity. The effect of this cannot be entirely discounted in the improving productivity figures of the era.

Eisenman differentiated between two periods of collectivisation: the 'Leftist Commune' (1965–1969) characterised by high savings rates and without increases of scientific investment in productivity and the 'Green Revolution Commune' (1970–1979) that had high savings and increases in productive investment.[51] He contends that under the Green Revolution Commune, total productivity and exports improved because of 'expanded use of hybrid seed varieties, chemical fertilisers, mechanisation and irrigation'.[52] After trying since the mid-1950s, China had found

---

[45] CIA-RDP78-02771R000300180002-5: 4.

[46] CIA-RDP78-02771R000300180002-5: 4.

[47] CIA-RDP79R00890A001100040029-0: 'Communist China' (1959), [J].

[48] CIA-RDP78-02771R000300120006-7: 'Collective Life and the Family in Chinese Communes' (by Nils Steffanson, *Backgrounder on Communism*, 1958), 1.

[49] CIA-RDP78-02771R000300120006-7, 1.

[50] CIA-RDP78-02771R000300120006-7, 5.

[51] Eisenman, 110.

[52] Eisenman, 119.

the optimum collective formula for scientific investment, feeding people and exporting grain that it had long sought.

After the Great Famine, the CCP's central committee instigated the 'agriculture first' policy which provided greater state investment in agriculture and industrial support for machinery.[53] By 1964, enough tools, water pumps and chemical fertilisers became available 'to make an impact on agricultural production'.[54] Nevertheless, due to China's growing population, efforts to stabilise and increase grain production had, by 1964, appeared to barely keep pace with demand. Availability of grain, on a per capita basis, had declined by 15% on the level in 1957.[55] Peking was forced ease restrictions on private vegetable plots for households and rely on import of 6.5 million tons of grain to supplement the paltry 170 million tons produced in 1964 due to bad growing conditions caused by excessive rainfall levels. By 1964, the commune system had been relaxed so that vegetables, fruit and meat could be produced on private plots and sold in free markets to supplement the rice and grain diet and this was credited as generally increasing food availability in net terms to levels much higher than those 1957 or 1961.[56]

Post-famine, Eisenman believed that the *Dazhai* workpoint remuneration system in the collective agriculture of the 1960s and 1970s raised China's national production by encouraging each household to 'vigorously monitor its workpoints and encourage others to contribute to increase collective income'.[57] Liberalisation by allowing private plots enabled a measure of food security to rural collectives to the extent that, by 1967, 20–25% of individual calorie intake was met but small private gardens and pig sties.[58]

The objective of the communes and collectivisation period was based quite faithfully on socialist theory. i.e. the purpose of increasing grain production was not personal profit but the domestic triumph of equality on the praiseworthy basis of 'from each according to his ability and to each on the basis of his needs'.[59] Underinvestment in scientific and technological methods of agricultural production stemmed in part from Mao's inordinate faith in the dam building and earth-moving capacity of China's 500 million strong masses. Underinvestment was also the result of difficult international trade conditions purposefully soured by the West. It is often the case that the CCP knew about the problem and made it clear in its policy position but its relentless leadership jockeying gave policies and their implementation an ad hoc character. Consider this from the *Resolution on the Establishment of People's Communes* (1958):

[53] CIA-RDP85T00875R001500020030-7: 'China Stays Even in Food/Population Race' (1970) 2.

[54] CIA-RDP85T00875R001500020030-7: 2.

[55] CIA-RDP78T05439A000400290077-6: 'Poor Autumn Harvest in Communist China Reduces Domestic Grain Supplies' (1964) 1.

[56] CIA-RDP78T05439A000400290077-6, 1.

[57] Eisenman, 179.

[58] CIA-RDP85T00875R001500020030-7: 3.

[59] CIA-RDP80M01009A001502550018-3: CCP Central Committee, 'Resolution on the Establishment of People's Communes in the Rural Areas' (1958), [5].

> After a number of years, as the social product increases greatly, the communist consciousness and morality of the entire people are raised to a much higher degree, and universal education is implemented and developed, the differences between workers and peasants, town and country, mental and manual labour – legacies of the old society that have been inevitably carried over into the socialist period, and the remnants of unequal bourgeois rights that are the reflection of these differences – will gradually vanish and the function of the state will be limited to protecting the country from external aggression but will play no role internally.[60]

One could chalk up such statements to the steady optimism of the Central Committee of the CCP during the 1950s; one could even laugh aloud at the folly of making orthodox Marxist doctrine and its prognosis into state policy. I contend that the 'increases' expected of 'social product' were presented as if it was an end, and the only end, of government policy, and as we have seen, the CCP leadership believed that it would emerge out of nowhere as the inevitable dividend of hard work and social progress. The people would become equal and the state would have no internal purpose; this would occur whether or not agricultural technology and science was used to increase the size of the pie.

Achieving large, year-on-year grain surpluses to fund industrialisation was not conformable to the equality principle. Economic conditions in the countryside were often impoverished to the point of starvation and called too heavily on the strained joints and flexed muscles of the peasants who were not given the scientific help needed to produce the expected yields. What the Central Committee of the CCP got 100% correct in the *Resolution* was that help for China was not going to come from the outside. The scientific exchange and transfer common to the relations between Western nations was not going to land on Chinese shores whilst the American trade blockade remained in place. Any hope in Russian partnership ended in 1956 with Mao's dismissive response to the Secret Speech of Khrushchev condemning Stalinism. The production revolution caused by hybrid seeds finally came to Chinese fields only in the 1970s.

6.5 Conclusion

This chapter has positioned the historical policy concerns of Chinese agriculture in relation to modern, middle class priorities of organic quality and environmental constraint. China's embrace of quality and sustainability in food production has been late because it was developmentally behind the rest of the world in hybridised seed production and other use of technology to improve crop yields. Only when these investment inputs in agriculture became widespread could productivity, domestic food security and export consistency be taken for granted. In this light, the modern Chinese consumer's concern for the quality, purity and novelty of food, and its truthful invigilation, sit atop decades of suffering, uncertainty and indomitable but rarely rewarded ambition. The CCP had no choice but to assume that feeding

---

[60] CIA-RDP80M01009A001502550018-3: [5].

hundreds of millions of people was more important than *what* was fed to them, or that help could come unconditionally from the West.

The production spurts of Chinese agriculture before and after the Great Famine were quite remarkable achievements considering it was locked out of many regional trading relationships and the opportunities for scientific exchange that would have given it a 'green revolution' sooner rather than later. The intervening era of the famine was a time when the rupture between the agendas of the Party leadership and rural people, saw the role of applied science in Chinese agriculture converted into a prop for foreign trade of rice and grain rather than in service to the food provision of the masses. The political sensitivity about food security in today's China is due to deep-rooted conceptions of agrarian sovereignty, and the vulnerability of the ruling party in recalling troubling history.

The effect of constantly playing catch up with population growth has been profound throughout China's twentieth century history. For most of its second half, China's communist leadership wrestled with maximising the quantity of food available to its burgeoning population, and crafting the socio-political institutions on the land thought able to achieve it. The environmental value put so strongly into the practice of sustainable agriculture these days, and at the heart of the Organic Guarantee, in the early days of communist China, was no more than a value in slight tension with other values.

China's agricultural story in the 1960s and 1970s came well before the realisation that industrial farming and irresponsible use of chemicals have resulted in large tracts of land becoming non-arable. Concerns about soil loss, land clearing, acidity, salinity and other types of water contamination due to modern agricultural practices were barely on the radar because of their endless belief in the adaptability of practices as though farmers were not, themselves, responsible for the environment. The main tension, between growing rice and grain sort-of scientifically for millions of people under the CCP's leadership and the political consequences of it failing to do so, masked many medium-term problems related to land degradation caused by what were thought to be practices conducted within environmental limits.

The modern day leadership of the CCP cannot ignore the land preservation imperative caused by China having less than 10% of the world's arable land and nearly 20% of its population. The big issues for China, include decreasing grain and rice import-dependency, increasing farm lots for a better scale of economy and increasing productive output without creating a generation of rural unemployment. The middle class consumers of the nation nevertheless make no trivial claim when demanding availability of, say, high quality Organic soymilk. The paradox of modern China is very much about having the wealth to buy the best but regularly being confronted with second-best in the market place. This is a historically unprecedented experience for the Chinese.

Joshua Eisenman's revisionist work, *Red China's Green Revolution,* concluded that 'Maoism was [an] essential source of commune productivity', and while acknowledging its problems, strongly credited consistent production yields after the famine to its principle of 'working harder and faster on collective undertakings and devoting less time to private affairs'.[61] Sigrid Schmaltzer's study *Red Revolution, Green Revolution* noted that the lesson to draw from China is that 'the people of the world still need to figure out how to do agricultural science differently' because 'we need to be able to trust that technologies ostensibly developed to increase our capacity to feed people are not in fact riding roughshod over social and environmental needs to profit private interests.'[62] Each author prompts reflection on trust in agriculture. The question of why food producers cooperated with each other in China has given way to the question of our age: why do consumers in the public at large trust agricultural producers? It is in this historical continuum that the momentum toward the Organic Guarantee, and environmentally sustainable production, must be seen.

The CCP's failure in the nation's food self-sufficiency during the Great Famine has left deep marks in the country's political culture. Food contamination and adulteration scandals, although occurring commonly in China's long history, these days require trenchant criminal justice responses because they highlight the recent re-evaluation of food quality as a national priority and recalls the historical sensitivity about the role of the state in food supply since the famine. The previous generation of Chinese were trapped in the living memory of scarcity, famine and under-investment in agrarian life. Times have changed greatly but the residues of the past are visible everywhere.

The Organics message, such as it is understood in China, relates to paying more as a consumer for lower yields of higher quality food produced within environmental constraints and freighted 'air fresh' to a supermarket. As middle class demands become ever louder, agricultural practices in China and around the world will change in response to this. Incorruptible methods of invigilating quality will become a Chinese requirement on domestic and imported foodstuffs alike. China has good historical reasons to be choosy about its consumption, and find a regulatory system to deliver it, now that it has the economic power to do so.

References

Articles

Becker J (2013) Review of Tombstone: the untold story of Mao's great famine and the great famine in China, 1958–1962: a documentary history. Asian Aff 42(2):340–346

Books

Eisenman J (2018) Red China's green revolution: technological innovation, institutional change and economic development under the commune. Columbia University Press, New York

Price R (2019) Resistance in colonial and communist China: anatomy of a riot. Routledge, Oxford

[61] Eisenman, 254–255.

[62] Schmaltzer, 26.

Schmaltzer S (2016) *Red revolution, green revolution: scientific farming in socialist China*. Chicago: Chicago University Press
Yang J (2012) *Tombstone*: the untold story of Mao's great famine. Allen Lane, London

Internet Sources

Ma C, Zinan C (2018) President Xi's quotes on food security. China Daily. http://www.chinadaily.com.cn/a/201810/16/WS5bc545baa310eff303282938.html. Accessed 10 June 2019
Macmillan T (2019). How China plans to feed 1.4 billion growing appetites. National Geographic. https://www.nationalgeographic.com/magazine/2018/02/feeding-china-growing-appetite-food-industry-agriculture/. Accessed 10 June 2019

Archival Sources

CIA-RDP82-00457R008400550011-1: 'Rice Production in Communist China' (27 August 1951)
CIA-RDP80-00809A000700010235-8: 'Reports of Bumper Harvests of Early Rice in Central and Southern China Region' (13 September 1951)
CIA-RDP80-00809A000700120362-5: 'East China Authorities Authenticate High Production Record' (31 July 1953)
CIA-RDP80-00809A000700170395-4: 'Improvements in Rice Growing Techniques in China' (23 April 1954)
CIA-RDP64-00014A000100130029-7: 'Asian Rice Problems and the Cold War' (18 January 1956)
CIA-RDP78-02771R000300180002-5, 'A Source Paper on Agricultural Cooperatives/Collectives in Communist China' (1 October 1957)
CIA-RDP80M01009A001502550018-3: CCP Report, 'Resolution on the Establishment of People's Communes in the Rural Areas' (29 August 1958)
CIA-RDP78-02771R000300120006-7: 'Collective Life and the Family in Chinese Communes' (by Nils Steffanson, *Backgrounder on Communism* (1 November 1958)
CIA-RDP79R00890A001100040029-0: 'Communist China' (29 April 1959)
CIA-RDP78T05439A000400290077-6: 'Poor Autumn Harvest in Communist China Reduces Domestic Grain Supplies' (30 December 1964)
CIA-RDP85T00875R001500020030-7: 'China Stays Even in Food/Population Race' (19 June 1970)
CIA-RDP04-01460R000100050001-2, CCP Report 'Cultivation of Chinese Paddy Rice' (1 June 1979)

Rohan Price A lecturer in the School of Law and Justice, Southern Cross University, Dr. Rohan Price has established a reputation as a foremost historian of the role of nationalism in the British/Chinese colonial encounter. He was awarded a Doctor of Philosophy from the University of New England (Australia) for his thesis on the use of property law to encourage civic loyalty to colonial Hong Kong between the world wars. His books, including *Reading Colonies: Property and Control of the British Far East* and *Resistance in Colonial and Communist China* (1950–1963), are based on extensive archival and digital repository research. His books have been described in reviews as "passionate", containing "argumentative strength and forthright originality" and "enormous attention to historic, theoretical and political detail". He has enjoyed lengthy stints as a Visiting Professor in three Chinese universities over the last decade, teaching in fields including common law history, maritime law and the law of trusts. His interest on Chinese food safety issues was prompted by a friend in Hong Kong who, in 2008, casually mentioned he should not buy the same brand of noodles every time he went to the supermarket. He has finished the year 2019 with a new monograph published by City University Press, Hong Kong, *Violence and Emancipation in Colonial Ideology: Hong Kong and British Malaya*. It has been described in an academic review as "a major contribution to the literature".

Chapter 7
Managing Food Waste, Improving Food Safety? The Case of Gutter Oil in China

Natalie W. M. Wong

Abstract In recent years there has been growing public concern in response to food safety problems in P.R. China. The discussion has focused on the widespread overuse of chemical fertilizers and pesticides, which has implications for food safety. However, there is a lack of knowledge about the impact of food waste on food safety. The increasing urbanization of mainland China and the rising affluence of its citizens has led to a change in dietary habits and consumption patterns. This in turn has led to an increase in food waste generation, greater contamination of the environment and more severe food safety risks. To review China's food waste management strategies, this chapter presents the case study of "gutter oil" (illegal cooking oil from leftovers or used oil and animal parts). It finds that the increase in food waste has had consequences for environmental pollution and health as a result of poor environmental decision making, insufficient food safety standards and non-transparent food labeling.

Keywords Gutter oil · Food safety · China · Food waste · Fragmented authoritarianism

7.1 Introduction

Regulating the quality of organic food is an important subject. However, in China the key issue to be tackled is the contamination found in everyday non-organic food. According to the World Bank, food waste is a challenging issue, posing a threat to food security, food safety, the economy and environmental sustainability.[1] Globally,

[1] Global Food Loss and Waste, World Bank (2015). Retrieved from: https://datatopics.worldbank.org/what-a-waste/global_food_loss_and_waste.html. [Accessed on 18 June 2019]

N. W. M. Wong (✉)
Department of Public Policy, City University of Hong Kong, Kowloon Tong, Hong Kong
e-mail: nataliew@cityu.edu.hk

GOH Bee Chen, R. Price (eds.), *Regulatory Issues in Organic Food Safety in the Asia Pacific*, https://doi.org/10.1007/978-981-15-3580-2_7

1.6 billion tons of food goes to waste each year.[2] Food waste refers to the discarding of food intended for human consumption.[3] Food waste contributes to an increase in greenhouse gas and methane emissions due to the decay of wasted food. Food safety is also a concern when disposing of food waste that goes to be incinerated or to landfills that leak chemicals into soil and water. For example, landfill leachate emits GHG that pollutes the soil.[4] Proper food waste management can improve the environment. However, a study found that in developing countries, mismanaged food waste systems, such as unsorted food waste, and unclear regulations for stakeholders to reduce food waste, create sanitary problems as well as increased risks to environmental and human health.[5] The P.R. China ('China' hereafter) generated 88.65 million tons of food waste in 2016.[6] Such a quantity of food waste not only affects the water and land resources but also impacts on human health.

As global interest in regulating the organic guarantee increases, it is important not to forget enforcement efforts aimed at food safety at the most basic level. This chapter briefly introduces the drivers of food waste and the food waste hierarchy in relation to food safety. Thereafter, an overview of food waste management in China is outlined by addressing the drivers of food waste and the food waste hierarchy. In particular, this section focuses on consumer food waste and its impacts on food safety in the country. The third part of this chapter presents the case of gutter oil to address the limitations of food waste management in the context of China. This section further highlights the constraints of environmental decision-making and insufficient food safety standards. Finally, the chapter concludes that limited departmental cooperation has weakened the capacity of food waste and food safety control and argues that there is a need for greater transparency with regards to food information to protect human health.

7.1.1 Drivers of Food Waste and the Food Waste Hierarchy

Food waste is a substantial contributor to the world's food system challenges, affecting natural resources and human security. Economic development and urbanization, as well as cultural practices, are the drivers of food waste generation.

[2] Hegnsholt, E., S. Unnikrishnan, M. Pollmann-Larsen, B. Askelsdottir, and M. Gerard, "Tackling the 1.6-Billion-Ton Food Loss and Waste Crisis", Boston Consulting Group, 20 August 2018. Retrieved from: https://www.bcg.com/publications/2018/tackling-1.6-billion-ton-food-loss-and-waste-crisis.aspx.[Accessed on 18 June 2019]

[3] Food and Agriculture Organization of United Nations, Food Loss and Food Waste. Retrieved from: http://www.fao.org/food-loss-and-food-waste/en/. [Accessed on 18 June 2019]

[4] Lin, A.Y, S. T. Huang, M.L. Wahlqvist, (2009), "Waste Management to Improve Food Safety and Security for Health Advancement", *Asia Pacific Journal of Clinical Nutrition*, 18(4):538.

[5] Ngoc B. D. T, G. Kumar, C. Lin, (2015), "An overview of food waste management in developing countries: Current status and future perspective", *Journal of Environmental Management*, volume 157, page 220, https://doi.org/10.1016/j.jenvman.2015.04.022. [Accessed on 24 June 2019]

[6] Envirunion, Food Waste Management Industry in the Thirteenth-Five Year, Beijixing. Retrieved from: http://huanbao.bjx.com.cn/news/20190211/961661.shtml. [Accessed on: 20 June 2019]

Economic growth and industrialization have contributed to the modernization of Chinese society, impacting the pattern of consumer behavior. Industrialized food systems provide consumers with diverse food choices, and economic globalization further encourages them to obtain food from different places, rather than selecting local food sources. Urbanization has also changed the people's dietary habits, with many seeking out an organic alternative flown into China from overseas or preferring to dine outside the home. In addition, cultural factors, such as luxurious banquet receptions, place increased strain on food resources and contribute to waste generation. Shopping practices also influence waste generation; for example, consumers in developing countries buy only small amounts of food at a time, which may reduce food waste.[7]

Food waste has implications for sustainable development in all societies. The food waste hierarchy provides a framework to identify the socio-economic impacts of food waste and delivers the best environmental solutions for reducing food waste.[8] The hierarchy sets guidelines for waste strategies.[9] It offers five options for managing food waste (from the most favorable to the least favorable option): prevention, reuse, recycle, recovery, and disposal. Preventing food loss and food waste is the most favorable option with substantial environmental, social and economic benefits. Disposal is the least favorable option, as it involves incinerating unavoidable food waste or sending it to landfill[10] In general, send to landfill or incinerator is the conventional practice for managing unavoidable food waste. However, improper food waste management may contribute to increased human health risks.

7.1.2 *Food Waste and the Relationship to Food Safety*

The public started to pay attention to gutter oil in 2003, when the Chinese government discovered 2.3 tons of gutter oil in Beijing.[11] The gutter oil was found in backstreet drainage near restaurants and was re-processed in small shops or factories in

---

[7]Thyberg K.L. and D. J. Tonjes, (2016), "Drivers of food waste and their implications for sustainable policy development", *Resources, Conservation and Recycling*, vol. 106, p. 12.

[8]Papargyropoulou, E., R. Lozano, J. K. Steinberger, N. Wright, Z. Ujang, (2014), "The food waste hierarchy as a framework for the management of food surplus and food waste", *Journal of Cleaner Production*, vol. 76, p.107, https://doi.org/10.1016/j.jclepro.2014.04.020. [Accessed on 22 June 2019]

[9]Rasmussen, C., Vigsø, D., Ackerman, F., Porter, R., Pearce, D., Dijkgraaf, E., et al., 2005. Rethinking the Waste Hierarchy. Environmental Assessment Institute, Copenhagen, p.1.; Porter, R.C., 2002. The Economics of Waste. Resources for the Future, Washington DC, p. 137; Price, J.L., Joseph, J.B., 2000. Demand management: a basis for waste policy: a critical review of the applicability of the waste hierarchy in terms of achieving sustainable waste management. Sustainable Development. Vol 8, Issue, 2, p. 97.

[10]Papargyropoulou, E., R. Lozano, J. K. Steinberger, N. Wright, Z. Ujang, (2014), "The food waste hierarchy as a framework for the management of food surplus and food waste", *Journal of Cleaner Production*, vol. 76, p.115, https://doi.org/10.1016/j.jclepro.2014.04.020. [Accessed on 22 June 2019]

[11]Li, J., Cui, N., & Liu, J. (2017). "Gutter oil: an overview of Chinese food safety issues and policies". *Global Health Promotion*, 24(3)75. https://doi.org/10.1177/1757975915623733

remote suburbs before being sold back to restaurants for a very low price.[12] Gutter oil refers to a type of recycled oil from the gutter, household drains, restaurants and animal grease. This recycled oil may contain other wastes, such as residues from kitchen fryers, animal fats and organs. After being collected, the gutter oil is turned into edible oil through a series of processes, such as mixing, filtering, boiling and refining. The edible oil refined from gutter oil is then sold to restaurants and school canteens. However, cooking with gutter oil poses great risks to human health as the oil contains dangerous bacteria and heavy metals.[13] The scandal of gutter oil is indicative of China's poor food waste management and raises broader concerns about food safety. Moreover, it reveals a lack of transparency surrounding food information in protecting human health and ensuring food safety. Hence, it is important to focus on improper food waste disposal to minimize the risks of environmental and human security first and then graduate to food quality claims made by those producers boasting an organic guarantee.

7.2 Food Waste Management and Food Safety Concerns in the Context of China

7.2.1 Drivers of Food Waste Generation and the Rise of Food Waste Concerns

China generates 60 million tons of food waste every year and the amount of waste grows by more than 10% annually.[14] In 2016, 88.65 million tons of food waste was generated nationwide.[15] The food waste figures indicate the increasing affluence and consumer behavior of the Chinese middle class as well as the skyrocketing food and beverage catering industry in China. According to the National Bureau of Statistics of China, the number of food catering service providers, including restaurants, increased from 3266 in 1999 to 25,884 in 2018.[16] Economic growth has changed the consumer and dietary patterns of contemporary Chinese society. Affluent consumers prefer to visit shops and restaurants, thus generating an increase in food waste.[17]

---

[12] Chen, S. "Chinese restaurant staff jailed for cooking with 'gutter oil', South China Morning Post, 27 April 2017. Retrieved from: https://www.scmp.com/news/china/society/article/2091125/chinese-restaurant-staff-jailed-cooking-gutter-oil. [Accessed on 20 June 2019]

[13] Li, J., Cui, N., & Liu, J. (2017). "Gutter oil: an overview of Chinese food safety issues and policies". *Global Health Promotion*, 24(3)75. https://doi.org/10.1177/1757975915623733

[14] The situation of food waste in China, China-Nengyuan.com. Retrieved from: http://www.china-nengyuan.com/baike/5371.html. [Accessed on 20 June 2019]

[15] China Biodiversity Conservation and Green Development Foundation, "The experiences of controlling food waste in other countries", Sohu.com, 10 October 2018. Retrieved from: http://www.sohu.com/a/258665185_100001695/ [Accessed on 20 June 2019]

[16] Number of catering service industry, National Bureau of Statistic of China. Retrieved on: http://data.stats.gov.cn/easyquery.htm?cn=C0. [Accessed on 20 June 2019]

[17] Bai, J. T.L. Wahl, B.T. Lohmar, and J. Huang. 2010. Food away from home in Beijing: Effects of wealth, time and "free" meals. *China Economic Review* 21(3):439.

This change in consumer and dietary patterns in China is a key factor contributing to food waste, and in turn, food safety.

The scandal of gutter oil was first highlighted in the media in 2003 when 2.3 tons of gutter oil was found in small shops and restaurants in Beijing. Other cities such as ones in Shandong, Jiangsu and Chongqing also discovered gutter oil between 2008 and 2014.[18] The waste cooking oil consisted of animal grease or organs and other chemical substances. Similar to the Beijing example, it was collected from street drainage and sold inexpensively to small shops and restaurants as edible cooking oil. Gutter oil may have multiple negative impacts on human health, such as diarrhea, indigestion, stomach and intestinal cancer. The scandalous processing of gutter oil reveals the problems of food safety and limitations of food waste management as China continues to modernise.

China's economic growth and economic globalization have reduced the cost of producing food. This generates food waste and raises concerns about food waste management, as improper food waste management may lead to unsafe food. Food safety concerns in China need firstly to be discussed before addressing the limitations of food waste management in the country.

The scandal of gutter oil processing shocked the nation, yet it was just the tip of the iceberg. Discoveries of poisonous Jinhua Ham in 2003, mushrooms in 2006 and fake eggs in 2008, for example, raised concerns about food safety in the country. In particular, the internationally reported scandal of melamine-tainted milk formula in 2008 showed that government regulation of food safety was ineffective. The scandal involved more than 300,000 victims in China[19]; 54,000 babies were hospitalized[20] and 6 died from kidney stones and other kidney-related illnesses. These food safety scandals are undermining consumers' trust in regulators' capacity to ensure food safety. A nationwide survey of 7500 people conducted from 2013 to 2016 asked participants whether they were satisfied with the level of food safety in China. The responses revealed an average food safety satisfaction score of 5.509 out of 10.[21] Interestingly, lower food safety satisfaction was found in provinces with higher GDP levels.[22] The low level of food safety satisfaction reflects a lack of trust in the government's efforts to manage food waste.[23]

The food safety issues also illustrate the limitations of food waste management in China. The gutter oil example in particular is indicative of how the Chinese government has failed to manage consumer food waste effectively.

---

[18] Chawang Zhongguo, Review of Food Safety Problem: Gutter Oil, Sohu.com, 26 August 2017. Retrieved from: https://www.sohu.com/a/167402318_425345. [Accessed on 20 June 2019].

[19] Branigan, Tania (2 December 2008). "Chinese figures show five fold rise in babies sick from contaminated milk". *The Guardian*. London.

[20] Scott McDonald (22 September 2008). "Nearly 53,000 Chinese children sick from milk". Google. Associated Press. Jane Macartney (22 September 2008). "China baby milk scandal spreads as sick toll rises to 13,000". *The Times*. London.

[21] He, C.; Han, G.; Liu, Y. (2019), "Food Safety Satisfaction in China and Its Influencing Factors: Empirical Study with a Hierarchical Linear Model". *Safety*, 5(1):17.

[22] Ibid, 18.

[23] Ibid, 19.

7.2.2 *Overview of Food Waste Management System in China and Its Implications for Food Safety*

Compared with municipal solid waste management, there are few regulations or policies on the management of food waste in China. Currently, the Chinese food waste management regulations categorize two types of management processes: food waste generation and food waste treatment.[24] At present, food waste ends up in landfills or incineration without proper treatment. In response to growing pressure, the Chinese government attempted to control food waste by encouraging local governments to implement regulations as early as 2007. The government also implemented the "Notification of Trial Points on Kitchen Waste Resources Recovery, Utilization and Harmless Treatment" (关于组织开展城市餐厨废弃物资源化利用和无害化处理试点工作通知) in 2010, which was the first regulation on food waste recycling.

In 2011, the government introduced more specific measures in a document titled, "Suggestions on Controlling Gutter Oil and Food Waste Management" (关于加强地沟油整治和餐厨废弃物管理的意见). The definition of food waste refers to food waste such as fruit, vegetables and meat, and other trash from restaurants, hotels and other catering industries. The plan aims to provide certain standards for minimizing food waste, recycling it and turning it into harmless resources. The Chinese government has further strengthened the food waste management regulations since 2011. With the financial support of five national level ministerial departments,[25] 100 cities including Dalian and Suzhou have launched pilot food waste management projects. These cities either implemented food waste regulations or built facilities for treating food waste.

Food waste has become an important social and political issue in China since 2012. During Xi Jinping's administration, campaigns were launched to reduce food waste, making food waste a priority issue for both government and society.[26] The central government set the target for managing food waste in the national city–town plan on building harmless facilities for municipal waste in the Twelfth Five Year Plan (2011–2015). According to the plan, 50% of Chinese cities should have implemented food waste separation by 2015 and invested RMB$10.9 billion on food waste management.[27] In the subsequent Five Year Plan (2016–2020) the Chinese government strengthened the regulations on food waste management. All

[24] Liu, G. (2014). "Food Losses and Food Waste in China: A First Estimate," OECD Food, Agriculture and Fisheries Papers 66, OECD Publishing, p.3.

[25] These are the National Development and Reform Commission, The Ministry of Housing, Urban-Rural Development, Ministry of Finance, Ministry of Ecology and Environment and the Ministry of Agriculture.

[26] Liu J., J. Lundqvist, J. Weinberg, and J. Gustafsson (2013), "Food Losses and Waste in China and Their Implication for Water and Land", *Environmental Science & Technology*,47 (18)10142. DOI: https://doi.org/10.1021/es401426b

[27] State Council, "Notification on building harmless municipal waste treatment facilities during the Twelfth-Five Year Plan", 19 April 2012. Retrieved on: http://www.gov.cn/zwgk/2012-05/04/content_2129302.htm. [Accessed on 20 June 2019].

cities are expected to establish food waste management systems and increase food waste management capacity to 34.4 thousand tons per day.[28] In addition, the government has planned a schedule to collect the food waste, begun to regulate food waste to turn it into fertilizer and separate food waste.[29] Most Chinese cities, in particular tier 1 and 2 cities, are still struggling to find a solution to food waste. Although they have implemented policies and technical solutions for food waste recycling and resource recovery, the overall treatment capacity has been overloaded by the increasing amount of food waste generated. Ineffective food waste control will continue to have disastrous consequences for the environment.

Informal waste collectors and traders use food waste to create potential economic profit. They collect leftovers from restaurants and sell them to pig farms as feed or to small workshops where the leftovers are processed into secondhand oils that are then sold on to markets. The gutter oil scandal has shown the limitations of both food waste and food safety management (Table 7.1).

Food safety management in China has evolved in four stages: 1949–1978, 1978–1992, 1992–2004 and 2004–2013[30] The current Food Safety Law of the People's Republic of China implemented several regulatory initiatives in 2018. For example, baby milk formula should be registered with the National Medical Products Administration, and food purchased online should be regulated.[31] However, the Food Safety Law has not yet introduced regulations to prevent food waste being turned into edible food. "Suggestions on strengthening gutter oil and food waste management" (关于加强地沟油整治和餐厨废弃物管理的意见), promulgated in 2010, is the only official document that provides guidelines on managing food waste, in particular gutter oil. In the document, the central government suggests that all illegal food processing workshops should be shut down and that the sale of gutter oil to restaurants and school canteens should be banned. In addition, local regulations on controlling food waste have initiated pilot projects on recycling food waste and turning it into harmless resources.[32] A compatible solution between food waste control and food safety is needed to make the food system less wasteful and more efficient.

[28] National Development and Reform Commission, "Plan of building harmless municipal waste treatment facilities during the Thirteenth-Five Year Plan", December 2016. Retrieved on: http://www.ndrc.gov.cn/zcfb/zcfbghwb/201701/W020170123357045898302.pdf. [Accessed on 21 June 2019]

[29] Ibid.

[30] Zhou, G. (2017), *The Regulatory Regime of Food Safety in China*, Studies in the Political Economy of Public Policy, Palgrave Macmillian, page 74, DOI https://doi.org/10.1007/978-3-319-50,442-1

[31] Five highlights on new food security law, Hualv.com, 28 June 2019. Retrieved on: https://www.66law.cn/special/shipinanq/. [Accessed on: 30 June 2019]

[32] State Council, "Suggestions on strengthening gutter oil and kitchen waste management", 13 July 2010. Retrieved from: http://www.gov.cn/zwgk/2010-07/19/content_1658092.htm. [Accessed on 21 June 2019]

Table 7.1 Milestones of food waste management in China (2007–2017)

Year	Milestones
2007	Requirement for centralized treatment of restaurant food waste
2009	Revised Food Safety Law
2010	100 pilot cities implement food waste recovery programme
	Food and Drug Administration strengthens "gutter oil" regulations
2011	12th Five-Year Plan targeted waste management
2012	Campaign on cutting official extravagance and government reception meals
2013	"Clear your plate" campaign to reduce food waste in catering industry
2014	State Council gives guidance to reduce food waste and overconsumption
2016	Mandatory Waste Sorting Plan in the 13th Five-Year Plan
2017	Transparent Kitchen programme to improve food waste management processes and reporting in restaurants

Collective Responsibility (n.d.), Sustainability Insights: Shanghai's Food Waste Management, page 4. Retrieved on: https://www.coresponsibility.com/wp-content/uploads/2018/11/Shanghai-China-Food-Waste-Management-Report.pdf. [Accessed on: 21 June 2019]

7.3 Discussion: Limitations of Food waste Management

The case of gutter oil reveals the failure of food waste management and how such mismanagement creates a risk to food safety. It also highlights how far China has to go before it can have confidence in the most basic standards of food safety. As mentioned previously, improper food waste management can be a risk to the environment too. For example, unsorted food waste and unclear regulations for stakeholders can yield sanitary problems,[33] as well as environmental risks and threats to human health. According to the food waste hierarchy, food waste management should utilize various options ranging from prevention to disposal, while China's regulations on food waste control have been in vain. Existing food waste management regulations are formulated and implemented at the local government level, such as in Shanghai and Suzhou. These local regulations, however, have been inconsistent in terms of compensation and recycling systems. It is also questionable whether any department takes responsibility for managing food waste.

In managing food waste in China, however, some obstacles have been identified. Several government ministries and agencies are linked with food waste generation and treatment, including the Ministry of Agriculture (for agricultural production), the Ministry of Commerce (for food marketing and consumption) and the Ministry of Ecology and Environment (for household waste treatment and collection). However, the new food safety standards system is only approved by the Ministry of Health and lacks the necessary mechanism to coordinate the food safety standards from multiple food regulators in the stages of food production and processing. As a

[33] Thi, N., G. Kumar, C. Lin (2015), An overview of food waste management in developing countries: Current status and future perspective. Journal of Environmental Management, vol. 157, p. 220.

result, the food regulators have not had enough authority over food safety issues beyond their regulatory authority.[34] As such, the coordination among these departments is weak[35] and lacks a comprehensive and integrated plan for addressing food waste and the associated risks to human health.[36]

Studies of policymaking in China frequently use the model of fragmented authoritarianism to explain policy as an outcome of negotiations and compromises among conflicting state agencies as well as the central and local governments.[37] Since the economic reforms of late 1978, national governing structures have been decentralized to give more power to local governments in establishing the conditions necessary for markets to take root.[38] The tiao-kuai (vertical and horizontal authorities) system has been implemented in China, with bureaucracies (tiao) vertically tied to the superior power at the center, while local administrations (kuai) tie them horizontally to the power at the local government level.[39] As such, the tiao-kuai relationship poses challenges to collaboration among the agencies, leaving the provincial governments to be perceived as "independent kingdoms" as they have significant decision-making power, are entitled to extract financial resources and are resistant to external influence from the central leadership.[40]

Through decentralization and growing regional autonomy, a variety of food waste management responses have emerged at the local level. The legal framework clearly describes the responsibilities for controlling food waste at the local level. The central government suggests that local governments have a city plan for manag-

---

[34] Zhou, G. (2017), *The Regulatory Regime of Food Safety in China*, Studies in the Political Economy of Public Policy, Palgrave Macmillian, page 98, DOI https://doi.org/10.1007/978-3-319-50,442-1

[35] Liu J., J. Lundqvist, J. Weinberg, and J. Gustafsson (2013), "Food Losses and Waste in China and Their Implication for Water and Land", *Environmental Science & Technology*,47 (18)1043. DOI: 10.1021/es401426b; Liu, G. (2014). "Food Losses and Food Waste in China: A First Estimate," OECD Food, Agriculture and Fisheries, OECD Publishing, p. 13; Sinkule, B. J. and L. Ortolano, (1995), Implementing Environmental Policy in China. U.S.: Praeger Publisher, p.6.

[36] Ibid, p. 13.

[37] Lieberthal, K. (1997). "China's Governing System and Its Impact on Environmental Policy Implementation", China Environment Series 1, Wilson Centre, http://www.wilsoncenter.org/publication/china-environment-series-11997, p.3 (Accessed 28 June 2019).

[38] Mertha, A. (2005). China's "Soft" Centralization: Shifting Tiao/Kuai Authority Relations. The China Quarterly, 184, 791. doi:https://doi.org/10.1017/S0305741005000500

[39] Lieberthal, Kenneth G., and David M. Lampton, editors (1992) Bureaucracy, Politics, and Decision Making in Post-Mao China. Berkeley: University of California Press, page 12. http://ark.cdlib.org/ark:/13030/ft0k40035t/

[40] Ross, L. (1988). Environmental Policy in China. U.S.: Indiana University Press, p. 199; Shapiro, J. (2012). China's Environmental Challenges. U.K.: Polity Press, p. 69; Morton, C. (2010). "Policy Case Study: The Environment". In W. A. Joseph (Ed.), *Politics in China: An Introduction*. N.Y.: Oxford University Press, p.284; Heilmann, S. (2011). "Policy-Making through Experimentation: The Formation of a Distinctive Policy Process", in S. Heilmann and E.J. Perry (Eds.), Mao's Invisible Hand: The Political Foundations of Adaptive Governance in China. U.K.: Harvard University Asia Center, pp. 63. https://doi.org/10.1016/j.resconrec.2015.11.016. [Accessed on 23 June 2019]

ing food waste as a goal in treatment and resource recovery. Within their jurisdictions, they may also establish their own controlling food waste standards based on the national plan. Cities such as Suzhou and Shanghai have implemented relevant regulations for managing food waste with a focus on collection, treatment facilities and resource recovery. This highlights how food safety management in China can, in some cases, be most effective on a local level. A case can be made that organic certification and the claims made for 'green' or 'pollution-free' supermarket food best belong on intermediate and national levels of governance.

The institutional limitations are further exacerbated by fragmented agency cooperation. Both bureaucracies and local governments play a role in influencing agenda setting, and particularly in furthering their respective institutional interests.[41] In managing municipal solid waste as well as food waste, the Ministry of Ecology and Environment (MEE) outlined a plan for household waste collection and treatment in the 12th and 13th Five-Year Plans for environmental protection. However, the Ministry is challenged by several constraints. Interdepartmentalism diffuses the ministry's authority; for instance, the Municipal Bureau of City Administration and Law Enforcement, instead of the MEP, is now responsible for municipal waste management and waste recycling. Other government agencies, such as the National Development and Reform Commission and the Ministry of Health, are also involved in food waste management. The policy-making process for waste-related laws and regulations in China is full of political interests among the ruling elites across national and local levels.[42] As a consequence, the government departments have little incentive to cooperate and there is a lack of explicit authority in addressing the problem of gutter oil. The protectionism of local government also hinders cooperation among government departments and accountability among restaurants. Local governments may turn a blind eye to illegal gutter oil trading as it drives large profits. In 2011, a licensed company in Beijing, which legally collected used cooking oil, was exposed for diluting the oil and reselling it to restaurants at a huge profit.[43] It was also revealed that the local government turned a blind eye to this illegal behavior in consideration of tax revenue and government performance.[44]

[41] Ross, L. (1988). Environmental Policy in China. U.S.: Indiana University Press, p. 190.

[42] Alpermann, B. (2010). "State and Society in China's Environmental Politics". In J.J. Kassiola and S. Guo, (Eds.). China's Environmental Crisis: Domestic and Global Political Impacts and Responses. N.Y.: Palgrave Macmillan, pp. 130; Wu, L., Ma, L. and Qi Y. (2012). "An Analysis of Policymaking Process of Provincial Environmental Policy: Case Study of the Ecological Compensation Policy for Conservation of Water and Soil in Shannxin and Yunnan", China Population, Resources and Environment, 22(3): 90.

[43] Zhao, X. (2012), "Investigation on profit chain of gutter oil", Tencents.com, 20 April 2012. Retrieved from: https://news.qq.com/a/20120420/001619.htm. [Accessed on 23 June 2019]

[44] Zhou, G. (2017), *The Regulatory Regime of Food Safety in China*, Studies in the Political Economy of Public Policy, Palgrave Macmillian, pp.231–232. DOI https://doi.org/10.1007/978-3-319-50,442-1

7.4 Conclusion

The gutter oil crisis highlights not only the failure of China's food safety regime, but also the limitations of food waste management. Lacking national regulations on food waste management, weak departmental cooperation has undermined the capacity of food waste control. Although China's political system is a highly centralized hierarchy, decentralization has to some extent provided autonomy at the local level. Local governments are entitled to determine public policies in their own jurisdictions, to approve the budgets of their administrative areas and to appoint or remove administrative functionaries. Moreover, local governments have autonomy in retaining financial resources for promoting economic efficacy, and the central government rewards local government on the basis of economic performance. As a consequence, food waste management is not at the top of local government agendas but city-based schemes offer the best hope for reforming restaurant practices. In addition, the absence of transparent food information for the general public contributes to the scale of the food safety crisis. Have good regulatory settings over organic food claims is important, but the case of gutter oil highlights a need for more basic priorities to be satisfied. Government should play an active role in regulating the food labelling system to protect customers and build consumer trust and credibility in enterprises.

References

Alpermann B (2010) State and society in China's environmental politics. In: Kassiola JJ, Guo S (eds) China's environmental crisis: domestic and global political impacts and responses. Palgrave Macmillan, New York, pp 123–151

Bai J, Wahl TL, Lohmar BT, Huang J (2010) Food away from home in Beijing: effects of wealth, time and "free" meals. China Econ Rev 21(3):432–441

Collective Responsibility (n.d.) Sustainability insights: Shanghai's food waste management. Retrieved on: https://www.coresponsibility.com/wp-content/uploads/2018/11/Shanghai-China-Food-Waste-Management-Report.pdf. Accessed on 21 June 2019

He C, Han G, Liu Y (2019) Food safety satisfaction in China and its influencing factors: empirical study with a hierarchical linear model. Safety 5(1):17

Heilmann S (2011) Policy-making through experimentation: the formation of a distinctive policy process. In: Heilmann S, Perry EJ (eds) Mao's invisible hand: the political foundations of adaptive governance in China. Harvard University Asia Center, London, pp 62–101. https://doi.org/10.1016/j.resconrec.2015.11.016. Accessed on 23 June 2019

Li, J., Cui, N., & Liu, J. (2017). "Gutter oil: an overview of Chinese food safety issues and policies". Glob Health Promot, 24(3)75–78. 10.1177/1757975915623733

Lieberthal, K. (1997). "China's governing system and its impact on environmental policy implementation", China environment series 1., Wilson Centre, http://www.wilsoncenter.org/publication/china-environment-series-11997 (Accessed 28 June 2019)

Lieberthal K, Lampton DM (1992a) Bureaucracy, politics, and decision making in Post-Mao China bureaucracy, politics, and decision making in Post-Mao China. University of California Press, Berkeley

Lieberthal KG, Lampton DM (eds) (1992b) Bureaucracy, politics, and decision making in post-mao China. University of California Press, Berkeley. http://ark.cdlib.org/ark:/13030/ft0k40035t/

Lin AY, Huang ST, Wahlqvist ML (2009) Waste management to improve food safety and security for health advancement. Asia Pac J Clin Nutr 18(4):538–545

Liu G (2014) Food losses and food waste in China: a first estimate. OECD Food, Agriculture and Fisheries Papers 66, OECD Publishing

Liu J, Lundqvist J, Weinberg J, Gustafsson J (2013) Food losses and waste in China and their implication for water and land. Environ Sci Technol 47(18):10137–10144. https://doi.org/10.1021/es401426b

Mertha A (2005) China's "soft" centralization: shifting Tiao/Kuai authority relations. China Q 184:791–810. https://doi.org/10.1017/S0305741005000500

Mertha A (2010) Society in the state: China's nondemocratic political pluralization. In Gries PH, Rosen S (eds) Chinese politics: state, society and the market. Routledge, New York, pp 69–69–84

Morton C (2010) Policy case study: the environment. In: Joseph WA (ed) Politics in China: an introduction. Oxford University Press, New York, pp 278–287

Ngoc BDT, Kumar G, Lin C (2015) An overview of food waste management in developing countries: current status and future perspective. J Environ Manag 157:220–229. https://doi.org/10.1016/j.jenvman.2015.04.022. Accessed on 24 June 2019

Papargyropoulou E, Lozano R, Steinberger JK, Wright N, Ujang Z (2014) The food waste hierarchy as a framework for the management of food surplus and food waste. J Clean Prod (76):106–115. https://doi.org/10.1016/j.jclepro.2014.04.020. Accessed on 22 June 2019

Porter RC (2002) The economics of waste. Resources for the Future, Washington, DC

Price JL, Joseph JB (2000) Demand management: a basis for waste policy: a critical review of the applicability of the waste hierarchy in terms of achieving sustainable waste management. Sustain Dev 8(2):96–105

Rasmussen C, Vigsø D, Ackerman F, Porter R, Pearce D, Dijkgraaf E et al (2005) Rethinking the waste hierarchy. Environmental Assessment Institute, Copenhagen

Ross L (1988) Environmental policy in China. Indiana University Press, U.S.

Shapiro J (2012) China's environmental challenges. Polity Press, Cambridge

Sinkule BJ, Ortolano L (1995) Implementing environmental policy in China. Praeger Publisher, Westport

Thyberg KL, Tonjes DJ (2016) Drivers of food waste and their implications for sustainable policy development. Resour Conserv Recycl 106:110–123

Wu L, Ma L, Qi Y (2012) An analysis of policymaking process of provincial environmental policy: case study of the ecological compensation policy for conservation of water and soil in Shannxin and Yunnan. China Popul Resour Environ 22(3):87–92

Zhou G (2017) The regulatory regime of food safety in China. Studies in the Political Economy of Public Policy, Palgrave Macmillian. https://doi.org/10.1007/978-3-319-50442-1

Natalie W. M. Wong currently is Visiting Fellow in the Department of Public Policy, City University of Hong Kong. She completed her PhD in Politics at the University of York, United Kingdom; and her dissertation mainly focuses on incinerator and waste policy in China. Despite her working experiences in academia, she was the researcher at a Hong Kong-based labour NGO which concerns the labour conditions in Pearl River Delta of China and at an international environmental NGO investigating the air quality of Beijing before commencement of the Olympic Games in 2008. She has published extensively on topics on environmental governance and environmental activism in *China Information*, *Voluntas* and *Journal of Environmental Planning and Management*.

Part III
National Regulatory Jurisdictions

CHINA – Chap. 8
INDIA – Chap. 9
SOUTH-EAST ASIA – Chaps. 10, 11 and 12
PACIFIC – Chaps. 13, 14 and 15

Chapter 8
Regulatory Issues on Organic Food in China

Hui Li

Abstract China is a growing market in organic food consumption. The new regulations and laws and their strong enforcement have created greater confidence and mutual trust among the stakeholders. After an overall review of the developing process of organic food law system in China, this article characterises the positions of different stakeholders in the entire system. It considers the opportunities, challenges and deficiencies in the design of the organic legal system, including insufficient punishment of those who violate the law, the need to strengthen collaboration and integration among the supervising authorities, and reinforce the public transparency. In total, there is still a relatively long way to go for the construction of a more effective organic food law system in China.

Keywords Regulatory issues · Organic food · China

8.1 Introduction

China has used practices associated with eco-systematic agriculture for thousands of years without knowing the concept of "organic farming". Over the past century, organic farming in Western countries can trace its roots to Rudolf Steiner's call for a differentiated agriculture in 1924, which was developed through the biodynamic agriculture movement in 1930s. Lord Northbourne invented the concrete concept of organic farming of 1940 and flowing from this was, the foundation of IFOAM (International Federation of Organic Agriculture Movements) in 1972, as well as the latest development in standards and certification.[1]

[1] John, Pawll. 2014. *Opportunities and Challengers for organic food and agriculture: China and Australia*. Chapter 3 in B. Mascitelli & B.O'Mahoney(Eds.), Good for All: Developing knowledge relationships between China and Australia (pp.50–80) Ballarat: Connor Court Publishing.

H. Li (✉)
School of Law, Dalian University of China, Dalian, China

GOH Bee Chen, R. Price (eds.), *Regulatory Issues in Organic Food Safety in the Asia Pacific*, https://doi.org/10.1007/978-981-15-3580-2_8

A U.S. professor, and agricultural scientist, Franklin Harim King, in his book *Farmers of Forty Centuries*, recounted his 9-month in-site investigations into what would now be called organic farming or sustainable agriculture of Asia in 1909. He described in his book the practices which had been developed in China over 2000 years. Many eco-cycles have been documented such as the typical fish-mulberry tree-silk cycle in Taihu Lake Basin. He praised that, "China …long ago struck the keynote of permanent agriculture …it remains for us and other nations to profit by their experience".[2] This work has been described by Lord Northbourne—the founder of organic agriculture—as a "classic" which "no student of farming or social science can afford to ignore".[3]

A leading anthropologist of China, Fei Xiaotong commented in *Farmers of Forty Centuries* that: Professor King recorded and documented Chinese agriculture on the basis of earth, he described the Chinese people as one part of entire ecological balance, which is the recycling between earth and humans. Humans are born from the earth, their excreta returns to earth, their body returns to earth after death, generation after generation, repeatedly. Depending on this recycling, humans have lived on this earth for many millennia, as part of this recycling system. Agriculture in this meaning is not opposite to earth; it is a harmonious agriculture.[4]

Although China has been a late-comer to the international organic movement, it has turned out to be the world's biggest potential consuming market[5] and the third largest country of organic agricultural land, with 3 million hectares, following Australia and Argentina in 2017.[6] According to the report of global organic market by the Institute of Germany Economic Research, in 2017 the consumption of organic food of Chinese market valued 85.1 billion EUR. Its organic food sales have been growing at around 10% annually in recent years. This constitutes a "high growth period" that exceeds the speed of the growth speed of U.S. market. That means greater attention will be given to the Chinese organic market.

The aim of this paper is to provide an overview of the Chinese regulatory framework concerning organic food. Section 8.1 reviews the developing process of organic food law system in China. Section 8.2 analyses the characters and positions of different stakeholders in the entire system. Section 8.3 discusses the opportunities, challenges and deficiencies in the design of the organic legal system. Section 8.4 makes conclusions.

---

[2] F.H.King(作者),程存旺(译者),石嫣(译者).2011.四千年农夫:中国、朝鲜和日本的永续农业.北京:东方出版社.

[3] Lord, Northbourne. 1940. Look to the land. London: Dent; and Paull, John. 2006. Permanent Agriculture: Precursor to Organic Farming. Journal of Bio-Dynamics Tasmania 83: 19–21.

[4] Fei Xiaotong. 1985. From 陈仁端.谈《四千年农夫》. http://wen.org.cn/modules/article/view.article.php/2680. Accessed 30 May 2019.

[5] 青木.德报告:中国人爱上有机食品,但"偏食". 环球日报. 2019-02-25; Also available on http://world.huanqiu.com/exclusive/2019-02/14397616.html?agt=15422. Accessed 27 May, 2019.

[6] IFOAM. The world of Organic Agriculture. https://www.ifoam.bio/en/news/2019/02/13/world-organic-agriculture-2019. Accessed 27 May 2019.

8.2 Overview of the Developing Process of Organic Food Law System

Many forms of sustainable agriculture in China are based on thousands of years of traditional practices including crop rotation and compost application with organic matter recycling. China also features traditional ecological systems in agriculture like mulberry trees—fish ponds, lotus root-fish ponds, canola oil-bee-fish & crab etc. Such use elements from natural environment to grow produce, help maintain soil fertility and manage ecosystems. On food product control, China has a long history of quality certification system, including the use of quality seals in trade process, which can be dated back thousands of years[7] and still exists in many agricultural products such as meat.

The recent development of certified organic agriculture in China has been based completely on the concept, standards, organization, accreditation, monitoring, and trade developed in the Western countries. In June of 1990, for the first time, organic tea from two tea plantations of Lin'an County of Zhejiang Province was exported to the Netherlands with SKAL certification, marking the start of organic certified production in China.[8] In May of 1990, the Ministry of Agriculture (MOA) initiated the Green Food Program. In 1992, the China Green Food Development Centre (CGFDC) was created to manage this programme specifically.[9] In 1994, the State Environment Protection Administration (SEPA) (restructured into MEE, Ministry of Ecology and Environment of PRC in 2018) set up the Organic Food Development Centre (OFDC) under its subordinated Nanjing Institute of Environmental Sciences, symbolising the national implementation of administration on development and certification of organic food. Thereafter, the OFDC successively established branches in 23 provinces, municipalities and autonomous regions such as Yunnan and Heilongjiang to be responsible for local quality control and management on organic food.[10] Since then, organic agriculture in China has been booming with international production and trade in organic foods developing rapidly. At the same time, the domestic market is also booming.

The CGFDC is specialized in responsible for Green Food labelling permissions, organic agricultural product certification, agricultural product geography labelling, including registration and protection, and coordination and guide local pollution-free agricultural product certification".[11] Therefore, the Green Food Program certifies voluntary quality systems that in that agricultural environment, production, pesticide and fertilizer usage, processing and package of the food products meet

[7] Biao Xie, Li Tingjou and Qian YI. 2011. Organic certification and the market: organic exports from and imports to China. British Food Journal 113 (10): 1200–1216.

[8] 纯天然无污染　我国有机茶渐成市场新宠, http://www.ofdc.org.cn/en/article_info.asp?n_id=679. Accessed 29 May 2019.

[9] http://www.greenfood.agri.cn. Accessed 29 May 2019.

[10] http://www.ofdc.org.cn/article.asp?m_id=1. Accessed 7 June 2019.

[11] http://www.greenfood.org.cn/jgjs/zxjj/201110/t20111018_5904305.htm. Accessed 30 May 2019.

relevant requirements. In a broad sense, organic food products and pollution-free food products are also declared included into the Green Food Program. Since 2001, MOA has carried out the nationwide Plan of Pollution-Free Agricultural Products. The system for pollution-Free Agricultural Products concerns food production in the particular area, Production processes and product quality are to meet relevant national standards. The certification process of Pollution-Free Agricultural Products is independent to the Green Food.[12]

From the beginning, the Green Food Program has enjoyed success. Green Food is popular in big supermarkets in China, enjoying a stable share in the retail market. Until 2012, 6622 Green Food companies have been certified, and 16,825 products were certified as Green Food products. From 2001–2011, the national sales of Green Food had increased from RMB 500 billion to 3134.5 billion. Exports had increased from 4 billion to 23.1 billion US dollar, the average annual growth rate had been around 20%, according to the report of the Vice Dean of Green Food Development centre in an international conference in 2014.[13] The Green Food certification is a certification trademark, and its issuance is governed by the Administrative Measures for Green Food Labelling (2012) issued by the MOA.[14]

The Pollution-Free Agricultural Products Plan was the first innovation in the development strategy of modern food safety in the agriculture of China. The Green Food Program was the second step, and the Organic Food Program was the third step. The Pollution-Free Agricultural Product Plan is the basic requirement of food safety. All ordinary food should meet this requirement. The Green Food program is the transitional process in the march toward China's Organic Food Program. Theoretically and practically, the certifying standards for Organic Food are much stricter than for Green Food and the certifying standards for Green Food are more onerous than those applying to Pollution-Free Food.

OFDC received the IFOAM recognition at the beginning of 2003. This was described as a milestone of the narrowing the gap in certificating administration between China and western countries.[15] The Chinese National Organic Standard (GB/T19630), first issued and implemented in 2005, was supplemented by the full regulation's implementation on 2006. The development of China's organic sector was consolidated by the issuing of this national regulation for certification and accreditation. The Standard was revised on 2011 (GB/T 19630.1-4—2011). The third version of the standard will be effective from 2019.[16] The regulation includes

[12] Alberto Scorzon, Bernd Van der Meulen, Li Jiao. 2014. Organics in Chinese Food Law. 9 Eur. Food & Feed L. Rev. 179.

[13] 王建平. 2012. 绿色食品认证管理制度与标准体系. http://www.mpoc.org.my/upload/P7_WangJianPing_POTSChina2012_chn.pdf. Accessed 20 May 2019.

[14] 绿色食品标志管理办法. http://jiuban.moa.gov.cn/zwllm/tzgg/bl/201208/t20120802_2814698.htm. Accessed 30 May 2019.

[15] 国家环保总局有机食品发展中心(OFDC)十年发展情况简介. http://trhj.mee.gov.cn/nchjgl/201604/t20160424_335971.shtml. Accessed 26 May 2019.

[16] Zhou Zejiang, senior advisor of Organic farming Development Centre and-MEP. IFOAM, the Research Institute of Organic Agriculture (FiBL). 2019. The world of Organic Agriculture. https://www.ifoam.bio/en/news/2019/02/13/world-organic-agriculture-2019. Accessed 27 May 2019.

requirements for imports to be certified by an approved certification body whose inspection personnel have specific qualifications and training in relation to the Chinese organic regulatory system.[17] The certificate must be obtained prior to the import, except the country that have signed mutual recognition arrangement with China for certified organic food.[18]

According to the Chinese National Organic Standard (GB/T 19630.1-4—2011), organic agriculture is an agricultural production method that follows a particular agricultural production principle. No genetically modified organisms and related products are permitted; nor are synthetic pesticides, fertilizers, growth regulator, feed additives allowed in production. Agricultural producers are required to respect natural rules and ecological principles. This includes coordinating the balance of crop farming and the breeding industry as well as maintaining sustainable agricultural technology to ensure a steady and stable agricultural production system. Organic products are defined as "products for human and animal consumption, which are produced, processed and sold according to this standard".[19]

In 2005, the issuance of the Chinese National Organic Standard included a national logo for organic products. This is a bilingual logo, in Chinese and English, and is comprised of two versions, "Organic" and "Conversion to Organic".

At present, organic claims are governed by the Administrative Measures for Organic Product Certification (AMOPC) (2014 came into force, 2015 modified)[20] issued by the AQSIQ (General Administration of Quality Supervision, Inspection and Quarantine). This body was restructured into SAMR, or State Administration for Market Regulation in 2018. Under its revised Measures, in response to the misuse of the logo "Conversion to Organic", only the use of the "Organic" mark was kept and the former was cancelled. "Organic" is a special term and it cannot be used to describe a food product unless it is certified by the required professional institution, including its Chinese "有机" (organic). Other words or similar logos or pictures that may mislead the public must be avoided. This requirement also applies to imported food.[21] Furthermore, every certification mark on a qualified organic product shall bear a unique number (organic numeric code), and take effective anti-fak.e and traceability technology to ensure that the product can be traced to its

[17] Article 10, article 19 and article 20 of Administrative Measures for Organic Product Certification, AMOPC.

[18] 王晓东,2016.中国与新西兰签署首份有机产品认证互认协议. http://cn.chinadaily.com.cn/2016-11/14/content_27368355.htm. Accessed 1 June 2019.

[19] The Chinese National Organic Standard (GB/T 19630.1–4—2011). Organic products—Part 1: Production. http://www.bcschina.com/UploadFiles/2018-06-07/15283703244196851.pdf. Accessed June 62,019.

[20] 有机产品认证管理办法(总局令第166号). http://www.cnca.gov.cn/bsdt/ywzl/flyzcyj/bmgz/201210/t20121024_36669.html. Accessed 30 May 30 2019.

[21] Article 32, article 35, chapter 3 of AMOPC. http://www.cnca.gov.cn/bsdt/ywzl/flyzcyj/bmgz/201210/t20121024_36669.html. Accessed 1 June 2019.

corresponding certificate, as well as the producer and processor of the certified product.[22]

The China online organic food product traceability system, The National Certification and Accreditation Information Public Service Platform, was set up in June of 2018.[23] It is similar to those operating in the other big organics consuming countries worldwide. If the customer needs to check the organic food at hand, one can log in the service platform, input the organic numeric code printed on the food product and then the specific details, including name of the enterprise, name of the certifying institution, packing specification, certificate number, category of certification, etc. are all presented.

In 2018, with the institutional restructure of the state council, the Certification and Accreditation of Administration of China (CNCA) was merged into the newly established "State Administration for Market Regulation". The organic sector is now mainly under the supervision of the new body.

8.3 Stakeholders Involved

Stakeholders involved in organic food industry include the main bodies who are responsible for enforcement of organic food laws (supervising authorities), the certifying institutions and food producers (processors included).

8.3.1 Supervising Authorities

Several organic food scandals were recently reported by the national media, CCTV and Xinhua Web.[24] The scandals related to the deceptive use of organic marks in food, in spite of the use of ingredients from conventional agricultural production. A need was felt to apply stricter standards and a more severe implementation of the rules and clauses. The Chinese National Organic Standard is undergoing a third modification. This version of thc standard will be effective in 2019.

The Certification and Accreditation Administration of China (CNCA),[25] a subordinate body to AQSIQ (now SAMR) is in charge of the national plan to promote organic farming in China. It exercises administrative responsibilities of undertaking

[22] Article 33 of AMOPC.

[23] 佟明彪，2018.全国认证认可信息公共服务平台"认e云"上线.经济日报-中国经济网. http://www.ce.cn/cysc/ztpd/2018/cdf_299595/dt/201806/08/t20180608_29379934.shtml.. Accessed 30 May 2019.

[24] 有机食品有玄机. Reported by Focus Report of China Central Television (CCTV) on 6 May 2018. 夏丹，2018.有机食品公司揭底:虚假标注+认证违规.Xinhuanet. http://www.xinhuanet.com/legal/2018-05/15/c_1122832164.htm. Accessed 28 May 2019.

[25] http://english.cnca.gov.cn/. Accessed 28 May 2019.

unified management, supervision and overall coordination of certification and accreditation activities across the country, among them including issuing the official organic certificate through the certifying institutions, granting licenses to individual organic inspectors employed by the certifying institutions.

The current Chinese certification and accreditation system has two governmental agencies involved, Certification and Accreditation Administration of China (CNCA), and China National Accreditation Service for Conformity Assessment (CNAS)[26] under the CNCA, who act as: an approval body; and a sole official accreditation body, respectively. Those certifying institutions should first register with and be approved by CNCA and then accredited by CNAS before their certification operation starts.

Administrative Measures for Organic Product Certification (AMOPC), which was issued in 2014, covers a broad range of aspects that include the competences of the certifying institutions as well as the responsibilities of the authorities on monitoring the functioning of the certification activities. Local branches of AQSIQ shall supervise foreign invested certifying institutions that deal with import and export business,[27]while the local quality and technical supervising departments shall supervise Chinese-funded certifying institutions dealing with the domestic market.[28]

8.3.2 *Certifying Institutions*

In 1994, Organic Food Development Center (OFDC) was founded to explore the internationalization road on organic products. OFDC is the first specialized organization in China who is engaged in organic agriculture research and certifying activity.[29] OFDC was accredited by IFOAM in 2003.

In 2002, the MOA appointed the Chinese Organic Food Certification Center (COFCC) to certify and promote the organic food sector. Since then COFCC is top-ranked of the organic production certifiers nationwide.[30]

To June 2019, China has 80 Certifying institutes all approved by CNCA and accredited by CNAS.[31] These institutions can be classified into three categories: subordinate organizations to MOA such as COFCC; affiliates to universities such as NAC (Northwest A&F University Certifier)[32]; and private ones, including a few international certifying institutions such ECOCERT (ECOCERT SA, France). A

---

[26] https://www.cnas.org.cn/english/lawsandregulations/12/718389.shtml. Accessed 28 May 2019.

[27] From institutional restructuring of State Council of 2018, merged into General Administration of Customs, GAC.

[28] Article 38 AMOPC.

[29] http://www.ofdc.org.cn/article.asp?m_id=1. Accessed 29 May 2019.

[30] http://www.ofcc.org.cn/index.php?optionid=669. Accessed 29 May 2019.

[31] http://ffip.cnca.cn/ffip/publicquery/certSearch.jsp. Accessed 17 June 2019.

[32] https://cszx.nwsuaf.edu.cn/rzzx/zxjj2/index.htm. Accessed 10 June 2019.

foreign-funded enterprise seeking to be qualified as a certifying institution shall, in addition to the requirements prescribed to the domestic ones, meet two more requirements, (1) be accredited by an accreditation body in its home country or region, and (2) have engaged in certification activities for no less than 3 years.[33]

AMOPC lays out the responsibilities of certifying institutions and procedures,[34] at the same time, the business activities of the institutions have to meet the stipulations of the Regulations on Certification and Accreditation. Among other responsibilities, when a certifying institution accepts entrustment, it shall assign certifying inspectors to conduct on-site inspection on production and processing of organic products, as well as entrust inspecting and testing institutions with qualifications to inspect and test the products to be certified.[35] A certifying institution is required to ensure the integrity, objectivity and authenticity of the certification process, and make a complete record for archive retention to ensure that the certification process and results are traceable.[36] The record-keeping period of all activities shall be 5 years.[37] Furthermore, the certifying institution shall conduct effective follow-up inspection to the products certified and their producing and processing processes, to ensure that the certification conclusions can meet the certification requirements continuously.[38] In cases of misconduct or fraudulent practice by the producer, the certifying institution shall suspend or revoke the certificate and announce it to the public.[39] The certifying institute can be fined, suspended from operation, or even disqualified when failing to do so.[40] When a certifying institution issues a false conclusion or the conclusion is highly inconsistent with the facts, its qualification shall be revoked and then be announced to the public. In such cases, both the management in direct charge and the certifying personnel shall be disqualified, and criminal liability can be investigated according to the law concerned.[41]

8.3.3 Producers

A Chinese organic farm Donghuang Farm (东荒农场) located in north-eastern China Dengta City of Liaoning Province was set up in 1996. It received the organic certification on June 13, 2017. The farm covers an area of 1200 Mu, around 80 Hectares. From 2001, the whole farm has exercised large-scale agricultural

[33] Article 11 of Regulations on Certification and Accreditation (2016 revised).

[34] See Chapter II and Chapter IV of AMOPC.

[35] Article 10 of AMOPC.

[36] Article 12 of AMOPC.

[37] Article 12 of AMOPC.

[38] Article 13 of AMOPC.

[39] Article 30 and 31 of AMOPC.

[40] Article 54 of AMOPC, article 60 of the Regulations on Certification and Accreditation.

[41] Article 62 of the Regulations on Certification and Accreditation.

mechanization.[42] As a successful conversion example to organic farming, it attests that the certifying process requires "over 20 indicators of agricultural pesticide residue assessment which are all required to be 0, and this criterion is harsh to the rice production business".[43]

Producers, especially the ones who have not yet built their own sale channel to international or national markets, unavoidably encounter significant difficulties including the certifying costs and the big investment during the 3-year conversion period. For such reasons, the small business food producers including the famous Sun Commune avoiding applying for an organic certificate despite having sustainable practices. The Sun Commune is described as "a thriving eco-farm" in the report of New York Times, located in Lin'an County, Zhejiang Province from where coincidently the first organic tea was exported. It has a well-known reputation as the most beautiful pig sty in China. The beautiful and humane pig barn of its organic pigs was designed and built up by a professor of the China Academy of Art,[44] which was reported by New York Times.[45]

Concerning the organic certification, the creator of the Sun Commune Cheng Wei said: "the truthfulness of organics is in the heart of people … our certification is on the 'face' of the farmers."[46] Cheng always illustrates the reliability of his business with these words. In Sun Commune, beside every piece of field or farming section, there is a post erected to show the details of the particular responsible farmer, including his/her name, picture, scope of responsibility, and even his/her motto. Furthermore, the farmers' pictures are printed on the package of organic products bought by customers of Sun Commune. To Cheng's point of view, this method of demonstration of the producers should be better than the official certification.

For Chinese producers, the choice to adopt organic agriculture was mainly market-driven. Many of the market-oriented organic agricultural enterprises are joint partnerships between trading companies and farmers like Sun Commune. This type of enterprise is much more common in the south-eastern coastal regions of China where the private economy is more energetic.

Three main types of ownership structure can be concluded among the organic sector. The state owned ones, a historical market presence, have export-oriented

---

[42] 夏晓峰, 2018. 东荒农场发展实现从量到质的转变. Liaoyang Daily. Also available Ly.nen.com.cn. Accessed 11 June 2019.

[43] 东荒农场微信公众号. Donghuang Farm Wechat.

[44] 李莉容. 2017. 一个有机农场, 它的生意从 "中国最美猪圈" 开始|有机这生意怎么样了①_商业_好奇心日报. https://www.qdaily.com/articles/40071.html. Accessed 10 June 2019 人民网. 2015. http://society.people.com.cn/n/2015/0414/c136657-26843237-6.html. Accessed 10 June 2019

[45] Amy, Chen. 2016. Architects Seize on Potential in China's Countryside – The New York Times. https://www.nytimes.com/2016/06/18/arts/design/architects-see-potential-in-chinas-countryside.html. Accessed 10 June 2019.

[46] 李莉容, 2017. 一个有机农场, 它的生意从 "中国最美猪圈" 开始|有机这生意怎么样了①_商业_好奇心日报. https://www.qdaily.com/articles/40071.html. Accessed 10 June 2019.

markets. There are also Chinese privately owned ones, such as family farms, which run on principles such as the updated version of household contractual management. Donghuang Farm is a typical Family Farm type. Sun Commune is another example of Chinese privately owned ones. Their products are usually sold to the domestic market or exported depending on the actual product, like the products of the Half Hectare Organic Farm, Xinjiang Uygur Autonomous Region.[47] Lastly, there are private foreign owned entities, established by foreign investors targeting the precise demands of international markets.

In Chinese organic markets, both domestic producers and foreign producers must meet the equivalent organic standards of China. As for imported food products, they have to meet the specific import rules for conventional agricultural products as well. Producers of organic products must voluntarily entrust and submit application materials prescribed to a certifying institution to obtain the certificate, organic mark and the subsequent sales permit.[48] The traceability system, as well as a complete record of all the operations, including procurement, storage, transportation and sales activities, must be set up for ensuring product quality and safety.[49] Furthermore, the organic certificate has to be renewed annually. These stipulations can be considered as effective efforts to build a high level of reliability and trust between customers and organic producers.

8.4 Opportunities, Challenges and Deficiencies in the Design of the Organic Regulatory System

From 2003 to 2019, Chinese authorities issued the first policy document at the beginning of each year, underscoring the importance of rural reform and agriculture modernization. The policy document is issued by the Central Committee of the CPC (Communist Party of China) and the State Council every year and has been dubbed the "No 1 Central Document". This is the 16th consecutive year in which the document has focused on the "three rural issues"—agriculture, rural areas and farmers. The document reinforced "vigorously developing the production in shortage and green high quality agricultural products, promoting agriculture to switch from quantity increase-oriented to quality improvement-oriented".[50]

From the No. 1 Central Document, the commitment to the development of green food plan (including organic farming) of the authorities is visible. It is obvious that the central government do pay attention to the general benefits of organic

[47] 杜源良，2018半亩园:用心种好有机菜，打造新疆—流有机品牌-要闻-南方农村报. http://wap.nfncb.cn/show-730-1164749-1.html. Accessed 10 June 2019.

[48] Article 8 and article 14 of AMOPC.

[49] Article 41 of AMOPC.

[50] 中共中央 国务院关于坚持农业农村优先发展做好"三农"工作的若干意见. http://www.moa.gov.cn/ztzl/jj2019zyyhwj/2019zyyhwj/201902/t20190220_6172154.htm. Accessed 12 June 2019.

agriculture in terms of conservation of resources. It is becoming common sense in China that organic agriculture is a concrete and applicable way to reduce the environmental contamination, to minimise the use of chemical synthetic substances, and to reduce the public health costs of pesticide residue poisoning.

In recent years, the unprecedented promotion of restoration of the natural environment facilitates the development of organic agriculture. It provides qualified natural resources that organic agriculture needs. A large number of projects relevant to environmental governance and ecological restoration are being successively implemented; at the same time, the government frequently has enacted a series of favourable policies, even through a Public Private Partnership (PPP) model to advance the development of the environmental restoration industry. The basic condition for organic agriculture is no pollution in soil, air and water. Restoration of the natural environment, not only can ensure product quality, but also can enhance production of organic agriculture, ultimately lowering production costs and increasing the profits.[51]

As a nation of a long history in the practice of ecological agriculture, the idea of natural ecological agricultural model is still kept deep in the mind of the ordinary people of China, which is proved by the existence of unique and visionary eco-cycles in some regions. It will translate into a natural tendency to transfer to organic agriculture (a more international modern concept), or "turn back" to this ecological agriculture in the future, from a historical perspective. This continuity can be seen as one of the big potential subjective advantages for China to develop organic agriculture.

Although objectively, the developments of legislation and the central government's increased awareness of the environmental issues in China both push forward the development of organic agriculture, the challenges are still arduous for the organic sector. Some of the remaining challenges will be discussed below.

8.4.1 Pollution

Environmental pollution is an unavoidable obstacle to developing organic agriculture. Compared to other parts of the world, China has a relatively serious pollution problem, with air pollution sometimes "at worrying level". The melamine in milk scandal, water pollution, heavy metals (lead, cadmium) discovered in grains and vegetables, mycotoxins in grains and nuts, and illegal use of veterinary drugs in livestock are causes for serious concern, as are illegal use of pesticides in vegetables and tea. Analysis has found that the main causes for these issues are environmental pollution. Small-scale and scattered farming (planting and breeding) are also implicated.[52] To effectively conquer the negative impacts of environmental pollution of

[51] 2017-2021年中国有机农业行业发展及预测分析-北京华夏元素有机农业产业联盟. http://www.choaia.com/newsitem/278038460. Accessed 12 June 2019.

[52] Joseph J. Jen, Junshi Chen. 2017. Food Safety in China: science, technology, management and regulation. New Jersey: John Wiley & Sons, Inc.

water, soil and air is the first challenge to the development of a sustainable and toxin-free organic agriculture sector in China.

8.4.2 Premium

In China, price premiums for certified organic food are excessive in some regions, even higher than in North America, Europe and Australia.[53] For example, organic rice commands premiums of ten to twenty times more expensive than non-organic versions in China.[54] Such high premiums push most potential customers out of the door of the organic food market and they lose interest in organic food completely. Practically, the reasonable and sustainable way for organic food should "charge what the market will bear". That also may be one aspect reflecting the immaturity of the organic market. It is reasonable to expect organic premiums in China to drop to a more manageable range that is witnessed elsewhere in the world, and, then we can expect to see increasing sales volumes at the same time.[55]

8.4.3 Consumer Awareness

A lack of consumer awareness of organic products, together with opportunistic practices by some producers and retailers, continues to provide obstacles to the development of an organic market in China. Large quantities of consumers including middle-class people are more committed than ever to healthier lifestyles but have no clue what "organic" means in concrete terms. Most people are still puzzled about the differences between pollution-free food, green food and organic food in the market, especially when organic production is a relatively new concept, as well as being a foreign one. For the promotion of organic food in China, just as suggested, there is a need to clearly explain the meaning of the concept "organic". Educational campaigns need to be jointly organized by authorities and industry to raise awareness of and clarify the term.[56]

---

[53] John, Pawll. 2014. *Opportunities and Challengers for organic food and agriculture: China and Australia*. Chapter 3 in B. Mascitelli & B.O'Mahoney(Eds.), Good for All: Developing knowledge relationships between China and Australia (pp.50–80) Ballarat: Connor Court Publishing.

[54] 李劼,2018.有机食品"真假"谁说了算?南方日报. http://www.xinhuanet.com/2018-03/14/c_1122533817.htm. Accessed 13 June 2019

[55] John, Pawll. 2014. *Opportunities and Challengers for organic food and agriculture: China and Australia*. Chapter 3 in B. Mascitelli & B.O'Mahoney(Eds.), Good for All: Developing knowledge relationships between China and Australia (pp.50–80) Ballarat: Connor Court Publishing.

[56] Chen, J. 2012. A study investigating the determinants of consumer buying behaviour relating to the purchase of organic food products in urban China. PhD thesis. Swinburne University OF Technology, Melbourne.

8.4.4 Effective Supervision and Transparency

As certifying process is essential to obtaining the organic certificate, the "paying too much attention to the certification, and no attention to supervision" is considered the prominent reason for the messy situation of the Chinese organic market.[57] According to the opinion of an industry expert, on the prominent problem of fraudulent advertisements in the organic food market: parties 'pay too much attention to the certification, no attention is paid to supervision".[58] Thus, certifiers are interested in the charges and ignores its supervisory responsibilities, therefore, 'the government authorities should strengthen supervision of the certifying institutions, to keep the entire chain run in an effective and transparent system.'[59]

8.4.5 Coordination of Authorities

Collaboration and cooperation among the supervising authorities should be further reinforced; it is vital to the development of the organic commerce in China. At present, different authorities with vary standards were currently involved in certification and identification of organic products in China, resulting in problems for consumers, making them confused and suspicious about the credibility of organic food. The authorities involved at least including SAMR (State Administration for Market Regulation) and its internal department CNCA and CNAS, MARA (Ministry of Agriculture and Rural Affairs) and its subordinate unit CGFDC (having an internal department Organic Center[60]), GAC (General Administration of Customs) and its subordinated local quality and technical supervision departments, and the MEE (Ministry of Ecology and Environment, before 2018 is Ministry of Environmental Protection) and its OFDC.[61]

[57] 夏丹, 2018.有机食品公司揭底:虚假标注+认证违规.Xinhuanet. http://www.xinhuanet.com/legal/2018-05/15/c_1122832164.htm. Accessed on 28 May 2019.

[58] 夏丹, 2018.有机食品公司揭底:虚假标注+认证违规.Xinhuanet. http://www.xinhuanet.com/legal/2018-05/15/c_1122832164.htm .Accessed on 28 May 2019.

[59] 夏丹, 2018.有机食品公司揭底:虚假标注+认证违规.Xinhuanet. http://www.xinhuanet.com/legal/2018-05/15/c_1122832164.htm. Accessed on 28 May 2019.

[60] http://www.greenfood.org.cn/jgjs/nscs/yjzx/. Accessed on 14 June 2019.

[61] 中心简介_生态环境部南京环境科学研究所. http://www.nies.org/jgsz/kjzx/hjbhbyjspfzzx/zxjj/. Accessed 15 June 2019.

8.4.6 Perfection of the Legislation

The penalty paid by those who violate the AMOPC is too weak. Transgressions need stronger vigilance such as those under article 48 (counterfeit, altering, taking off, illegal trade, transfer, amending the organic certificate), article 50 (label non-certified products as "有机" or "organic" or other description or picture that will likely misleading the public) and article 53 (the certifying institution certifies and issues an organic certificate to those processed products that the amount of organic ingredients of which is less than 95%). All of the activities above-mentioned will be fined to up to 30000RMB (about US$4285). The cost for violation of the law is too low. Without powerful sanctions in the cases of producing and marketing fake and shoddy goods in organic market, how can it be stopped effectively?

8.5 Conclusion

As a nation who has a great depth of agricultural expertise, and whose population has a deep food culture, China has farmers who can draw on a history of millennia of sound ecological agriculture. China can lead the world in organic agriculture if develops the right will.[62] The Chinese government has been making effort to make more reliable and stricter organic law and standards during the past years and has achieved impressive outcomes, however, the growth of the organic sector remains affected by serious regulatory issues. Environmental pollution, reasonable premiums, consumer awareness, effective supervision and transparency, the coordination of authorities, and perfection of the legislation, are some of the challenges lying ahead.

References

Articles

Paull J (2006) Permanent agriculture: precursor to organic farming. J Bio Dyn Tasmania 83:19–21

Scorzon A, Van der Meulen B, Jiao L (2014) Organics in Chinese Food Law 9. Eur Food Feed L Rev 179:179–186

Xie B, Tingjou L, Qian YI (2011) Organic certification and the market: organic exports from and imports to China. Br Food J 113(10):1200–.1216

[62] John, Pawll. 2014. *Opportunities and Challengers for organic food and agriculture: China and Australia*. Chapter 3 in B. Mascitelli & B.O'Mahoney(Eds.), Good for All: Developing knowledge relationships between China and Australia (pp.50–80) Ballarat: Connor Court Publishing.

Books

Jen JJ, Chen J (2017) Food safety in China: science, technology, management and regulation. Wiley, Hobokeb

John P (2014) *Opportunities and challengers for organic food and agriculture: China and Australia*. In: Mascitelli B, Mahoney BO (eds) Chapter 3: Good for all: developing knowledge relationships between China and Australia. Connor Court Publishing, Ballarat, pp 50–80

King FH (作者),程存旺(译者),石嫣(译者) (2011) 四千年农夫:中国、朝鲜和日本的永续农业.北京:东方出版社

Lord N (1940) Look to the land. Dent, London

Internet Sources

Amy C (2016) Architects seize on potential in China's countryside – the New York Times. https://www.nytimes.com/2016/06/18/arts/design/architects-see-potential-in-chinas-countryside.html. Accessed 10 June 2019

http://english.cnca.gov.cn/. Accessed on 28 May 2019

http://www.greenfood.org.cn/jgjs/nscs/yjzx/. Accessed on 14 June 2019

http://www.ofcc.org.cn/index.php?optionid=669. Accessed 29 May 2019

http://www.ofdc.org.cn/article.asp?m_id=1. Accessed 7 June 2019

https://cszx.nwsuaf.edu.cn/rzzx/zxjj2/index.htm. Accessed 10 June 2019

IFOAM. The world of Organic agriculture. https://www.ifoam.bio/en/news/2019/02/13/world-organic-agriculture-2019. Accessed 27 May 2019

The Chinese National Organic Standard (GB/T 19630.1-4—2011). Organic products—Part 1: Production. http://www.bcschina.com/UploadFiles/2018-06-07/15283703244196851.pdf. Accessed 6 June 2019

Zhou Zejiang, IFOAM, the Research Institute of Organic Agriculture (FiBL) (2019) The world of organic agriculture. https://wwwifoambio/en/news/2019/02/13/world-organic-agriculture-2019. Accessed 27 May 2019

2017-2021年中国有机农业行业发展及预测分析-北京华夏元素有机农业产业联盟. http://www.choaia.com/newsitem/278038460

国家环保总局有机食品发展中心(OFDC)十年发展情况简介. http://trhj.mee.gov.cn/nchjgl/201604/t20160424_335971.shtml. Accessed 26 May 2019

陈仁端 (2005) 谈《四千年农夫》, 节选自《关于太湖流域的水环境与生态农业的若干思考》. http://wen.org.cn/modules/article/view.article.php/2680. Accessed 30 May 2019

IFOAM. The world of Organic Agriculture. https://www.ifoam.bio/en/news/2019/02/13/world-organic-agriculture-2019. Accessed 27 May 2019

纯天然无污染 我国有机茶渐成市场新宠. http://www.ofdc.org.cn/en/article_info.asp?n_id=679. Accessed 29 May 2019

http://www.greenfood.agri.cn. Accessed 29 May 2019

杜源良 (2018) 半亩园:用心种好有机菜, 打造新疆一流有机品牌-要闻-南方农村报. http://wap.nfncb.cn/show-730-1164749-1.html. Accessed 10 June 2019

杭州现全中国最美猪圈. 人民网. http://society.people.com.cn/n/2015/0414/c136657-26843237-6.html. Accessed 10 June 2019

李劼 (2018) 有机食品"真假"谁说了算?南方日报. http://www.xinhuanet.com/2018-03/14/c_1122533817.htm. Accessed 13 June 2019

李莉容 (2017) –个有机农场, 它的生意从"中国最美猪圈"开始|有机这生意怎么样了①_商业_好奇心日报. https://www.qdaily.com/articles/40071.html. Accessed 10 June 2019

绿色食品标志管理办法. http://jiuban.moa.gov.cn/zwllm/tzgg/bl/201208/t20120802_2814698.htm. Accessed 30 May 2019

2017–2021年中国有机农业行业发展及预测分析-北京华夏元素有机农业产业联盟. http://www.choaia.com/newsitem/278038460. Accessed 12 June 2019
佟明彪 (2018) 全国认证认可信息公共服务平台"认e云"上线.经济日报-中国经济网. http://www.ce.cn/cysc/ztpd/2018/cdf_299595/dt/201806/08/t20180608_29379934.shtml. Accessed 30 May 2019
王建平 (2012) 绿色食品认证管理制度与标准体系. http://www.mpoc.org.my/upload/P7_WangJianPing_POTSChina2012_chn.pdf. Accessed 20 May 2019
王晓东 (2016) 中国与新西兰签署首份有机产品认证互认协议.http://cn.chinadaily.com.cn/2016-11/14/content_27368355.htm. Accessed 1 June 2019
夏丹 (2018) 有机食品公司揭底:虚假标注+认证违规.Xinhuanet. http://www.xinhuanet.com/legal/2018-05/15/c_1122832164.htm. Accessed on 28 May 2019
夏晓峰 (2018) 东荒农场发展实现从量到质的转变. Ly.nen.com.cn. Accessed 11 June 2019
有机产品认证管理办法(总局令第166号). http://www.cnca.gov.cn/bsdt/ywzl/flyzcyj/bmgz/201210/t20121024_36669.html. Accessed 30 May 2019
中共中央 国务院关于坚持农业农村优先发展做好"三农"工作的若干意见. http://www.moa.gov.cn/ztzl/jj2019zyyhwj/2019zyyhwj/201902/t20190220_6172154.htm. Accessed 12 June 2019
中心简介_生态环境部南京环境科学研究所. http://www.nies.org/jgsz/kjzx/hjbhbyjspfzzx/zxjj/. Accessed 15 June 2019

Other

Chen J (2012) A study investigating the determinants of consumer buying behavior relating to the purchase of organic food products in Urban China. PhD thesis. Swinburne University of Technology, Melbourne
青木 (2019) 德报告:中国人爱上有机食品, 但"偏食". 环球日报

Hui Li PhD, is Associate Professor in the School of Law, Dalian University, China. The projects she presided include Research on Allocation and Coordination Mechanism of Authorities and Functions Relating to Banking Supervision—Based on Deposit Insurance Ordinance of China, a project supported by the National Social Science Foundation of China, 2015 (Grant No. 15BFX174), and Research on Banking Deposit Insurance Legislation, a project supported by Humanities and Social Science Foundation for Young Scholars, Ministry of Education of the People's Republic of China, 2014 (Grant No. 14YJC820029).

Chapter 9
Regulatory Framework for Organic Food Safety in India

Suhail A. Nathani and Mehfuz Mollah

Abstract The regulation of organic food safety in India can be classified under three broad heads – universal regulation applicable for all food, specific regulation applicable to organic food and certification systems. India follows a combination of "principle-based" and "rule-based" approaches to safety regulation of any food item, with the *Food Safety and Standards Authority of India* having power to impose penalty, compensation and terms of imprisonment. Regulations specifically targeted for organic food mandates certification of organic food before being offered for sale, with relaxation given to certain small traders. It further provides detailed guidelines which covers the labelling, traceability, transport, storage, distribution, retail and import of the certified organic food. India currently allows certification either (i) under the *National Programme on Organic Production* that provides accreditation to various third-party certification bodies and lays down the standards for organic production; or (ii) through the *Participatory Guarantee System for India,* a quality assurance system that is established on active participation of local stakeholders and shared responsibilities between farmers and traders.

Keywords Organic food · Food safety · Labelling · Traceability · Certification · Participatory certification

Glossary of Acronyms

APEDA	Agricultural and Processed Food Products Export Development Authority
FSS Act	Food Safety and Standards Act, 2006
FSSAI	Food Safety and Standards Authority of India
GGC	Grower Group Certification

S. A. Nathani (✉) · M. Mollah
Economic Laws Practice, Mumbai, India
e-mail: suhailnathani@elp-in.com

GOH Bee Chen, R. Price (eds.), *Regulatory Issues in Organic Food Safety in the Asia Pacific*, https://doi.org/10.1007/978-981-15-3580-2_9

INR	Indian Rupee
MT	Metric Tonnes
NAB	National Accreditation Body
NAC	National Advisory Committee
NCOF	National Centre of Organic Farming
NITI	National Institution for Transforming India
NOP	National Organic Programme
NPOP	National Programme for Organic Production
NSC	National Steering Committee
NSOP	National Standards for Organic Production
PGS-India	Participatory Guarantee System for India
PMKSY	Pradhan Mantri Krishi Sinchai Yojana
QMS	Quality Management System
USD	United States Dollar
USDA	United States Department of Agriculture

9.1 Introduction

India is primarily an agrarian economy, with 70%[1] of its rural households still depending mainly on agriculture for their livelihood. Given India's population, in order to eliminate external food dependencies, the Government of the day ushered in a 'green revolution' in the sixties which was hugely successful, and India became a net exporter of food. However, this came at a cost – all traditional methods of agriculture gave way to the liberal use of chemicals and fertilizers to spur yield.

With the rise of spending power of the urban population and increasing global demand, the production of organic food is increasing year on year. In 2017–2018, India produced around 1.70 million Metric Tonnes ("MT") of certified organic products which includes all varieties of food products. India's exports of organic products increased by 17% between 2015–2016 and 2016–2017.[2]

Along with the rise of consumption of organic food, however, there was a need for regulations to address concerns of safety and standards of organic food. Though the Indian government tried to give an impetus to the organic food sector, the pace of evolution of an enabling system of laws and regulations was slow. This changed in 2017 when the government enacted organic food specific safety and standards regulations. The following sections detail various laws and regulations that are applicable from the safety perspective of organic food.

---

[1] http://www.fao.org/india/fao-in-india/india-at-a-glance/en/

[2] APEDA. http://apeda.gov.in/apedawebsite/organic/Organic_Products.htm

9.2 Organic Food Safety Laws in India

India is a federation of states and while agriculture is a State governed subject, food safety comes under both the Central and the State Governments.[3] Consequently organic food standards in India are regulated by administrative bodies – both under the Central government and the State government.

The regulation of organic food safety can be classified into three heads:

(i) Universal Regulations – applicable to all food (including organic food) wherein all food for human consumption is required to follow the safety and standards as mandated by *the Food Safety and Standards Act, 2006,* which is administered by the Food Safety and Standards Authority of India.

(ii) Specific regulations for organic food – *the Food Safety and Standards (Organic Food) Regulation, 2017* administered by *the Food Safety and Standards Authority of India.*

(iii) Certification Regime – adopting either one of two methods of organic food certification:

 (a) *National Programme on Organic Production regulated by Agricultural and Processed Food Products Export Development Authority*;

 (b) *Participatory Guarantee System in India* introduced by *Department of Agriculture Co-operation and Farmer's Welfare.*

Apart from regulating the sector from a food safety and security perspective, these bodies also promote, encourage and incentivize organic farming and ecologically sustainable agricultural practices. The subsequent sections will examine the above regulatory framework in greater detail.

9.3 The Food Safety and Standards Act, 2006

Prior to the implementation of the Food Safety and Standards Act, 2006 ("FSS Act") on August 24, 2006, numerous food safety laws and authorities were simultaneously operating in India.[4] The FSS Act repealed the multiplicity of such laws and consolidated these into a single statute with an overarching policy framework for food safety and nutrition.

---

[3]The seventh schedule of the Indian Constitution allocate subject matters between the Centre (List I), the states (list II) and jointly by Centre and the states (List III called the Concurrent List). *Adulteration of foodstuffs and other goods* is listed as entry 18 in the Concurrent List.

[4]For example, as the Prevention of Food Adulteration Act, 1954; the Fruit Products Order, 1955; the Meat Food Products Order, 1973 etc. and any other order issued under the Essential Commodities Act, 1955 relating to food.

- **Objective**

The FSS Act is created to establish science-based standards for articles of food and to regulate their manufacture, storage, distribution, sale and import and thereby "*to ensure the availability of safe and wholesome food for human consumption.*"

- **Enforcement of the FSS Act: The Food Safety and Standards Authority of India**

The FSS Act also established the Food Safety and Standards Authority of India ("FSSAI") under the aegis of the Ministry of Health and Family Welfare to enforce its provisions. The FSSAI is assisted by scientific committees and panels in setting standards.

- **Salient Features**

The FSS Act mandates the FSSAI to ensure prevention of unsafe or contaminated or sub-standard food while framing regulations and specifying standards under the Act and takes a combination of "principle-based" and "rule-based" approaches to regulation. Section 18 of the FSS Act lays down certain general principles to be followed while implementing the provisions relating to the FSS Act (Box 9.1).

In addition to the above general principles, the FSS Act also provides for the following rules:

- Section 19: Use of food additive or processing aid – *not allowed unless in accordance to the provisions of the FSS Act.*
- Section 20: Contaminants, naturally occurring toxic substances, heavy metals, etc. – *presence in excess of specified quantity is not allowed.*
- Sect. 21(1): Pesticides, veterinary drugs residues, antibiotic residues and microbiological counts – *not allowed in cases where there is an excess of specified tolerance limit.*
- Section 21(2): *No insecticide can be used directly on article of food except certain fumigants.*

Box 9.1: Guiding Principles Under FSS Act

- To achieve appropriate level of protection of human life and health and the protection of consumers interests.
- To carry out risk management before implementation of any measure
- If scientific uncertainty persists with respect to any situation, provisional risk management measures necessary to ensure appropriate level of health protection to be adopted
- Measures adopted should be proportionate and should be least trade restrictive
- Review of measures should be undertaken in a reasonable period of time
- Dissemination of information regarding risk to health from food

- Section 22: Genetically modified foods, organic foods, functional foods, proprietary foods, etc. – *restriction on any person to manufacture, distribute, sell or import these categories of food unless in accordance with regulations.*
- Section 23: All packaged food should be marked and labelled as prescribed.
- Section 24: Restrictions of misleading or deceiving advertisement and prohibition of unfair trade practices.

- **Penalty Provisions under the FSS Act**

Section No.	Offence	Fine/penalty (in INR and USD)	Imprisonment, if any
50	Selling food not of the nature or substance or quality demanded	Penalty not exceeding INR 5 lakh (USD 7000)	–
51	Sub-standard food	Penalty which may extend to INR 5 lakh (USD 7200)	–
53	Misleading advertisement	Penalty which may extend to INR 10 lakh (USD 14400)	–
54	Food containing extraneous matter	Penalty which may extend to INR 1 lakh (USD 1400)	–
59	Punishment for unsafe food	No injury – fine which may extend to INR 10 lakh (USD 14,400)	Imprisonment for a term which may extend to **6 months**
		Non –grievous injury – fine which may extend to INR 3 lakh (USD 4300)	Imprisonment for a term which may extend **to 1 year**
		Grievous Injury – fine which may extend to INR 5 lakh (USD 7200)	Imprisonment for a term which may extend to **6 years**
		Death – fine which shall not be less than INR 10 lakh (USD 14,400)	Imprisonment for **life**
63	Punishment for carrying out a business without licence	Fine which may extend to INR 5 lakh (USD 7200)	Imprisonment for a term which may extend to **6 months**
65	Compensation in case of injury or death of consumer	Not less than INR 5 lakh (USD 7200)	–
		Not exceeding INR 3 lakh (USD 4300) in case of grievous injury	
		Not exceeding INR 1 lakh (USD 1400), in all other cases of injury.	

9.3.1 Important Regulations Under FSS Act

Apart from the provisions discussed above, the FSSAI has promulgated several regulations detailing various aspects of food safety and standards. Discussed below are some of the important regulations notified by FSSAI pertaining to food safety.[5]

- **The Food Safety and Standards (Food Products Standards and Food Additives) Regulations, 2011**

  It deals with the implementation of various standards of food. It provides detailed mandatory standards of various food products and prescribes the limits of various food additives used across different food groups and lays down microbiological requirement for different foods. Those food products which do not have requisite standards as per these regulations must apply for product approval from the FSSAI.

- **The Food Safety and Standards (Contaminants, Toxin and Residues) Regulations, 2011**

  It deals with the compliance of various contaminants, toxins and residues standards prescribed in food. The regulation provides detailed standards of various contaminants, naturally occurring toxic substances and residues of insecticides in different food articles.

  The term "crop contaminant" has been defined in this regulation as any substance not intentionally added to food, but which gets added to articles of food in the process of their production (including operations carried out in crop husbandry, animal husbandry and veterinary medicine), manufacture, processing, preparation, treatment, packing, packaging transport or holding of articles of such food as a result of environmental contamination.

  This regulation prescribes that the chemicals as described in the monograph of Indian Pharmacopoeia (autonomous institution under the Ministry of Health and Family Welfare) when used in food shall not contain metal contaminants beyond the limit prescribed therein.

  The regulation also prescribes the quantity in ppm (per parts million) per weight of various metal contaminants across different article of food. Further, a total of 149 insecticides have been covered in this regulation whose tolerance limit have been prescribed in chapter two of the regulations.

9.4 Regulations Pertaining Specifically to Organic Food

9.4.1 Provisions Relating to Organic Food in FSS Act

The FSS Act provides for the general rules and regulations to ensure the safety and standards of all types of articles of food.

[5] All regulations, including the detailed standards can be accessed at: https://www.fssai.gov.in/cms/food-safety-and-standards-regulations.php

Section 22, however, makes a special reference to "organic food" and states that no person shall manufacture, distribute, sell or import any organic food unless in accordance to the FSS Act and the regulations made under it.

The term "organic food" is explained under the FSS Act as food products that have been produced in accordance with specified organic production standards.

9.4.2 Food Safety and Standards (Organic Food) Regulations, 2017

The FSSAI promulgated *the Food Safety and Standards (Organic Food) Regulations*, 2017 ("Organic Food Regulations") which covers the labelling, transport, storage, distribution, retail and import of certified organic foods. The Organic Food Regulations also prohibit any non-compliant person from engaging in any organic food related business activities.

- **Salient Features**

Certification System

The Organic Food Regulations mandate that any organic food that is offered or promoted for sale must also comply with any one of the following certification systems:

1. National Programme for Organic Production ("NPOP") under the Agricultural and Processed Food Products Export Development Authority ("APEDA")
2. Participatory Guarantee System for India ("PGS-India")
3. Any other system approved and notified by the FSSAI.[6]

It is important to note that the Organic Food Regulations do not attempt to create any new standards for the production of organic food. Instead, it provides clarity and uniformity by approving and mandating the adoption of these certification systems as it lays down the standards for organic farming.

A detailing of the certification system is provided in subsequent sections of this chapter.

Labelling and Packaging

There are mandatory labelling requirements for organic products under the Organic Food Regulations, for the purpose of providing consumers with appropriate distinctions between organic food products and other food products.

The seller, manufacturer or distributor must necessarily put a certification or quality assurance mark of one of the above-mentioned certification systems on the packaging.

[6] The FSSAI has not notified any new certification system under Sub-Clause (1) Regulation 4 of the Organic Food Regulations.

Such products also need to comply with the packaging and labelling requirement specified under the Food Safety and Standards (Packaging and Labelling) Regulations, 2011 and this labelling must convey full and accurate information.

In addition to these, the Regulatory Compliance Division of the FSSAI has mandated the use of FSSAI's Organic Logo or the '*Jaivik Bharat*' Logo with effect from April 1, 2019 for all organic products. The logo was introduced with the purpose of providing the consumers and other stakeholders with a unified logo to help distinguish organic products from non-organic products.[7] The FSSAI has directed the State Food Safety Departments that commodities which are "*in conversion*" under NPOP or PGS-India scheme may write "*In conversion to organic*" on the labels.[8]

Traceability

The Organic Food Regulations mandate that traceability should be established up to the producer level. This is to ensure the complete organic integrity of the food product. There are various mechanisms which ensure the traceability of the product.

- Under PGS-India, the system ensures the traceability only if it stays in the custody of the PGS-India group and once the product leaves this group, the PGS-India group or system does not have any control on the integrity of the organic product.
- Under the NPOP system, the APEDA (being the regulatory body) has set up a traceability mechanism. This ensures that the genuineness of the organic product certified under the NPOP scheme can be checked by stakeholders, even from international markets. *Tracenet* is a web-based traceability system implemented by the APEDA for organic products in line with the NPOP system. The *Tracenet* system currently covers the certification of all horticultural and agricultural products with the plan being to eventually include the database of organic livestock products.[9]
- The FSSAI introduced a web-based portal, the *Indian Organic Integrity Database*, to help consumers verify the authenticity of organic food. This Portal allows the consumers to access all information with respect to the producer, the certification system and the availability of certified organic products in specific markets. The portal was developed in collaboration with APEDA and PGS-India.[10]

Import of Organic Food

The FSSAI, through the Organic Food Regulations, has relaxed the norms for import of organic food provided that there is equivalence of standards between

[7] Details of the logo can be accessed at: https://jaivikbharat.fssai.gov.in/

[8] Direction under 16(5) of FSS Act, 2006 regarding enforcement of Food Safety and Standards (Organic Foods) regulations, 2017, dated April 2, 2019.

[9] APEDA's initiatives in Traceability in India, available at http://www.apeda.gov.in/apedawebsite/TracenetOrganic/TraceNet.htm

[10] The database can be accessed at: https://jaivikbharat.fssai.gov.in/

National Programme for Organic Production (NPOP) and the organic standards of the respective exporting countries. These imported organic products, if compliant with the exporting countries' certification rules and with a transaction certificate issued by the appropriate body, are not required to be re-certified on import to India.

Relaxation for Small Traders

The FSSAI has provided a breather to smaller original producers or producer organizations involved in direct sales of their produce, and has exempted them from the mandatory compliance of certification systems.

The FSSAI has notified that a small original producer/producer organization with an annual turnover of organic produce of not more than INR 12 lakh (USD 17,000) is allowed to sell organic products without any certifications. Similarly, aggregators/intermediaries who collect organic food from small producers and sell it to end consumers directly, having a turnover of organic produce not exceeding INR 50 lakhs per annum (USD 70,000) are permitted to sell without any certifications.

Recognizing the cumbersome requirements of the NPOP and PGS-India schemes and with the intention of providing a producer friendly environment, the FSSAI has directed the State Food Safety Departments to consider the Organic Food Regulations as "*enabling regulations*" and not as "*provisions for prosecution*", especially for small original producers during the initial phase of the implementation of the regulations.

This is a welcome move on the part of the FSSAI as it not only attempts to ease compliance requirements on small producers but also eases the entry barriers for new small producers who are attempting to get into this space.

9.5 Certification Systems Applicable to Organic Food

9.5.1 National Programme on Organic Production (NPOP)

The Ministry of Commerce and Industry under the Government of India implemented the NPOP, which involves the accreditation programme for Certification Bodies, standards for organic production and the promotion of organic farming. In order to get a better understanding of the NPOP it is important to discuss it against the context of the Agricultural and Processed Food Products Export Development Authority (APEDA), the regulatory body which is responsible for laying down the rules and standards for certifying an organic product.

- **Agricultural and Processed Food Products Export Development Authority (APEDA)**

APEDA is mandated with the responsibility of promoting and developing the export of agro-products from India.

Its main functions includes promoting the export-oriented production and development of scheduled products (including fruits, vegetables, cereals and rice), fixing of standards and specifications for the scheduled products for the purpose of exports, carrying out inspections of meat and meat products for quality purposes and other matters.

- **Objective and Scope**

The Central Government, through the APEDA, has been implementing standards for organic farming and promoting the same through a third-party certification programme under the NPOP. Another intent of being in compliance with the NPOP is to be at par with the stringent standards on export of organic food products.

The NPOP has the following objectives and principles:

- To provide the means of evaluation of certification programmes for organic agriculture and products as per approved criteria
- To accredit certification programmes
- To facilitate certification of organic products in conformity to the National Standards for Organic Products
- To encourage the development of organic farming and organic processing

The scope[11] of NPOP includes the following:

- Policies for development and certification of organic products as notified by the Department of Commerce from time to time
- National standards for organic products and processes
- Accreditation of certification programmes to be operated by Certification Bodies
- Certification of organic products

The NPOP provides the institutional mechanism for implementation of the National Standards for Organic Production ("NSOP"). The NSOP sets out the standards to be followed in cultivation/harvest/production/processing and trading of organic products. The NSOP has been inspired by the International Federation of Organic Agriculture Movements ("IFOAM") – Basic Standards, where 'general principles', 'recommendations', 'basic standards'and 'derogations' are established for individual aspects of organic production, handling and processing.[12]

In December 2008, India's organic certification process under the NPOP had been granted equivalence with the European Union.[13] It has also been recognized

[11] Section 2.1, Chapter 2, National Programme for Organic Production, Seventh Edition, 2014.

[12] Role of Indian Government in the Development of Organic Agriculture, Archana, K *p-ISSN: 2319–2372. Volume 2, Issue 6 (Mar. – Apr. 2013), PP 32–39*

[13] Commission Regulation (EC) No 1235/2008 of 8 December 2008 laying down detailed rules for implementation of Council Regulation (EC) No 834/2007 as regards the arrangements for imports of organic products from third countries

for conformity assessment by the National Organic Programme ("NOP") of the United States Department of Agriculture ("USDA").[14]

- **Organizational Structure**

The National Steering Committee ("NSC") is the apex policy formulation committee, responsible for the implementation and administration of NPOP and lays down the NSOP. The National Accreditation Body ("NAB") is the body which is responsible for the drafting the procedure for evaluation and accreditation of Certification Bodies and the accreditation of Certification Bodies. The APEDA functions as the secretariat for the implementation of the NPOP.

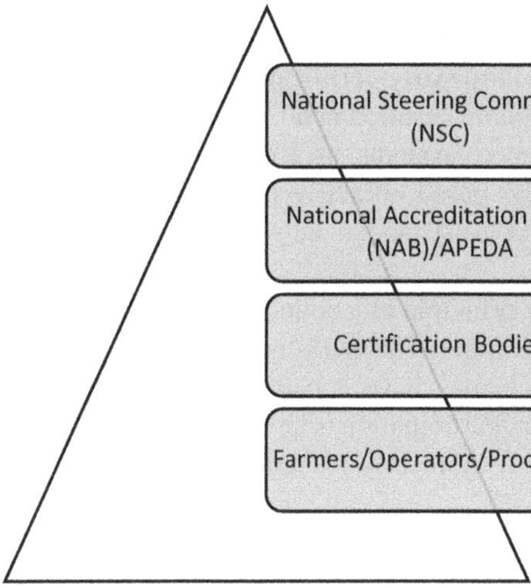

- **Certification Process**

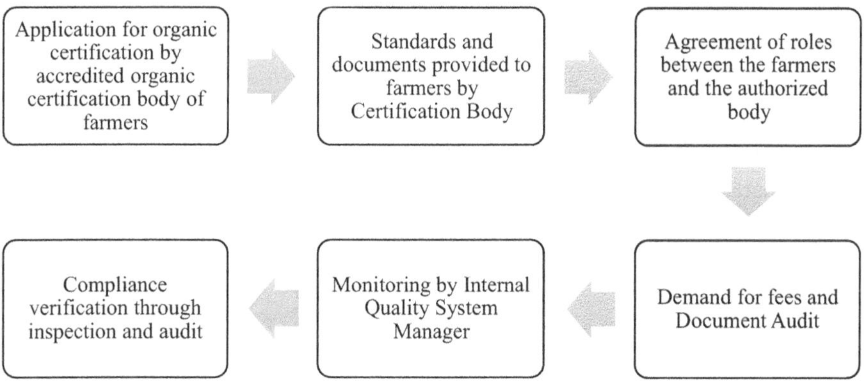

[14] https://www.ams.usda.gov/services/organic-certification/international-trade/India

9.5.2 Participatory Guarantee System for India (PGS-India)

PGS-India is a process in which people in similar situations (in this case smallholder producers) assess, inspect and verify the production practices of each other and make decisions on organic certification. It is a quality assurance initiative to guarantee the organic integrity of products that has been developed by small domestic producers. This programme was introduced by the Ministry of Agriculture and Farmer's Welfare under the Government of India.

In contrast to the previously discussed framework of third-party certification, this scheme emphasizes the participation of stakeholders, including producers and consumers. Third-party certification schemes are essential for ensuring the integrity of the organic product in world trade. However, the cumbersome procedure (inherent expense and paperwork required) of the third-party certification schemes discourages most small organic producers from being certified at all.[15]

IFOAM defines Participatory Guarantee Systems as locally focused quality assurance systems. They certify producers based on active participation of stakeholders and are built on a foundation of trust, social networks and knowledge exchange".[16]

The idea of participation embodies the principle of a collective responsibility for ensuring the organic integrity of the PGS-India. The PGS-India system comprises of certain basic elements which are inclusive, have a participatory approach, a shared vision, transparency as well as trust. Participation is an essential and dynamic part of PGS-India system. This collective responsibility is reflected through:

1. Shared ownership of the PGS-India
2. Stakeholder engagement in the development process
3. Understanding of how the system works
4. Direct communication between producers, consumers and other stakeholders.

PGS-India controls the quality system through an internal Quality Management System ("QMS") through a Farmer Grower Group Certification ("GGC"). The role of GGC is as follows:

- Act as an internal quality control system for small farmer group and cooperatives;
- Producers who share common products are organized under one management and marketing system;
- Collective marketing of the organic products;
- Centralized processing, marketing, and distribution system; and
- Internal control and supervision system to ensure compliance with organic certification.

[15] 'Participatory Guarantee System for India [PGS-India] Operational Manual for Domestic Organic Certification' [National Mission for Sustainable Agriculture] – Department of Agriculture

[16] https://www.ifoam.bio/en/organic-policy-guarantee/participatory-guarantee-systems-pgs

- **Operational Structure**

The National Advisory Committee ("NAC") is the apex body which is responsible for policy making. The National Centre of Organic Farming ("NCOF") is the secretariat (executive authority) of the PGS-India programme. Under the NCOF are the Zonal Councils and Regional Councils. At the root level are the Local groups or the Farmer groups. Local group is the main functional and decision-making body under the Participatory Guarantee System.

- **Advantages of the PGS-India System**

PGS-India is a process-based certification system wherein a group certifies that its members are growing food organically. This certification process has been made available at no cost to the farmers, cost other than a nominal cost as set by the farmer community. According to the Ministry of Agriculture, Government of India,[17] the PGS-India system has other advantages over the NPOP scheme and it includes:

- The procedure is simple and in a local language, thus more comprehensible to farmers
- All the members are local and known to each other
- Peer appraisers are amongst the group and live in the same village, therefore have better access to surveillance
- Peer appraisal instead of third-party inspections reduces cost
- Mutual recognition and support between Regional PGS-India groups ensures better networking for processing and marketing
- Empowers farmers with increased capacity building
- Bring consumers to the farm without the need of middleman
- Unlike grower group certification system, PGS-India offers every farmer with an individual certificate and each farmer is free to market its own produce independent of group
- Consumers and buyers are often involved in production and verification process
- Random residue testing at regular intervals ensures the integrity and increases the trust

- **Limitations of PGS-India system**

While it provides an alternative for local and small producers, there are certain inherent limitations of the PGS-India system. These are:

- Individual farmers or groups of farmers consisting of fewer than five members are not covered under PGS-India. Such parties will have to go under the NPOP system.
- Inputs/activities not directly provided by the farmer or group cannot be covered under the PGS-India Scheme.

[17] Frequently Asked Questions (FAQs) – On Participatory Guarantee Systems for India, https://pgsindia-ncof.gov.in/pdf_file/faq/Faq1.pdf

- Traceability only if it stays in the custody of the PGS-India group. Once the product leaves this group, for example by sale to aggregators or intermediaries who are not part of the PGS-India group, the PGS-India group or system does not have any control on the integrity of the organic product.

9.6 Other Programmes, Schemes & Initiatives

Generation of food and food products is deeply influenced by several other factors. Therefore, safety of organic food products cannot be solely ensured by food regulators or certification schemes. In order to buttress the above schemes, the Government of India has introduced several other schemes and initiatives which include:

- National Mission for Sustainable Agriculture – It is an initiative taken upon by the Department of Agriculture Cooperation and Farmers Welfare under the National Action Plan on Climate Change. The aim of this programme is to promote sustainable agriculture through climate change adaption measures.
- Pradhan Mantri Krishi Sinchai Yojana ("PMKSY") – It is an initiative taken upon by the Department of Agriculture Cooperation and Farmers Welfare for providing end-to end solutions in irrigation supply chain, including, water sources, and distribution network and farm level applications. PMKSY not only focuses on creating sources for assured irrigation, but also creating protective irrigation by harnessing rainwater at micro level. PMKSY adopts State level planning and execution that allows States to draw up their own irrigation development based on District Irrigation Plans and State Irrigation Plans.
- To facilitate organic farming, 11 state governments in India (viz. Kerala, Karnataka, Andhra Pradesh, Sikkim, Mizoram, Nagaland, Himachal Pradesh, Madhya Pradesh, Gujarat, Rajasthan and Odisha) have created their own State Organic Farming Policies, and Sikkim became the first state to be declared as Organic State with 75,000 hectares under organic cultivation, thereby providing impetus to other states to pursue similar objectives. Meghalaya, for instance, aims to make the switch by 2020. Many other Indian states such as Chhattisgarh and Uttarakhand are also promoting an organic marketplace wherein producers and consumer can directly interface.
- NITI Aayog[18] has pitched for completely removing agricultural commodities from the Essential Commodities Act and initiated a shift towards organized trading wherein lower number of traders with enough capital will dominate the market. The NITI Aayog is of the view that removing stock restrictions from agriculture commodities will lead to organized trading, improve scale and logistics benefit and bring about more capital into trade with handful of big traders competing with each other.

---

[18] The National Institution for Transforming India, also called NITI Aayog, is the premier policy 'Think Tank' of the Government of India, providing both directional and policy inputs.

- The Directorate General of Foreign Trade ("DGFT") liberalized the quantitative restrictions on the export of organic products. The export of organic wheat, non-basmati rice, edible oils and sugar have been exempted from all the annual quantitative ceiling limits. Additionally, the quantitative ceiling on pulses and lentils has been increased from 10,000 MT to 50,000 MT.

9.7 Conclusion

Organic farming brings better economic value to the farmer, especially where farms are small. Conversion of farms and adoption of organic farming is not high yielding in the initial years of implementation, and this is an impediment to switching over to organic farming methods. Farmers need ample scientific and financial support to convert from conventional farming methods to organic farming methods.

With growing awareness and demand for organic foods, the market is becoming increasingly lucrative. The regulatory framework is encouraging small farmers to opt for organic farming. With e-commerce market places in India potentially connecting farmers directly to consumers, and large retailers offering these products to discerning this segment is posed for exponential growth. In the words of Dr. M.S. Swaminathan, the Father of Green Revolution in India, "India must focus on improvement of productivity without harming ecology and thereby push for an 'Evergreen Revolution'".[19] India is embarking on that phase now, at least with organic farmers.

References

Archana K (2013) Role of Indian Government in the development of organic agriculture. IOSR J Agric Veter Sci 2(6):32–39

Commission Regulation (EC) No 1235/2008 of 8 December 2008 laying down detailed rules for implementation of Council Regulation (EC) No 834/2007 as regards the arrangements for imports of organic products from third countries. [online] Available at: https://eur-lex.europa.eu/legal-content/EN/ALL/?uri=CELEX%3A32008R1235. Accessed 18 Jul 2019

Department of Commerce, Government of India(2014) National Programme for Organic Production (NPOP), 7th edn. [online] Available at: http://apeda.gov.in/apedawebsite/organic/organic_contents/national_programme_for_organic_production.htm. Accessed 3 Jul 2019

International Federation of Organic Agriculture Movements. Participatory Guarantee Systems (PGS). [online] Available at: https://www.ifoam.bio/en/organic-policy-guarantee/participatory-guarantee-systems-pgs. Accessed 18 Jul 2019

[19] 'M.S. Swaminathan calls for sustainable farm practices' – The Hindu – https://www.thehindu.com/news/national/other-states/ms-swaminathan-calls-for-sustainable-farm-practices/article5497311.ece

Ministry of Agriculture, Government of India. Frequently asked questions (FAQs) – on participatory guarantee systems for India. [online] Available at: https://pgsindia-ncof.gov.in/pdf_file/faq/Faq1.pdf. Accessed 18 Jul 2019

National Mission for Sustainable Agriculture – Department of Agriculture, Government of India (2015) Participatory Guarantee System for India [PGS-India] – operational manual for domestic organic certification. [online] Available at: https://pgsindia-ncof.gov.in/pdf_file/PGS-India%20Operational%20Manual.pdf. Accessed 3 Jul 2019

The Agricultural and Processed Food Products Export Development Authority (APEDA), Government of India. Organic products. [online] Available at: http://apeda.gov.in/apedawebsite/organic/Organic_Products.htm. Accessed 3 Jul 2019

The Agricultural and Processed Food Products Export Development Authority (APEDA), Government of India. Tracenet. [online] Available at: http://www.apeda.gov.in/apedawebsite/TracenetOrganic/TraceNet.htm. Accessed 3 Jul 2019

The Food and Agriculture Organization of the United Nations. India at a glance, FAO in India. [online] Available at: http://www.fao.org/india/fao-in-india/india-at-a-glance/en/. Accessed 3 Jul 2019

The Food Safety and Standards Authority of India. Food safety and standards regulation. [online] Available at: https://www.fssai.gov.in/cms/food-safety-and-standards-regulations.php. Accessed 3 Jul 2019

The Food Safety and Standards Authority of India. Jaivik Bharat. [online] Available at: https://jaivikbharat.fssai.gov.in/. Accessed 3 Jul 2019

The Hindu (2013) M.S. Swaminathan calls for sustainable farm practices. [online] Available at: https://www.thehindu.com/news/national/other-states/ms-swaminathan-calls-for-sustainable-farm-practices/article5497311.ece. Accessed 3 Jul 2019

United States Department of Agriculture, the Agricultural Marketing Service. International trade policies: India. [online] Available at: https://www.ams.usda.gov/services/organic-certification/international-trade/India. Accessed 18 Jul 2019

Suhail A. Nathani is the Managing Partner of Economic Laws Practice and has over 25 years of experience spanning corporate and commercial laws, international trade and customs and competition law and policy. He has been widely recognised as a leading lawyer and has been named as a top 30 International Trade Practitioners in the World by the Best of the Best Expert Guides 2016 and India Business Law Journal's A List as India's Top 100 Lawyers in 2017 and 2018. He has appeared before the WTO Panel and Appellate Body on behalf of the Government of India. He earned his master's degree from Cambridge University, England, and has also received his LLM from Duke University, USA. Apart from India, he is also admitted to the State Bar of New York. He is an Adjunct Professor at the Jindal Global Law School in India.

Mehfuz Mollah is a lawyer with a background in economics. At Economic Laws Practice, he regularly advises clients on regulatory and policy issues. He completed his degree in Law from Delhi University, his master's degree in Economics from the Delhi School of Economics and his LLM from the London School of Economics. He is also a Visiting Professor at NMIMS University, India.

Chapter 10
Organic Food Policy and Regulation in Malaysia: Development and Challenges

Mohammad Firdaus Bin Abdul Aziz, Muhamad Shakirin Bin Mispan, and Febri Doni

Abstract The organic food industry is relatively new in Malaysia and the government has recognised its potential to contribute to economic growth. Consumers' interest in organic food has increased due to local demand. This has resulted from the image associating organic food as natural and chemical-free making it wholesome and safe for health. Malaysia has introduced national agriculture policies since before Merdeka (independence from colonial rule in 1957). Recently revised government policy and regulation aims to foster further development. Nonetheless, there are some issues and challenges, for instance, a lack of incentives provided for farmers to adopt organic farming and lack of public trust towards the quality of organic products. There have also been concerning questions about the quality of the accreditation system. This chapter explores the development of the organic agriculture industry in Malaysia using its policy and regulation as a framework to examine challenges facing the industry that need addressing for the country to become a globally recognised exporter of organic food.

Keywords Organic food · Consumer perception · Agriculture policy of Malaysia · Implementation of organic standards · Sustainable agriculture

M. F. B. A. Aziz (✉)
Centre for Law and Ethics in Science and Technology, Faculty of Law, University of Malaya, Kuala Lumpur, Malaysia
e-mail: firdausaziz@um.edu.my

M. S. B. Mispan · F. Doni
Institute of Biological Sciences, Faculty of Science, University of Malaya, Kuala Lumpur, Malaysia
e-mail: shakirin@um.edu.my; febridoni@um.edu.my

GOH Bee Chen, R. Price (eds.), *Regulatory Issues in Organic Food Safety in the Asia Pacific*, https://doi.org/10.1007/978-981-15-3580-2_10

10.1 Introduction

Organic food production is always being associated with a healthy agriculture ecosystem, which includes rich biodiversity, dynamic organic cycles, and active soil biotic activities (Lin et al. 2011; Somasundram et al. 2016; Jambon et al. 2018). Extensive use of synthetic chemicals in conventional agriculture practices has been commonly known to exert harmful effects to the environment including soil degradation, water pollution, and direct toxic effect. The use of such chemicals can also cause medical conditions among farmers such as dermatological, gastrointestinal, neurological, carcinogenic, respiratory, reproductive, and endocrinal diseases (Pimentel et al. 2005; Busby et al. 2017; Doni et al. 2019).

Substituting synthetic chemicals with agroecology-based input using cultural, biological, and mechanical practices is the main mantra in organic farming as an alternative approach to produce agro-food in a more sustainable way (Ahmad 2001; Tiraieyari et al. 2014). Organic agriculture promotes the health of soil, water, and air by minimising all forms of pollution from overly used synthetic inputs (e.g. fertilizer, mineral compounds and pesticides) and maximising the natural biological activities in the agro-ecosystem (Uphoff 2017; Doni et al. 2019). Henceforth, organic agriculture is regarded as an integrated farming approach that gives importance to every aspect of technology, economic efficiency, and human health. Consumers can benefit from high quality food and a synthetic chemicals-free approach also safeguards farmers against toxicity-related occupational health problems (Somasundram et al. 2016).

Organic farming is one of the environmentally friendly production methods that have been practised for many years ago among the local community in Malaysia. Local farmers would use compost from agricultural waste and animal manure for their crops. Traditionally, the local community also consumed organic foods, which they would collect from nearby forests or naturally grow on their land (Suhaimee et al. 2016). In recent years, the organic agriculture industry in Malaysia has been driven mostly by non-governmental organisations (NGOs) and the private sector.

This industry has also received increasing support from the government through various efforts in encouraging farmers to adopt organic approach, particularly among the small-scale farmers, as well as recognising the importance of this industry in national policies, and introducing regulatory frameworks to regulate this area to ensure high quality of produce. Nevertheless, there are challenges facing this industry in Malaysia. This article aims to provide an overview of the development of the commercial organic food industry in Malaysia, the background of its recognition as an important area in the national policies, its current regulatory system, and the difficulties facing the key players including small-scale farmers.

10.2 Organic Food Production in Malaysia

A well-established academic literature depicts the commercial organic agriculture or production of Malaysia in its infancy (Dardak et al. 2009; Suhaimee et al. 2016). The Malaysian organic industry contributes less than 0.01% of organic shares of total agriculture land in the world and relatively very small in terms of acreage and percentage of organic land use compared to other South-East Asian countries such as the Philippines, Thailand, and Indonesia (Willer and Lernoud 2019). Similarly, organic food comprises only a small fraction of the food market in the country. Nonetheless, its rapid growth has drawn a lot of interest among diverse parties from consumers, researchers, businesses, and the government (Suhaimee et al. 2016).

10.2.1 Current Development of Commercial Organic Agriculture

At present, the organic food production industry in the country is estimated to be worth more than USD 1 billion (Loh 2019). In general, organic market approaches in Malaysia are divided into three categories, namely: (1) direct home-distribution; (2) through specialist organics retail/supermarkets; and (3) through organic farmers' associations (Somasundram et al. 2016; Suhaimee et al. 2016). The emergence of commercial organic farms in Malaysia were initiated by the upsurge of concern of many non-profit organisations (NGOs) and private companies in the 1990s about food safety and the environmental impacts of conventional farming or agriculture practices (Partap 2010; Suhaimee et al. 2016).

The Centre for Environment, Technology and Development Malaysia (CETDEM) has played a crucial role in highlighting the issues of conventional agriculture including environmental degradation, occupational hazard from pesticide usage, and food safety, which has urged by CETDEM to look for alternative food sources in organic agriculture (Somasundram et al. 2016; Suhaimee et al. 2016).

This effort received positive response from members of the public and eventually created demands for organic products in major supermarkets, especially in Kuala Lumpur (Suhaimee et al. 2016). Following this, many pioneering organic farms were then established in other States including Johor, Penang, Pahang, Perak, Melaka, and Negeri Sembilan (Somasundram et al. 2016). These organic farms are established through the Malaysian Organic Scheme Certification (myOrganic) program introduced by the Department of Agriculture (DOA) Malaysia.

There were 142 farms that have received a myOrganic certificate between 2003 and 2014 covering a total area of 1757 ha. The DOA is working closely with farmers to bring about more than 240 farms to receive myOrganic certification by 2020 (DOA Malaysia 2019a). Currently, there are less than 100 certified organic farms in the country, and they are mostly located in Peninsular Malaysia with products spanning from mushrooms, herbs, vegetables, coffee, fruits, and rice (Table 10.1).

Table 10.1 Certified organic farms in Malaysia

States	Agricultural products	Total farms
Kedah	Rice, mushrooms and fruits	6
Pulau Pinang	Vegetables and fruits	2
Perak	Mushrooms, vegetables and herbs	2
Selangor	Mushrooms, vegetables, herbs and fruits	12
Negeri Sembilan	Vegetables and herbs	2
Melaka	Mushrooms	1
Johor	Mushrooms, vegetables and herbs	10
Pahang	Potato, vegetables and mushrooms	11
Terengganu	Mushrooms	2
Kelantan	Vegetables, mushrooms and fruit	5
Sabah	Coffee, fruits, vegetables and mushrooms	15
Sarawak	Vegetables, fruits and herbs	3

Source: DOA Malaysia (2019b)

It has been reported that vegetables are the most consumed organic foods in Malaysia followed by fruits and rice (Suhaimee et al. 2016). Majority of the organic farms are vegetables farms with increasing production and diverse products such as spinach, broccoli, asparagus, okra, and water spinach (DOA Malaysia 2019b). It is also worth noting that there is a certified organic farm in the State of Kedah that can produce more than 2 tonnes of organic red ruby, white and brown rice on a monthly basis. The farm is applying an environmentally and economically tenable way in growing their rice, namely the System of Rice Intensification (SRI), with less water, zero chemicals, and lower production cost (Doni et al. 2019). This development is important because of the history of ammoniated and salinated land and water due to rice production in South East Asia. Besides rice and vegetables, there are a significant number of organic farms that are producing mushrooms and fruits such as papaya, avocado, guava, and dragon fruits (DOA Malaysia 2019a). In spite of this, the existing organic farms are unable to meet the growing demand for organic foods in the market. This, along with export hopes, has been challenged by the number of Malaysian organic farms decreasing in the year 2019 (DOA Malaysia 2019b).

10.2.2 *Public Perception of Organic Food Products*

It has been reported that public acceptance and positive consumer perception of organic foods and products are proliferating in the country since its market began to boom in the early 2000s (DOA Malaysia 2019a). The demand for organic products in Malaysia is projected to grow more than 12.4% a year with the financial value of more than RM20 million a year [USD 4,800,000] (Suhaimee et al. 2016). It was argued that the development and growth of organic market and organic-based business are well supported by the consumers due to the increasing awareness and the standard of life in Malaysia (Tiraieyari et al. 2014).

A number of studies have been conducted in the country to show the level of awareness as well as different attitudes and opinions shared by members of the public. In a nationwide survey in 2008, it was reported that more than 90% of 625 respondents were aware and understood about organic food with 55% of the respondents have consumed organic products (Dardak et al. 2009). The respondents also associated the term 'organic' with chemical-free, healthy and natural products. This showed that the popularity and demand for organic food were established mainly as part of the promotion of a healthy lifestyle.

The increase of awareness on a green environment and sustainable agriculture production also has added to the swift consumption pattern towards organic products in the country (Ahmad and Juhdi 2010; Thambiah et al. 2015; Lian 2017). A survey of 625 consumers in selected regions in Peninsular and East Malaysia has identified three major factors influencing the consumption of organic food products, which include: (1) organic products as a safe and healthy option, (2) being better quality, and (3) a more environmentally friendly way to produce food (Dardak et al. 2009). The survey also found that the respondents were willing to spend up to a 25% higher price for organic food products than the conventional ones.

Another study demonstrated that the majority of organic consumers believed that organic products are healthier, tastier, and also better for the environment as compared to the non-organic or industrially farmed counterpart (Ahmad and Juhdi 2010). Similar to the previously mentioned study, this study also indicated that the main influences that shaped the intention in purchasing organic products were the perception of organic products as worthy of purchase and the belief in the safety and health aspects of the product. However, in contrast, a study on consumers' purchasing behavior in Penang (the second largest population centre in Malaysia) showed that health consciousness, environmental concerns and price do not have any significant relationship with consumers' behavior towards organic foods (Hossain and Lim 2017).

A more recent study conducted in Kuala Lumpur, however, demonstrated a different attitude among the urban consumers in that particular State, whereby the findings shown that health consciousness does play a significant role in consumers' decisions (Lian 2017). Consumers' experience and personal preference were also reported as the other motivational factors that have some influences on purchasing attitude. In addition, it was also argued that consumers' behavior towards organic food was influenced by other factors such as government support and policy, individual beliefs and attitudes, knowledge level on organic food, and the availability of the product.

A similar pattern was also observed among younger Malaysians whereby knowledge of organic food, familiarity with such products, and price consciousness were the main factors influencing the consumption of organic products among generation Y (Thambiah et al. 2015). Price consciousness as one of the influencing factors was also captured by study conducted by Ahmad and Juhdi (2010) showing that Malaysian buying power of organic food was limited due to its price, which is 50–300% higher than conventional food. The price difference between organic and conventional food in Malaysia is particularly substantial, compared to the U.S. and

E.U. where, the price gap is only 25–30% (Teh 2012). According to Ahmad (2001), the high price is due to the labor-intensive approach adopted in the industry in Malaysia.

10.3 Concerns of Organic Food Safety Among Malaysian Consumers

It has been widely reported that the increase in demand for organic products was due to a widespread belief that anything labeled 'organic' is perceived as natural, healthier, and safer in comparison with the conventional agriculture products (Doni et al. 2018). This general perception is strongly associated with the notion that organic food production is free from synthetic chemicals and used environmentally sound techniques. Nonetheless, it is worth noting that many organic advocates are heavily depending on testimonial experiences and beliefs that have allured them to be receptive to the idea of organic food is superior than the conventional alternative (Morkerberg and Porter 2001; Magkos et al. 2003).

This is also due to the fact that scientific discussions and evidence on organic food and its association with better health is fairly limited and unclear. At the same time, the available information was rarely discussed with the public and it was only discussed within the academic realm (Magkos et al. 2003). Therefore, it has been argued that the effect of organic food on health is not conclusively proven so as to be scientifically recommended and advertised to the public based on the current knowledge of the available research studies (Rock et al. 2017).

The safety issues of the organic food were also only discussed occasionally, especially on the zoonotic pathogens in organic produce and mycotoxin risk, due to high usage of biofertiliser and the absence of synthetic fungicide, respectively. It has been speculated that keeping livestock outdoors, and the use of animal excreta for fertilization, may result in elevated levels of microbiological hazards on organic foods. However, there is no evidence to support the allegations that organic fertiliser in organic crops increases the prevalence of bacterial pathogens and/or of antibiotic-resistant bacteria on fresh produce (Lücke 2017).

Organic food production can be perceived as safer than conventional farming merely because of fewer expected agrochemical residues and a lower level of nitrate in the produce (Doni et al. 2019). However, there is no solid evidence and clear conclusion to differentiate conventional farming from organic production with regards to environmental contaminants and food hazards (Magkos et al. 2003; Rock et al. 2017). The safety of the products from both origins are indecisively dependent on good agriculture practices and maintaining hygienic conditions from the farm to the plate.

As previously discussed, a number of studies have shown that the majority of the members of the public still believe and always consider 'organic' as equal to 'safe' due to strong attitudes and beliefs instilled since the early introduction of organic farming in Malaysia (Dardak et al. 2009; Ahmad and Juhdi 2010; Thambiah et al. 2015; Lian 2017). The concerns that the public have are only in regards to the credibility of organic certification and labels awarded by a local authority and not at the concept of 'organic' food as a whole (Saleki and Seyedsaleki 2012). This is further discussed in Sect. 10.6. However, there are limited discussions regarding safety issues in organic food products in Malaysia and there are no official reports on this matter. Therefore, based on the current scenario, it can be said that organic food safety has not yet been identified as a prime concern among the consumers in Malaysia.

The following section provides a brief background on the development of national policies on agriculture in Malaysia and how organic agriculture was introduced as one of the key areas identified as crucial to pursue the national interests.

10.4 National Policy on Agriculture

British Malaya introduced agricultural policy during the colonial twilight years between the Malayan Emergency (1948) and Merdeka (1957). This policy constituted an official document formulated as a set of strategic directions for the agricultural industry in the country as it moved toward decolonisation. Given the fact that the formulation of the first policy was carried out during the British colonisation, it can be said that the objectives of the policy were aligned strongly with the interests of the British administration. Emphasis was given on plantation crops such as rubber, oil palm, and cocoa.

During this era, Malaysia was known as the top rubber producer across the globe. The rubber production has significantly contributed towards the national income. Countries in the West such as the United Kingdom and the United States of America were the major importing countries. They imported raw materials to be used for industrial and manufacturing sectors in their countries and hit its peak during the Korean war. Meanwhile, agricultural activities for food commodities such as rice, fruits, and vegetables were left to small producers such as traditional farmers. The British policy was for rice production in the country to maintain the Tamil rubber tapping communities and the Chinese workers in the tin mines. The focus of the early agricultural policies was to develop agricultural commodities including latex, copra, and palm oil for export markets.

After successfully sought independence from the British administration in 1957, new national agricultural policies were formulated for the country. The policies introduced in the post-independence period represented the concerns of the local government with a focus on strategies to reduce poverty in the agricultural sector yet benefited substantially from British initiatives in public housing and land redistribution to discourage transient Chinese rural populations from supporting the armed insurgency of the Malaysian Communist Party (MCP) that flared between 1948 and 1953. The government also introduced agricultural policies in the First Malaya Plan (1956–1960), the Second Malaya Plan (1961–1965), and the First Malaysia Plan (1966–1970).

Being a former British colony, there were still influences from the colonial period. Agricultural activities were consistently focusing on industrial crops for export markets, whereby large forestland was cleared to make way for palm oil and rubber plantations that were protected by long leasehold periods of up to 999 years. One could argue that it was only comprehensible and expected given the fact that this industry was and continued to be the main contributor of the country's economy. However, the newly established government also incorporated a new focus and strategies. The combination of old and new approaches has resulted in the country benefiting from foreign exchange income, capital creation, and the agricultural industry was able to provide employment for the local people. The following sections provide an overview of the development of national policies on agricultural industry and how organic agriculture was then given emphasis and recognition as an important area to address national concerns such as the need to improve the living standard of small farmers, to boost agricultural produce for local consumption and export markets, and to foster sustainable agriculture, which are highlighted as some of the objectives of national policies as discussed in the following sections.

10.4.1 The First National Agricultural Policy (NAP1)

Significant changes were made to the focus of the country's policy on agriculture industry when the first national policy on agriculture (National Agriculture Policy 1984–1991) was introduced in 1984. Its introduction aimed to address the issues of rural poverty and the gap identified between commercial and traditional farmers, to improve the performance of the agriculture sector, and to increase the value of the agricultural produce for export markets. To implement the policy, new lands were cultivated to create more opportunities for employment, especially in the rural areas, and also to increase the production of agro-food products for local consumption.

10.4.2 The Second National Agricultural Policy (NAP2)

In 1992, the second national policy on agriculture was introduced, having an eight year span. The focus to meet the demand for agro-food products for local consumption identified in the NAP1 was extended to include the global markets in NAP2. Therefore, greater emphasis was given on the productivity and efficiency of this industry. The objectives of this policy were to increase competitiveness and production as well as to foster sustainable production. The government also realised the need to put in place measures to ensure this area is developed sustainably.

Incentives were also given by the government to facilitate the development of new infrastructures and subsidy programs were introduced to benefit traditional farmers. This new policy maintained the focus of the pre-independence policy, which was to emphasise industrial crops as a targeted area to boosting national income albeit that the British administration had, throughout the 1950s, increasingly opened up the rubber estates to small Malayan producers. More arable lands were created under this policy to further increase the productivity of rubber and palm oil and to encourage the plantation of agro-food such as paddy rice, pineapple, and vegetables. In addition, five strategies were adopted in order to achieve the objectives stated above:

- Maximise participation of the private sector to transform the sector into a more competitive and efficient sector;
- Optimise resource utilisation to diversify export crops cultivation into other activities;
- Enhance human-resource development through training;
- Accelerate agro-based industrial development that generates more off-farm opportunities for small holders to earn additional income; and
- Enhance R&D and promote 'Western' technology transfer to overcome the limitations on production processes caused by Malaya not industrializing under the British administration, labor shortage and other constraints in the agricultural sector.

10.4.3 The Third National Agricultural Policy (NAP3)

As has been highlighted previously, one of the concerns identified by the government in the previous policies was to address the issues of food availability for domestic consumption. This issue continued to be a growing concern and it was given further emphasis after the incident of Asian Financial Crises (AFC) from 1997 until 1998. After the crisis, the government realised that, after a period of currency devaluation and price inflation, there was a need to recognise the importance of agriculture as the food providers for the people. Among other identified concerns were such as the need to utilise resources efficiently in order to improve competitiveness, the lack of arable land, and to address the competition posed by other sectors, which has resulted in a shortage of labour for the agriculture sector. The revision led to the introduction of the third national policy on agriculture spanning from 1998 to 2010.

The policy aims at the maximisation of income including increasing the income for farmers, breeders and fishermen while ensuring measures are in place to conserve and utilise natural resources on a sustainable basis. This shows that sustainability continues to be one of the aims of the national policy. The objectives of the NAP3 are identified as follows:

- To enhance food security;
- To increase productivity and competitiveness of the sector;
- To deepen linkages with other sectors;
- To create new sources of growth for the sector; and
- To conserve and utilise natural resources on a sustainable basis.

The policy also introduced two strategic approaches namely agroforestry and product-based approaches. Whilst the former addresses the issue of scarce resources such as land, labour, and raw materials, the latter focuses on identifying key market and products by taking into account determining factors namely market demand, market potential, and consumer preferences. The NAP3 also shows the recognition given by the government to the organic agriculture industry and the commitment shown by the government in commencing the efforts in developing it. The following two sections provide a brief overview of how NAP3 set forth the importance of organic agriculture and this recognition is continued in the current and latest national policy namely the National Agro-Food Policy (NAP).

10.4.4 Organic Agriculture in NAP3

The Malaysian government recognised organic agriculture industry as a niche market and an opportunity for traditional or small-scale farmers to increase their income. It can also be seen as a sustainable measure to protect the environment and promote the country's exports. Over the years, the government has actively encouraged small-scale farmers to invest in the organic agriculture industry.

Organic agriculture is particularly identified in the third national policy (as a market opportunity especially for vegetable and fruit producers (Ahmad 2001). This can be seen from the formulation of the policy, which is based on an integrated agriculture that focuses on organic farming and recycling of organic wastes among other elements such as mixed farming, rehabilitation of marginal and contaminated land, ago-forestry, and composting.

Paragraph 34 of the policy states, "vegetables production will be expanded to meet domestic and export demand…The Government will continue to provide institutional support, infrastructure and incentives to encourage the private sector to venture into large scale commercial production…Quality will also be emphasised through product differentiation and product brand names such as pesticide-free vegetables and organic farm produce."

Under the NAP3, the Malaysian Government highlighted the opportunity in the organic farming sector that could bring more income to the country through organic food-export. This signals that the government no longer primarily relies on industrial crops to increase the income of the country. This move is in line with the new lifestyle of local and global consumers, which are increasingly concerned about the need to have a healthy lifestyle. This includes consuming organic food, which is associated with healthy food that is produced without chemicals.

Given the existing demand at the local and global level, the Malaysian Government implemented certain strategies in order to encourage small-scale farmers to involve in organic farming (Somasundram et al. 2016). The Malaysian Government has planned to increase organic food production from 1900 tons in 2016 to 6700 tons in 2020, as well as an increase in farmers' income from USD 1500 per farmer per month in 2016 to USD 1900 per farmer per month in 2020. In order to achieve these goals, Malaysian Government has launched several initiatives such as improvement of agricultural infrastructure, providing machinery and equipment, and supplying biofertiliser and biopesticides (DOA Malaysia 2019b).

10.4.5 The National Agro-Food Policy (NAP)

On the 28th of September 2011, the Malaysian Cabinet approved the 10-year national policy on agro-food (National Agro-Food Policy 2011–2020), which effectively replaced the National Agriculture Policy. The NAP identified agriculture as a National Key Result area. This policy was introduced as the guiding principle for Malaysia to stay competitive in this industry while ensuring sustainable production to address food security and food safety and nutrition aspects along its value chain and to address the challenges posed by domestic and global markets. The main objectives of the National Agro-food Policy are:

- To address food security and safety to ensure availability, affordability and accessibility;
- To ensure the competitiveness and sustainability of the agro-food industry; and
- To increase the income level of agri-based entrepreneurs.

As with the previous policies, the focus of this latest policy is to further address the issues of food supply for domestic consumption. The Malaysian government has consistently aspired to self-sufficiency in food and fiber supply and this can be seen from the policies discussed earlier. It is also in line with the national desire to work with the agriculture sector to facilitate small-scale farmers and agro-entrepreneurs to increase their income and to decrease rural poverty in Malaysia – an explicit aspiration of UMNO/Alliance since 1948 and emphasized in its formulation of the NAP1 in 1984.

It is obvious in its formulation that the government also wants to highlight the issues of food security and the importance to address it. The world is seeing a sharp rise in global population, whereby the number is projected to reach 9.9 billion in 2050, and this is 29% more than the estimated 7.6 billion people now (Population Reference Bureau 2018). It is therefore crucial for Malaysia to be part of the global efforts to ensure food security not only at the national level, but also at the international level. There is also an issue of climate change that affects every nation. Scientists have demonstrated that climate change will affect the production of food commodities.

It was reported by Badgley et al. (2007) that organic farming could produce enough food supply for the current human population without having to cultivate more land for agriculture. This is echoed by Rundgren (2006), who also argued that organic farming is an agro-ecological approach that can be adopted to ensure food security, whereby ample food can be produced for the increasing population. Ann (2007) stated, "organic farming can yield up to two or three times as much food as conventional farming on the same amount of land".

Scholars have also argued that organic farming approach is in line with the global effort to combat climate change. The reduction in usage of chemicals in conventional agriculture can prevent soil and water pollution, which can lead to the loss of biodiversity (Sadati et al. 2010). It can also be seen from the strategies adopted in the policies, whereby lands are created for agricultural activities and it can be argued that it is not a sustainable approach to keep on clearing land to increase the size of plantation land as this means that natural habitats will be affected and lead to many other negative consequences (Sadati et al. 2010).

All these reasons support the need to develop organic agriculture in the country given this industry is believed to be able to facilitate to increase crop yield, and does not need chemicals, which is a safer option not only to the consumers but also to the environment. In 2006, the American Chemical Society reported that developing countries have suffered from food shortages and environmental degradation as a result from unsustainable agriculture (American Chemical Society 2006). This validates the move taken by developing countries such as Malaysia to incorporate organic food industry in the national policy. It is a positive move for Malaysia to change its direction in agriculture industry.

In so doing, the NAP incorporated programs of which implementation is aimed at increasing food production through sustainable agricultural activities by setting a target for the production of agro-food commodities to increase at around 4% a year, upgrading agriculture infrastructure, and improving the quality and safety of food by expanding the compliance of standards. In the Ninth Malaysian Plan (2006–2010), the government targeted organic farming as one of the key areas to be developed and estimated a potential value of more than USD 200 million over 5 years. The Ministry of Agriculture also planned to expand organic farms to 20,000 ha by 2010 and to increase production by 4000 ha per year (Tiraieyari et al. 2014). The following section introduces the development of the standard introduced by the Malaysian government in regulating agriculture industry in general and an accreditation system specifically established to ensure the safety and quality of organic food produced in the country.

10.5 Accreditation and Standard for Organic Agriculture

10.5.1 Standard for Malaysian Farm Certification Scheme for Good Agricultural Practice (SALM)

To govern the safety of agricultural practice in Malaysia, in general, the government has introduced Good Agricultural Practice through the Standard for Malaysian Farm Certification Scheme for Good Agricultural Practice (SALM 2005). It is an integrated system to "manage the hazards associated with the elements of land, input, processes and output of agricultural production, within an agricultural spatial entity, in a good way to achieve productivity, sustainability, quality and safe produce in a socially acceptable worker friendly and environmentally safe way" (SALM 2005). It is based on the Malaysian Standard MS1784 for Crop Commodities and encompasses national laws to control hazards that have an impact on the environment, food safety and workers' health and safety.

The Standard is an important document, which can be used as a reference by key players such as farmers. The guiding rules in the Standard are presented in two categories namely 'MUST' and 'ENCOURAGED'. The former includes rules that are mandatory in nature, and critical to be adhered to, in order to ensure the objective of the Good Agricultural Practice is met with emphasis on sustainability, environmental protection, products that are safe to be consumed, and workers' health and safety protection. The MUST rules are further subdivided into two types; MAJOR MUST and MINOR MUST. The MAJOR MUST rules are rules that need to be strictly met by farmers and, in case of non-compliance, may result in a severe impacts on the food safety, environment, workers safety, and sustainability. The MINOR MUST rules also need to be complied with, however, the impacts of non-compliance are predicted not to be as severe as the MAJOR MUST rules.

Rules that fall within the latter category, 'ENCOURAGED', are referred to rules that are not mandatory in managing farm and crop production process. These rules are recommended for adoption in good agricultural practices. If the rules are not complied with, the impacts are predicted not to be hazardous or severe in their effect on the environment, sustainability of the agricultural practice, the safety of workers, and/or food safety. This Standard is applicable to all agricultural practices including organic agriculture.

10.5.2 Malaysian Organic Certification Program

Malaysia also introduced regulation specifically to regulate organic farming in the country known as the Malaysian Organic Certification Program. The Ministry of Agriculture, and Agro-based industry mandated the Department of Agriculture Malaysia (DOA) to set up the certification program. The certification program was launched in 2002 by the DOA and a total of 581 ha of vegetable farms were accredited (Ramli 2002). This certification program aims to facilitate the development of organic farming in Malaysia, and it is based on the Malaysian Standard MS1529:2001.

The Malaysian Standard MS1529:2001 was then updated in 2014 to MS1529:2015, which covers plant-based organically produced food. The certification requirements cover all stages of organic food value chain, which includes production, processing, handling, labelling, and marketing. The aims of the standards are as follows:

- To protect the producers and consumers against the deception in the market place and fraud such as unsubstantiated claims;
- To ensure all stages of production, preparation, storage, transportation and labelling are in compliance with the standard; and
- To be on par with international standards to facilitate international trade.

The Standard also aims to control hazards that have effects on the environment, food safety, and workers' health and safety. The DOA functions as a monitoring body to implement the organic certification scheme. As part of the monitoring process, the DOA has assigned a number of agricultural officers who are given the mandate to carry out field inspections to ensure that farm operations are in compliance with the organic standards. Certified organic farms have to be examined annually. The government also made a revision to the accreditation scheme to myOrganic in March 2015 to be consistent with the changes made to the Malaysian Standard.

Notwithstanding the policies and standards introduced to regulate organic agriculture in Malaysia, a number of studies have reported that there are challenges facing one of the key players in this industry, namely, the small-scale farmers. The following section discusses the challenges that exist within the current regulatory system and the importance to address them in order to facilitate the development of organic agriculture in Malaysia.

10.6 Challenges Associated with the Certification Process

As mentioned in the NAP, its objectives are; (1) to address food security and safety to ensure availability, affordability and accessibility; (2) to ensure the competitiveness and sustainability of the agrofood industry; and (3) to increase the income level of agri-based entrepreneurs. However, a number of studies conducted involving local farmers have shown that there are challenges identified and it can be argued that the objectives of the NAP may be compromised by these challenges and therefore, needed addressing.

10.6.1 Competitiveness Between Small-Scale Farmers and Commercial Entities

Based on an empirical study carried out in Cameron Highlands (Pahang) among the local organic farmers, it was found out that the farmers had financial difficulties to venture into organic food industry (Tiraieyari et al. 2017). They expressed dissatisfaction with the current regulatory system. In particular, the certification process is said to be highly costly and there is a lack of financial support. This financial burden is compounded by the fact that organic farming requires significant on-farm and off-farm investments such as the overall production costs, labour costs, certification costs, soil conservation, as well as packing and storage facilities. In contrast to the strategies stipulated in the national policy, whereby the government aimed to provide incentives to encourage farmers to adopt organic farming, they argued that they mostly receive assistance from the non-governmental organisations (NGOs) (Tiraieyari et al. 2017).

There were also issues reported in the study involving land especially among farmers, who are not landowners. This has a colonial history whereby small Malay farmers did not produce to optimum levels due to fear that their land and produce would be confiscated by their own aristocracy. They still need to rely on a temporary land tenure and are required to obtain Temporary Occupation Licenses (TOLs) (Tiraieyari et al. 2017). The TOLs need to be renewed on an annual basis. They argued that the government can reclaim their temporary land and this is discouraging especially when they need to invest a lot of money in cultivating the land. This means that farmers who work on a temporarily licensed land are in an uncertain situation. There is no guarantee for them to receive an expected return from their investment since they do not know for how long they can use the cultivated land.

They did not see any incentives for them to venture into organic agriculture industry, especially when they were also having difficulties in securing financial assistance. It was also argued that the TOLs system was regarded by local banking institutions as an adverse factor should they wish to apply for a bank loan.

Despite the government's concern in increasing the income of small-scale farmers' has been consistently highlighted in the national policies introduced earlier, and specifically to encourage them to adopt organic farming in the current policy, the findings of this study do not reflect such aspiration. Based on the findings reported, one could argue that a future in organic agriculture can only be more certain for landowners and commercial entities that have the financial ability to meet the costs required than that of small-scale farmers with a temporary land.

Small-scale farmers are at a disadvantage in order to compete with commercial and industrial entities that have better financial capabilities. For instance, the latter can offer higher returns for the use of land in nearby areas, which means small-scale farmers would not be able to compete especially without having financial assistance. This means that they would have to move to areas far from their own houses or require a lot more money to cultivate less suitable lands, which will incur higher cost production. This does not reflect the objective of the national policies to ensure competitiveness in the industry.

Also, the current procedure to obtain organic certification could be lengthy, whereby it could take roughly between 2 and 3 years (Tiraieyari et al. 2017). In addition, there is a grace period imposed on farmers before they could be awarded the certificate. Farmers are required to practice organic farming for 2 years before the certificate is granted and within this period, farmers cannot use 'organic' labels to market their produce. Despite selling the produce without organic' label, farmers are still required to strictly maintain the quality of the produce as organic products. This means that not only the cost for the production is high to meet the requirement for organic farming, without the labels, farmers have to sell their produce at a similar price to conventional produce, and not at a premium price.

Based on this scenario, it clearly shows that only farmers who are equipped with financial capabilities would be able to adopt organic farming as it does require money and time before they could begin to make a profit. This suggests that small-scale farmers could perceive such a system as not only discouraging, but also unfair. Without having a proper system in place for small-scale farmers to seek financial assistance, or secure land tenure to encourage personal investment, they would never be able to compete with commercial and industry entities. The organic food industry is undermined by the failure of national policies to address challenges or encourage small land holders to adopt organic farming.

10.6.2 Food Security and Sustainability

The organic food business is comfortably established in the market, and it can be said that the demand for organic products in Malaysia is rising due to the increasing standard of life especially among the health conscious population. Organic food is distributed through different types of organic marketing in Malaysia, namely: direct home-distribution, organic outlets, conventional supermarkets, wholefood supermarkets, and through organic farmers associations. This industry is instrumental in achieving the objective of the national policy, which is to ensure food security and affordability at the national and international level, especially after the AFC in 1997–1998. It has also changed the agricultural practice from primarily focusing on the export markets, whereby a significant amount of pesticides are used to increase production, which causes damage to the environment, to a more environmentally friendly and sustainable practice using biopesticides.

Nonetheless, it was reported that the organic producers are currently unable to commit to supply organic produce to local markets, especially from retail outlets, which need a fixed supply on a monthly basis. Since they could not meet the demand for a fixed supply, it is challenging for them to make a business deal with local distributors such as supermarkets and outlet retailers that need to set a certain amount of quantity in advance and require a continuous supply. Therefore, Malaysia needs to import organic food from other countries such as the United States, Japan, Australia, and New Zealand (Dardak et al. 2009). Malaysia also receives a high demand for organic food products from Singapore, however, the capacity to meet the demand is limited due to a shortage in supply and market access issues at the Causeway.

One could argue that the lack of ability for the organic food industry to meet the local and international demand is due to the lack of number of organic producers in the country. One of the contributing factors could be the difficulties facing small-scale farmers in venturing into this industry. Therefore, it is vital for the government to revisit the implementation of the national policy in facilitating and providing incentives to attract more farmers to adopt organic farming with the hope that this could then ensure the organic food industry will meet the demand in future.

10.6.3 Food Safety and Quality

It was reported that the public is not entirely confident about the quality of organic produce even though with the ‘organic’ labels are granted by the local authority (Saleki and Seyedsaleki 2012). This shows a vital issue that needs addressing as it reflects that the public do not trust the current regulation. One of the serious uncertainties revealed was about the genuinity of the organic produce. The quality of organic food in Malaysia is questioned by the public and there were views suggesting that organic food producers are not honest in selling their produce by mixing them with inorganic products. This further questions the credibility of the new regulatory system, myOrganic.

The skeptical view towards the reliability of myOrganic scheme among the local population is compounded by the fact that it has not received recognition by the international bodies. This suggests that Malaysia’s aspiration to boost its income by leveraging on organic food industry can be jeopardised given that many countries would not accept organic products from Malaysia without the international organic certification, or acceptance that equivalence with European or U.S. standards has been achieved. To convince both local and international consumers of the quality and safety of Malaysian organic products, it is vital for the authority to ensure that the current regulation is recognised internationally. Without doing so, local producers will need to rely on the international certification schemes if they wish to export their products internationally.

The negative views regarding the organic food regulations could be a signal that revision and improvements are needed and that there is a need to deepen nationwide research and politically commit to transitional subsidies for organic producers in order to support consumer confidence in the growing organic food market.

10.6.4 Regulatory Awareness

According to Saleki and Saleki (2012), there is a lack of awareness among the key actors in this industry including consumers, retailers, and even producers, of the wider extent of the implementation of the standards in the country. This is echoed in a finding from the study conducted in Cameron Highlands, Pahang, Malaysia whereby small-scale farmers viewed the system as too complex and has caused some confusion among them. To overcome this issue, the farmers created an association to build a support system among themselves and use it as a platform to spread information and educate themselves to address the complexity of the process.

It is important to have a system that is easy to be understood by the regulated parties, in this case, the farmers and retailers, so that they would be able to meet the policy requirements in order to ensure food quality and safety. One could argue that it is the responsibility of the authorised authority to ensure that the rules and regulatory system as a whole facilitate regulated parties to understand the system as rules are not self-explanatory, or capable of being implemented without significant income expenditures by small land holders. Appropriate measures are needed to be introduced by the government to address such issues. The DOA especially, needs to be more active to engage with the regulated parties, especially small-scale farmers, who may require attention and tangible assistance to understand the system.

10.7 Conclusion

The formulation of the national policies on agriculture in Malaysia has gone through a journey that reflected changing ideas of the national interest before and after Merdeka. Throughout its development, Malaysia maintains its interest recognizing agricultural industry as one of the main contributors towards the national income and also to benefit the local population by alleviating rural poverty. After a few revisions made to the national policies, organic agriculture eventually surfaced as one of the key areas identified by the government as instrumental to achieve the national aspirations, namely, to increase the income level of farmers or agri-based entrepreneurs, to ensure food security, and to foster sustainable agriculture practice in the country.

The discussion above shows that despite the existence of national policies on organic food industry in Malaysia, the regulation does not adequately reflect the spirit of the national agenda in developing this industry. There are challenges that need addressing before the objectives of the national policy can be achieved. This includes aspirations:

(1) to convince consumers by addressing their skepticism as to the credibility of the regulation system in ensuring the quality and safety of organic products in Malaysia, as well as to increase the level of awareness of the implementation of the standards;
(2) to create a more just and competitive environment to attract more farmers to adopt organic farming by revisiting the current regulation that is perceived as financially burdensome among small-scale farmers, in order to increase their income level; and
(3) to facilitate the effort to ensure food security, affordability and to meet the demand for organic food locally and, eventually, internationally.

Overall, even though organic food industry, in particular organic agriculture, is relatively new and still considered at its infancy in Malaysia, it has been shown that much progress has been made by the government toward consumer recognition of the organic ideal. However, in further developing this area, there is a need to revisit the regulatory system to strengthen the current settings in order for the government to achieve the objectives of the national agriculture policies.

References

Ahmad F (2001) Sustainable agriculture system in Malaysia. In: Regional workshop on Integrated Plant Nutrition System (IPNS), development in rural poverty alleviation. United Nations Conference Complex, Bangkok, pp 18–20

Ahmad SNB, Juhdi N (2010) Organic food: a study on demographic characteristics and factors influencing purchase intentions among consumers in Klang Valley, Malaysia. Int J Bus Manag 5:105–118

American Chemical Society (2006) Sustainable farm practices improve third world food production. ScienceDaily. Retrieved July 28, 2019 from https://www.sciencedaily.com/releases/2006/01/060123163315.htm

Ann A (2007) Organic farming can feed the world, study suggests. ScienceDaily. Retrieved July 28, 2019 from www.sciencedaily.com/releases/2007/07/070711134523.htm

Badgley C, Moghtader J, Quintero E, Zakem E, Chappell MJ, Aviles-Vazquez K et al (2007) Organic agriculture and the global food supply. Renew Agric Food Syst 22:86–108

Busby PE, Soman C, Wagner MR, Friesen ML, Kremer J, Bennett A, Morsy M, Eisen JA, Leach JE, Dangl JL (2017) Research priorities for harnessing plant microbiomes in sustainable agriculture. PLoS Biol 15:e2001793

Dardak RA, Zainal Abidin AZ, Ali AK (2009) Consumers' perceptions, consumption and preference of organic product: Malaysian perspective. Econ Technol Manag Rev 4:95–108

DOA Malaysia (2019a) Department of agriculture strategic plan 2016–2020. http://www.doa.gov.my/index/resources/aktiviti_sumber/sumber_awam/penerbitan/pelan_strategik_doa_2016_2020.pdf

DOA Malaysia (2019b) List of certified organic farms in Malaysia. http://www.doa.gov.my/index/resources/perkhidmatan/skim_pensijilan/penerima_sijil_myOrganic_dis2018.pdf

Doni F, Zain CR, Isahak A, Fathurrahman F, Anhar A, Mohamad WN, Yusoff WM, Uphoff N (2018) A simple, efficient, and farmer-friendly *Trichoderma*-based biofertilizer evaluated with the SRI rice management system. Org Agric 8:207–223

Doni F, Mispan MS, Suhaimi NSM, Ishak N, Uphoff N (2019) Roles of microbes in supporting sustainable rice production using the system of rice intensification. Appl Microbiol Biotechnol 103:5131–5142

Hossain MTB, Lim PX (2017) Consumers' buying behavior towards organic foods: evidence from the emerging market. Malays Manag Rev 51:7–25

Jambon I, Thijs S, Weyens N, Vangronsveld J (2018) Harnessing plant-bacteria-fungi interactions to improve plant growth and degradation of organic pollutants. J Plant Interact 13:119–130

Lian SB (2017) What motivates consumers to purchase organic food in Malaysia? Asian Soc Sci 12:100–109

Lin BB, Chappell MJ, Vandermeer J, Smith G, Quintero E, Bezner-Kerr R et al (2011) Effects of industrial agriculture on climate change and the mitigation potential of small-scale agroecological farms. CAB Rev Perspect Agr Vet Sci Nutr Nat Resour 6:1–18

Loh PQ (2019) The determinants of attitudes towards organic food in Malaysia. Master Thesis, UTAR, Malaysia

Lücke FK (2017) Microbiological safety of organic and conventional foods. FOODBALT-2017 https://doi.org/10.22616/foodbalt.2017.003

Magkos F, Arvaniti F, Zampelas A (2003) Organic food: nutritious food or food for thought? A review of the evidence. Int J Food Sci Nutr 54:357–371

Malaysian Farm Certification Scheme for Good Agricultural Practice (2005) Department of Agriculture, Malaysia

Malaysia, National agro-food policy (2011–2020), Ministry of Agriculture and Agro-based Industry

Malaysia, National agriculture policy (1984–1991), Ministry of Agriculture. Executive summary

Malaysia, Second national agriculture policy (1984–1991), Ministry of Agriculture. Executive summary

Malaysia, Third national agriculture policy (1984–1991), Ministry of Agriculture. Executive summary

Mørkeberg A, Porter JR (2001) Organic movement reveals a shift in the social position of science. Nature 412:677

Partap T (2010) Emerging organic farming sector in Asia: a synthesis of challenges and opportunities. In: Organic agriculture and agribusiness: innovation and fundamentals. Asian Productivity Organization, Tokyo

Pimentel D, Hepperly P, Hanson J, Douds D, Seidel R (2005) Environmental, energetic, and economic comparisons of organic and conventional farming systems. BioSci 55:573–582

Population Reference Bureau (2018) World population data sheet. Retrieved July 28, 2019 from https://www.prb.org/2018-world-population-data-sheet-with-focus-on-changing-age-structures/

Ramli B (2002) The development of organic farming in Malaysia. Paper presented at workshop on green productivity and natural farming, Seoul. Asian Productivity Organization (APO)

Rock B, Suriya J, Vijay B, Thalha N, Elango S, Rajajeyakumar M (2017) Organic food and health: a systematic review. J Community Med Health Edu 7:1–7

Rundgren, G (2006) Organic agriculture and food security, International Federation of Organic Agriculture movements, Retrieved August 1, 2019 from https://www.ifoam.bio/sites/default/files/organic_agriculture_and_food_security_printcopy.pdf

Sadati SA, Shaabanali Fami H, Asadi A, Sadati SA (2010) Farmer's attitude on sustainable agriculture and its determinants: a case study in Behbahan County of Iran. Res J Appl Sci Eng Technol 2(5):422–427

Saleki ZS, Seyedsaleki SM (2012) The main factors influencing purchase behaviour of organic products in Malaysia. Interdiscip J Contemp Res Business 4:98–116

Somasundram C, Razali Z, Santhirasegaram V (2016) A review on organic food production in Malaysia. Horticulturae 2(3):12

Suhaimee S, Ibrahim IZ, Wahab MAMA (2016) Organic agriculture in Malaysia. http://ap.fftc.agnet.org/files/ap_policy/579/579_1.pdf

Teh C (2012) Organic agriculture and food in Malaysia. http://www.christopherteh.com/blog/2012/02/organic-agriculture/

Thambiah S, Khin AA, Muthaiyah S, Yen YY (2015) Organic food consumption among generation Y in Malaysia: a conceptual. J Appl Sci 15:570–575

Tiraieyari N, Azimi H, Bahaman AS, Jegak U (2014) Knowledge and perceptions of extension workers on sustainable agricultural practices. Am J Environ Sci 9:45–50

Tiraieyari N, Hamzah A, Samah BA (2017) Organic farming and sustainable agriculture in Malaysia: organic farmers' challenges towards adoption. Sustain Dev Org Agric Hist Perspect:135

Uphoff N (2017) SRI: an agroecological strategy to meet multiple objectives with reduced reliance on inputs. Agroecol Sust Food 41:825–854

Willer H, Lernoud J (2019) The world of organic agriculture. Statistics and emerging trends. Research Institute of Organic Agriculture FiBL and IFOAM-Organics International. https://shop.fibl.org/CHde/mwdownloads/download/link/id/1202/?ref=1

Mohammad Firdaus Bin Abdul Aziz is a senior lecturer in the Faculty of Law, University of Malaya (UM), Kuala Lumpur, Malaysia. He is a coordinator for a number of courses, namely, Medical Law and Ethics, Biosafety Law, Biodiversity Law, Law and Society and Ethical Issues of Emerging Sciences. He completed his BSc (Hons) in Genetics and Molecular Biology from the University of Malaya, his MA in Biotechnological Law and Ethics from the University of Sheffield and his DPhil in Public Health from the University of Oxford. For his doctoral research, he examined the regulatory techniques adopted by selected Commonwealth countries to regulate human stem cell research and proposed an appropriate regulatory framework for Malaysia. He is also currently a researcher and production editor in the Centre for Law and Ethics in Science and Technology (CELEST) at the University of Malaya; an Academic Affiliate in the Centre for Health, Law and Emerging Technologies (HeLEX) at Oxford; and an Affiliate Member of Young Scientist Network, Academy of Sciences Malaysia, working on policy and governance. His research interests are in the legal and ethical issues of emerging technologies. He is also actively involved with UNESCO Bioethics as a trainer for the UNESCO Ethics Teachers' Training Course (for higher education) in the region.

Muhamad Shakirin Bin Mispan is currently a senior lecturer at the Institute of Biological Sciences, University of Malaya, Malaysia, teaching bachelor degree courses on Ecology and Management of Weeds, Soil Ecology and Management and Quantitative Ecology and Applied Weed Science and Pest Management in Organic Crops for Master of Science in Crop Protection. He is also the coordinator of GlamiLemi Biotechnology Research Centre, University of Malaya, where part of the Centre's responsibilities are to promote sustainable agriculture and develop new agriculture technologies for local farmers. He is a PhD holder from South Dakota State University working on QTL mapping for weedy adaptive traits in rice, haplotype analysis for a major seed dormancy QTL region in rice and seed longevity and photoperiod sensitivity in weedy rice. His current research focuses are mainly on ecological genetics of weedy rice focusing on adaptation to various ecological systems, seedbank dynamics and longevity, environmental impact from various weed managements, integration of sustainable agriculture for rice production and precision agriculture in rice farming.

Febri Doni is a Research Fellow at the Institute of Biological Sciences, Faculty of Science, University of Malaya. He obtained his PhD in Microbiology from the School of Biosciences and Biotechnology, Faculty of Science and Technology, National University of Malaysia. His research interest includes plant-microbe interactions, agroecology and plant pathology. He published dozen of papers focusing on microbe-rice plant interactions, System of Rice Intensification (SRI), organic agriculture and agroecology. He is the Executive Secretary of the Malaysian Agroecology Society (SRI-Mas) and is involved in various social activities in Southeast Asia in promoting agroecology and sustainable agriculture.

Chapter 11
Understanding Production and Safety Situations of Organic Food in Thailand

Thararat Chitov

Abstract The rising demand for safe food in Thailand has led to the increasing popularity of organic food. This calls for a significant transition from conventional agriculture, in which agricultural chemicals are employed to protect plants and animals and to enhance yields of agricultural produce, to organic agriculture. Such transition, however, is a complex process that requires farmers' commitment, support from governmental and non-governmental organisations through the production chain and marketing, and supportive national policy and regulations. This chapter presents on how organic food in Thailand is produced, and addresses issues related to food production and safety that organic farmers and organic food manufacturers are facing or can anticipate. Management of organic production, using both traditional knowledge and modern technology, and how they can affect food safety, are discussed. It also raises key issues about existing regulatory systems and mechanisms that can support organic agriculture and organic food safety control measures in the Thai context.

Keywords Organic agriculture · Organic food production · Regulatory system

11.1 Transitional Phases of Agriculture in Thailand

11.1.1 "Traditional" Agriculture to Conventional Agriculture: The Problem of Pesticide Residue

There is a need to assess the contrast between conventional and organic farming in Thailand, including the role of excessive pesticide residues in bringing safe food to the public's attention. Then the regulatory system currently in place in the country can be examined.

T. Chitov (✉)
Microbiology Division, Department of Biology, Faculty of Science, Chiang Mai University, Chiang Mai, Thailand

GOH Bee Chen, R. Price (eds.), *Regulatory Issues in Organic Food Safety in the Asia Pacific*, https://doi.org/10.1007/978-981-15-3580-2_11

Thailand has always been known as a land of Agriculture. In the Thai context, rice farmers are considered the "Backbone of the Thai Nation." The saying about Thailand that describes the natural abundance of food in the land, translated as "fish in every water body and rice in every paddy field," is well-known through the generations. Even in the midst of the economic transitions, this saying is still well-recognised among the Thais. In the past, the traditional farming in Thailand relied on sufficient agriculture, as the Thai society was based on the concept of Sufficient Economy (or subsistence farming) which has strong associations with organic farming techniques. It was very common that a Thai household owned a piece of land on which a rice field was located or an integrated orchard with various edible plants was cultivated, and on which animals or aquatic animals were raised. As a result, the family had sufficient supplies of food, grown or raised naturally, almost throughout the year. Having been blessed with such an abundance, however, did not prevent Thai society urbanising or the traditional Thai agriculture of sufficiency transforming into economically-driven industrial agriculture. Agricultural pieces of land were sold, ponds were filled, and animal labours were replaced by machinery. Integrated fields and orchards were replaced by a single crop or single plant farms. Nowadays, small farming of free-range chicken raised under the Thai houses on stilts and pigs in the backyards has been replaced by massive chicken and pig farming that holds large number of animals in high densities. This new form of farming is vulnerable to pests and the rapid spread of plant and animal diseases, causing massive destruction to the produce, if no control measure is applied. As a result, the chemical-free traditional farming that had long been part of the Thai society has dispersed into what is now known as "conventional agriculture," in which chemicals are heavily applied to control pests and diseases and to increase production yields. The data from the Office of Agriculture Regulation (OAR) showed that in 2018 Thailand imported 170,826,647.30 kg of potentially hazardous pesticides, with the total value of 27,982,162,163.89 Thai Baht (slightly under one billion USD), with herbicides making up the largest proportion of the chemicals, followed by fungicides, insecticides, and other groups of pesticides (Fig. 11.1).[1] Surveys have shown that significant proportions of vegetables and fruits in Thai markets contained unsafe levels of pesticide residues (Table 11.1).

Production under a conventional agriculture system in Thailand is comprised of a complex socio-economic system in transition which is characterised in recent times by changes of ownership of land, the promotion of single crop planting or single animal raising, the production, supply, and extensive use of agricultural chemicals, the marketing and encouragement of use of these products. Post-production, there exists a multi-party distribution chain of agricultural produce, enlarged scales of processed food production, and the changing expectation of consumers on produce appearance. Conventional agriculture undoubtedly plays a significant role in the socio-economic development of Thailand. Although it offers

[1] Office of Agriculture Regulation (OAR), Department of Agriculture (2018) Summary of imported hazardous agrochemicals 2018 Available at: http://m.doa.go.th/ard/

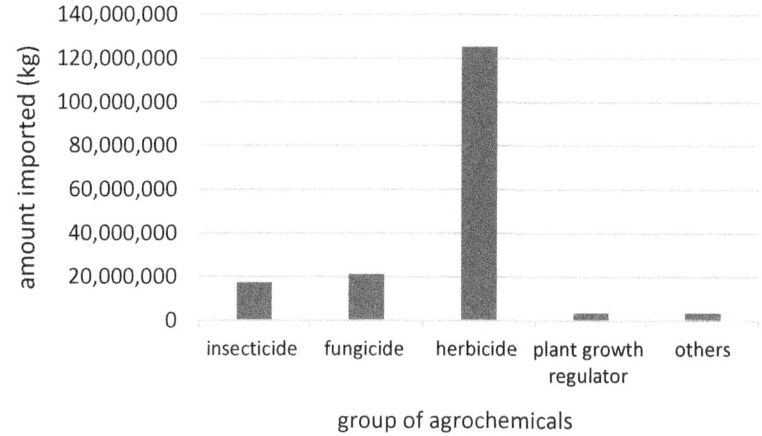

Fig. 11.1 Imported hazardous pesticides in Thailand in 2018. What is shown as 'others' include acaricide, rodenticide, mollussicide, fumigants, and nematocide. (Data from OAR 2018)

Table 11.1 Pesticide residues in agricultural produce available in Thai markets[a, b]

Produce group	Produce with pesticide residue levels exceeding MRL[c]	Percentage of samples contaminated with pesticide(s) exceeding MRL
Vegetables	Chinese kale (from local markets)	48%
	Chinese kale (from supermarkets)	35%
	Pakchoi (from local markets)	71%
	Pakchoi (from supermarkets)	55%
	Morning glory (from local markets)	42%
	Morning glory (from supermarkets)	49%
	Red chillies	100%
	Basil	66.7%
	Long beans	66.7%
	Chinese kale	55.6%
	Chinese cabbage	33.3%
	Morning glory	22.2%
	Tomatoes	11.1%
	Cucumbers	11.1%
Fruits	Oranges	100%
	Guava	100%
	Dragon fruit	71.4%
	Papaya	66.7%
	Mango	44.4%

[a]Wanwimolruk S, Phopin K, Boonpangrak S, and Prachayasittikul V. (2016) Food safety in Thailand 4: comparison of pesticide residues found in three commonly consumed vegetables purchased from local markets and supermarkets in Thailand. PeerJ. 4: 2432.
[b]Bio Thai (2016) Thai PAN has published the results of their second round of testing on chemical pesticide residues for 2016. Available at: https://www.biothai.org/node/1427
[c]Maximum residue limit

many immediate advantages, especially in increasing productivity, it did not come without a cost. It has taken a considerable period of time for governments, farmers, and consumers, to recognise many negative effects it has brought to the Thai society, economy, and environment. For example, the massive, single crop plantation system rapidly reduces the availability of nutrients and upsets the balance of nutrients and natural properties of soil.[2] It uses large amount of water and causes water resource to become drastically limited, affecting small farming and village life.

Quick but destructive methods, like burning, have been applied in post-harvesting treatments of plantation areas, contributing to air pollution; this problem is increasingly apparent in recent years, especially in the north and northeastern parts of Thailand.[3] Excessive uses of chemical fertilisers and chemical pesticides in plant production negatively affect ecosystems. Studies have shown that soil and water in many agricultural areas in Thailand are polluted with toxic agrochemical-related substances, including nitrate, pesticides, and heavy metals.[4,5] The study by Tirado (2007) pointed to a direct relationship between the amount of nitrogen fertiliser applied in an agricultural area and nitrate concentration in groundwater within the area.[6] These pollutants can also accumulate in plants and animals that are part of the food chains, leaving chemical residues in agricultural produce. They can cause immediate poisoning to human and animals and can potentially have a prolonged, chronic effects to health.[7]

Overuse of chemicals can cause environmental problems, such as algal bloom due to excessive amounts of nitrogen in water bodies, affecting aquatic ecosystems, and can potentially cause algal toxin accumulation in the food chain and eventually result in intoxication upon consumption of aquaculture products. Production and use of agrochemicals can also result in large-scale greenhouse gas emission and is considered one of the contributing factors to climate change.[8]

[2] Food and Agriculture Organization of the United Nations (FAO) (n.d.) AGP-Agriculture and soil biodiversity. Available at: http://www.fao.org/agriculture/crops/thematic-sitemap/theme/spi/soil-biodiversity/agriculture-and-soil-biodiversity/en/

[3] Junpen A, Jirataya Pansuk J, Kamnoet O, Cheewaphongphan P, and Garivait S. (2018) Emission of air pollutants from rice residue open burning in Thailand, 2018. Atmosphere 9: 449; doi:https://doi.org/10.3390/atmos9110449

[4] Tirado R, Englande AJ, Promakasikorn L, and Novotny V. (2008) Use of agrochemicals in Thailand and its consequences for the environment. Greenpeace Research Laboratories Technical Note 03/2008.

[5] Jaipieam S, Visuthismajarn P, Sutheravut P, Siriwong W, Thoumsang S, Borjan M, and Robson M. (2009) Organophosphate pesticide residues in drinking water from artesian wells and health risk assessment of agricultural communities, Thailand. Human and Ecological Risk Assessment 15: 1304–1316.

[6] Tirado R (2007) Nitrates in drinking water in the Philippines and Thailand. Greenpeace Research Laboratories Technical Note 10/2007.

[7] Nicolopoulou-Stamati P, Maipas S, Kotampasi C, Stamatis P, and Hens L. (2016) Chemical pesticides and human health: the urgent need for a new concept in agriculture. Frontiers in Public Health 4: 148–155.

[8] International Atomic Energy Agency (IAEA) (2019) Greenhouse gas reduction. Available at: https://www.iaea.org/topics/greenhouse-gas-reduction

11.1.2 Transition from Conventional Agriculture Back to Organic Agriculture

The inter-relatedness of human health, animal health, and environmental health has been increasingly recognised worldwide and in Thailand, especially after the reintroduction of the One Health Concept.[9] The public realisation of the contamination of chemicals in food of plant and animal origins has led to increasing demands among consumers in Thailand for safer food. In response to greater awareness about the problem of pesticide residues, many farmers and supporting agencies have made a step towards safer food production by operating "Pesticide-Safe Agriculture." In the Pesticide-Safe Agriculture concept, agricultural production may employ pesticides, but the types of pesticide used are limited to those permitted by the country's authority, and the use of them must comply with the recommendation of the Department of Agriculture (DOA), Ministry of Agriculture and Cooperatives or with instruction on the labels that are registered with the DOA. Application of pesticides must also be withdrawn before the harvest, leaving a certain period of time according to the manufacturer's or official recommendations for specific pesticide types. Pesticide residues in the produce must also be monitored through laboratory analyses and the residues must be kept under the maximum residue limits (MRL) under relevant standards. With these steps of control, the produce is expected to be "safe" for consumption, i.e. having pesticide residue levels that are understood to be not harmful to health.[10]

The fact that the agricultural produce that was believed to be pesticide-safe or labelled as pesticide-safe but was found to have pesticide residues, or have the residues exceeding the MRL limits, has made consumers lose their trust in "pesticide-safe" food.[11] Some educated consumers, moreover, still were not satisfied with other chemicals besides pesticides that may be used in the production of pesticide-safe produce. Moreover, a number of farmers who had felt the negative effects of chemicals used in agriculture on their health, crops, and land started to change their attitude towards conventional agricultural production.[12] These factors contribute to establishment of organic agriculture, or in a more accurate term for Thailand, re-establishment of it, thus making another transitional phase in agriculture in Thailand.

---

[9] Destoumieux-Garzón D, Mavingui P, Boetsch G, Boissier J, Darriet F, Duboz P, Fritsch C, Giraudoux P, Le Roux F, Morand S, Paillard C, Pontier D, Sueur C, and Voituron Y. (2018) The One Health Concept: 10 years old and a long road ahead. Frontiers in Veterinary Science 5: 14; doi: https://doi.org/10.3389/fvets.2018.00014

[10] National Bureau of Agricultural Commodity and Food Standards, Ministry of Agriculture and Cooperatives (2013) Thai Agricultural Standard TAS 9001–2013: Good Agricultural Practices for Food Crop.

[11] Bio Thai (2016) Thai PAN has published the results of their second round of testing on chemical pesticide residues for 2016. Available at: https://www.biothai.org/node/1427

[12] Boonchai K. (2016) Management of agricultural and food systems that are safe from agrochemicals: lessons from modelled areas. A publication of National Health Commission Office [article in Thai].

Organic agriculture, as defined by the Organic Agriculture Certification Thailand (ACT), is "farming system without the use of artificial fertilisers and synthetic pesticides and in compliance with ACT Organic Agriculture Standard".[13] It includes natural farming and ecological farming which embraces principles of promoting/maintaining the diversity of plants and animals and ecosystems, preserving natural resources and the environment, exercising the production system and management with care for humanity, while providing safe food. It is necessary for farmers to understand that conversion from conventional agriculture to organic agriculture is not just simply withdrawing chemical use from farming, but it is an overall system of farm management that integrates best environmental practices with the production of safe produce.

This transition, reverting back to where Thai agriculture had been, therefore, is not as easy as picking up a system from where it had been left off. It involves settling the impacts left by what came in between, including the social and environmental impacts of conventional agriculture. It is a complex transitional process that requires farmers' commitment, social support, science and technology support, and support from the national and local governments by means of policy and regulatory systems. At present, the organic form of agricultural production takes place in small scale compared to the conventional form, but it has high economic and environmental values. It is necessary that the mechanisms that support this transition are well established, for organic agriculture deserves to be a long-lasting trend.

11.2 Thai Regulatory Systems Relating to Organic Farming and Safety of Organic Food

11.2.1 General Regulatory Systems Relating to Agricultural Production

Regulation of agricultural production in Thailand is largely done by the Ministry of Agriculture and Cooperatives (MoAC).[14] Food safety is mainly regulated by MoAC and the Ministry of Public Health (MPH).[15] In the governmental level, several pieces

[13] Organic Agriculture Certification Thailand. (2017a) ACT Organic Standards 2016 (Revision 2017) Available at http://actorganic-cert.or.th/wp-content/uploads/2017/10/act_standards_2016_v4_revision24-02-17.pdf

[14] Ministry of Agriculture and Cooperatives (2016) Vision and mission page. Available at: https://eng.moac.go.th/ewt_news.php?nid=101

[15] Thailand National Food Committee (TNFC) (2012) Strategic framework for food management in Thailand (2012–2016). Available at: https://www.foodsafety.moph.go.th/document/Info_general/Strategic_Framework.pdf. This article gives an example of the active collaboration between these two main governmental agencies to form food safety policy.

of legislation form the basis of control mechanisms of agricultural production and food safety[16]:

Ministry of Public Health

- Food Acts B.E. 2522 (1979)

Ministry of Agriculture

- Plant Quarantine Act B.E. 2507 (1964)
- Hazardous Substances Act B.E. 2535 (1992)
- Plant Varieties protection Act B.E. 2542 (1999)
- Animal Epidemic Act B.E. 2499 (1956)
- Feed Quality Control Act B.E.2525 (1982)
- Animals and animal Carcasses Act B.E 2544 (2001)
- Fisheries Act B.E 2490 (1947)
- Agricultural Standards Act B.E. 2551 (2008)

The Food Act (1979) covers such diverse topics as food establishment registration, establishment of regulations on specific foods, control of food safety and food labelling. For agricultural production, it can be seen from the list above that different Acts were issued to cover plant, animal, and fishery production.[17] Another legal instrument that is very important for the regulation of safety in agricultural production and environment in Thailand is the Hazardous Substance Act B.E. 2535 (1992) amended B.E. 2544 (2001) and B.E. 2551 (2008), which includes pesticides in its scope.[18] However, there is a gap in managing pesticides at the farm level, compounded by other factors including the structure of regulatory agencies, marketing systems, and education levels of farmers, which make it almost impossible to control the level of pesticide used and the levels of pesticide residues left in agricultural produce.[19]

Agricultural Standards Act B.E. 2551 (2008) was issued to provide for the making of agricultural standards in order to facilitate the export of Thai agricultural and food products.[20] According to this law, the National Bureau of Agricultural

[16] Naewbanij JO. (2005) Food safety overview in Thailand. E-document of the presentation at SELAMAT EU-Asian Food Safety Platform Lisbon, Portugal, April 11, 2005. Available at: https://selamat.net/upload_mm/7/5/3/6be94190-ef9f-4753-9c80-126732c76da7_p11.pdf

[17] Food and Drug Administration. (n.d.) Unofficial translation of Food Act B.E. 2522. Available at: http://food.fda.moph.go.th/law/data/act/E_FoodAct2522.pdf

[18] Office of the Council of State (2008) Unofficial translation of Hazardous Substance Act, B.E. 2535 (1992) https://www.jetro.go.jp/thailand/e_activity/pdf/hazsubact2535.pdf

[19] Panuwet P, Siriwong W, Prapamontol T, Ryan PB, Fiedler N, Robson MG, and Barr DB (2012) Agricultural Pesticide Management in Thailand: Situation and Population Health Risk. Environmental Science and Policy 17: 72–81.

[20] Office of the Council of State of Thailand's Law for ASEAN project. (n.d.) Agricultural Commodity Standard Act, B.E. 2551 (2008) - initial version. Available at: http://web.krisdika.go.th/data/outsitedata/outsite21/file/AGRICULTURAL_COMMODITY_STANDARDS_ACT,_B.E._2551_(2008).pdf

Commodity and Food Standards (ACFS) under the Ministry of Agriculture and Cooperatives sets standards for agricultural systems, commodity and food items, as well as food safety. The standards are implemented by different responsible departments: for plants, by the Department of Agricultural Extension (DOAE) and the Cooperative Promotion Department (CPD), for livestock, by the Department of Livestock Development (DLD), and for fisheries, by the Department of Fisheries (DOF). The Department of Agriculture (DOA) plays an important role in inspection and certification of plant production, while DLD and DOF carry these responsibilities for livestock production and fishery production, respectively.

In recent years, the Thai government has taken many major steps to promote food safety. In 2004, the government launched its national policy on food safety to improve the quality of food. The year 2004 was announced as a Food safety year and the "Road Map of Food Safety" was approved to be implemented from that year onwards. The "road map" adopts the Farm-to-Table concept and provides a guideline for quality control strategies in each stage of food production, including the stages of imported raw materials and inputs, food production, food manufacturing, produce and food products, and the market (domestic and export). As a result, many programmes have been implemented, which may be considered milestones in Thai Food Safety Regulatory Systems. These include major programmes such as Good Agricultural Practice (GAP) Implementation and Certification (at farm level), GMP/HACCP Implementation and Certification (at manufacturing level), and product registration/certification and issuing of the 'Q' Mark.[21]

The legislation above and the recently introduced programmes, although are not directly related to organic food, form a basis of quality control of agricultural production, agricultural commodities, and food products. The rise in Thai organic standards, certification and enforcement will now be addressed.

11.2.2 Development of Organic Food Standards and Certification

As for production of organic commodities and organic food products, establishment of regulatory systems took place in a unique way. Perhaps, one of the most significant characteristics of the regulatory mechanisms of organic food in Thailand is the fact that much of it is done not by governmental agencies but by non-governmental organisations. These organisations have a long history of supporting organic conversion, organic production and marketing of organic products since the 1980s, before their major move to call for an official form of organic certification. In 1995, with the cooperation and initiative of sustainable agriculture advocates led by

[21] Thailand International Development Cooperation Agency (n.d.) Thailand's best practices and lessons learned in development, Volume 1. Available at: https://www.undp.org/content/dam/thailand/docs/TICAUNDPbpVol1.pdf

Alternative Agriculture Network (AAN), NGOs, academics, consumer networks, media representatives, and eco-friendly businesses, the Organic Agriculture Certification Thailand (ACT), was established.[22] ACT's legal status was gained with the registration of Foundation of Organic Agriculture Certification Thailand in 2001. ACT is an independent, non-profit certification body that works with inspection and certification of organic agriculture in Thailand and Asian regions.[23] ACT has its own standards, known as "ACT Organic Standards," which has been held as one of the key standards for organic agricultural production and processing.

Notably, ACT offers organic certification services which are recognised at the international level. ACT is accredited by the National Office of Agricultural Commodity and Food Standards (ACFS) and International Federation of Organic Agriculture Movements (IFOAM). The products certified by ACT are recognised at national and international level. Several foreign certification bodies are also available for certification of Thailand organic products which need to comply with or will be distributed in specific countries.[24] There are also several other certification bodies that certify organic products on the regional level. A well-established one among these is the Northern Organic Standards Organisation (NOSO), which is a local organic certification in Northern Thailand.[25]

As for the part of governmental organisations, certification of organic commodities is offered by different organisations, as listed below[26]:

- Department of Rice offers certification for organic rice.
- Organic Crop Institute under the Department of Agriculture offers certification of crops other than rice.
- Organic Aquaculture Farm and Product Certification Center under the Department of Fisheries offers certification for aquaculture.
- Department of Livestock offers certification for livestock.

The National Office of Agricultural Commodity and Food Standards also issued a national agricultural standard for organic agriculture: "Thai Agricultural Standard TAS 9000–2003: Organic Agriculture Part 1: The Production, Processing, Labelling and Marketing of Organic Agriculture".[27] This added to the regulatory systems in

[22] Thai Organic Trade Association (2011) Overview of organic agriculture in Thailand. Available at: http://www.thaiorganictrade.com/en/article/442

[23] Organic Agriculture Certification Thailand. (2019) bout ACT. Available at: http://actorganic-cert.or.th/about/

[24] Thai Organic Trade Association (2011) Overview of organic agriculture in Thailand. Available at: http://www.thaiorganictrade.com/en/article/442

[25] Nandwani D. and Nwosisi S. (2016) Global trends in organic agriculture. *In* Nandwani D (editor) Organic Farming for Sustainable Agriculture. Springer International, Switzerland. p. 22.

[26] Thai Organic Trade Association (2011) Overview of organic agriculture in Thailand. Available at: http://www.thaiorganictrade.com/en/article/442

[27] National Bureau of Agricultural Commodity and Food Standards, Ministry of Agriculture and Cooperatives (2003) Thai Agricultural Standard TAS 9000–2003: Organic Agriculture Part 1: The Production, Processing, Labelling and Marketing of Organic Agriculture. Available at: https://www.acfs.go.th/standard/download/eng/TAS-9000.pdf

Thailand to provide a mechanism to control safety of organic commodities (crops, livestock, and aquaculture) and organic products. Since ACFS also acts as an accreditation body, this standard also serves for national organic accreditation.

Recently, the Thai government launched the National Organic Agriculture Development Strategy, 2017–2021, to serve as a framework to develop organic agriculture for the purpose of increasing organic agricultural productivity and establish them on both local and international consumer markets.[28] It is hoped that this positive, proactive movement from the government side will bring significant positive impacts on organic production and organic food safety as a whole.

There are two prominent forms of organic farming in Thailand: integrated farming and mono crop/monoculture farming.[29] The first form, "integrated organic farming," is the form in which farmers grow plants or raise animals that are indigenous to their geographical areas. Most of these plants and animals are parts of the traditional Thai diet, and this form of farming mostly takes place on a small scale. It can take place in such a simple way. For example, a farmer has a plot of ground next to his house, and within the farming ground, native plants and herbs are cultivated in different divisions. The fence may be made of edible plants such as tamarind or betel plants. Within the farming ground, there might be root plants such as ginger, galangal, or turmeric. At one or a few locations within the plot of land, there might be bamboo trees that give bamboo shoots, or a few local fruit trees that give seasonal fruits. Some areas in the plot of land might be filled with different types of kitchen herbs such as basil, chili, lemon grass, or coriander. Farmers might grow some plants on a commercial basis, such as French beans, pumpkins, cabbage, or tomatoes, but not in large amounts. There might be a water well as a water source or a fish pond. Some farmers might raise free-range animals such as local-breed chickens, ducks, pigs, or cattle. Composts made within the farm from green and animal manures are the source of additional plant nutrients. The produce of the small farm is normally sufficient for consumption within the household and the excess may be sold in a local market. This type of farming can be naturally organic because of the repellent properties of some herbs, the diversity of native plants cultivated, or the free-range form of animal husbandry, which make them resist pests and diseases.[30] Some farmers choose to distribute their produce in ordinary markets or under the ordinary category of farm produce, but not necessarily classify them as "organic." However, some are part of organic farmer groups or associations and find some specific local markets or a special section in a market to distribute their "organic" produce.

[28] FAO (2018a) Country fact sheet on food and agriculture policy trends: Thailand. A publication of Food and Agriculture Organization of the United Nations. Available at: http://www.fao.org/3/I8683EN/i8683en.pdf

[29] Win HE. (2017) Organic agriculture in Thailand. FFTC Agricultural Policy Article. Available at: http://ap.fftc.agnet.org/ap_db.php?id=734&print=1

[30] Shelef O, Weisberg PJ, and Provenza FD. (2017) The value of native plants and local production in an era of global agriculture. Frontiers in Plant Science 8: 2069; doi: https://doi.org/10.3389/fpls.2017.02069

The second form of organic farming in Thailand is "monoculture organic farming," in which farmers grow certain types of plants or raise certain types of animals on a larger scale. The plants or animals are not necessarily indigenous to the geographical areas, but have high commercial values or are in high demand, especially among local urban and overseas communities. Such produce may include vegetables that are popular among city people, like broccoli, carrot, cabbages, tomato, salad leaves, beans, peas, cauliflower, Chinese kale, cucumber, etc. Monoculture organic fruit orchards can also be found, such as orchards of banana, mango, pomelo, avocado, durian, longan, lychee, etc. A structure that allows control of environmental factors might be used, such as greenhouses, mushroom chambers, and hen houses. Advanced technology can be adopted by such enterprises. With this design of farming, it is natural for main crops or main produce to be prone to pests and diseases. The non-indigenous or non-local breeds might not be the most pest- and disease-resistant breeds. Therefore, this form of farming would naturally require more care and effort by farmers to control pests and diseases. The produce is normally labelled or advertised as "organic," and the target are the high-end markets, which can be organic markets, organic or health stores, or the organic section in supermarkets. Organic produce and organic food products in Thailand's domestic market are mainly distributed in modern trade shops (supermarkets), accounting for approximately 60% of total domestic sales. The rest is distributed among green shops, food establishments, and farmer's markets.[31]

What both forms of farming share in common is the promise of safe, organic produce to consumers. It is certain that not all organic food production is covered by the support and control activities of governmental or non-governmental bodies, especially with the production by individual organic producers who are not part of an organic farmer group or by unregistered producers (which are the majority of organic producers). Mechanisms to allow involvement of such organic food producers in an organic network, or in a regulated system, are needed for more efficient protection of organic food consumers and of the environment.

11.2.3 Description and Labelling of Organic Food

Since there is a high demand for safe food and organic food, labels or indications of the quality of food or how food has been produced are becoming more important in Thai green and safe-food markets. Labels offer the opportunity to inform consumers and they are also used for marketing purposes. Organic food and safe food generally have a higher price than food produced through the conventional farming. Surveys

[31] Kongsom C and Panyakul V. (2016) Production and market of certified organic products in Thailand. International Journal of Economics and Management Engineering 10: 2723–2727.

have shown that there are consumers who are willing to pay for food that is safe and of high quality.[32]

Some farmers or manufacturers who distribute produce or products of animal origins use their self-designed labels besides "organic" or "safe" to distinguish their produce from "conventional" produce. These labels include, for examples, "grass-fed," "free-range," "naturally raised," "hormone-free," and "antibiotic-free." These labels do not mean to indicate the produce is organic, but they are intended to inform consumers of the production methods or processes which would normally result in safer produce/food products than those produced using the conventional method. With these labels, educated consumers may be able to distinguish the different degree of safety of such produce from organic produce, but many consumers cannot tell the differences or are unaware of the organic term.[33,34]

Agricultural commodities and food products with plant origins sold in Thai markets, especially in supermarkets, are marketed by use of at least four categories: conventional, safe, pesticide-free, and organic.[35] They have their specific production process and product characteristics described as follows (see also Table 11.2).

Conventional agricultural commodities. These are produced under a conventional farming concept. There is no control system over on-farm production, thus no certification nor label is required.

Safe produce and safe food products. The agricultural commodities and products of this category are those normally produced under the conventional agricultural system. However, the production steps are controlled and monitored or quality of products examined. So, in principle, the commodities or food products of this category should have agrochemical residue levels within the limits that are considered "safe" for consumption. Certification is granted by the Ministry of Agriculture and Cooperatives (MoAC), and the "Q" label is given to indicate quality and safety of the products.[36] Another certification, "Food Safety," is granted by the Ministry of Public Health, upon the satisfactory results from laboratory examination. The scope of this certification covers fresh food, cooked food, and processed food produced in small scale or sold in markets.

---

[32] Pracharuengwit, P and Chiaravutthi, Y. (2015) Consumer willingness to pay for organic food in Thailand: evidence from the random *n*th-price auction experiment. Business Administration Journal 146: 52–70.

[33] Roitner-Schobesberger B, Darnhofer I, Somsook S, and Vogl CR. (2008) Consumer perceptions of organic foods in Bangkok, Thailand. Food Policy 33: 112–121.

[34] Sangkumchalianga P and Huang W-C. (2012) Consumers' perceptions and attitudes of organic food products in Northern Thailand. International Food and Agribusiness Management Review 15: 87–102.

[35] Wanwimolruk S, Phopin K, Boonpangrak S, and Prachayasittikul V. (2016) Food safety in Thailand 4: comparison of pesticide residues found in three commonly consumed vegetables purchased from local markets and supermarkets in Thailand. Peer J. 4: 2432.

[36] National Bureau of Agricultural Commodity and Food Standards (2018) Manual of use and display of standard certification under the Agricultural Standard Act B.E. 2551 (2008). [article in Thai]

Table 11.2 Labels for non-conventional agricultural commodities and food products in Thai markets

Category/ production concept/product characteristics	Certification authority	Label	Remark
Safe: Normally produced in conventional agriculture system, but production steps are controlled, or products examined, so that the commodities or food products will have agrochemical residue levels within the safe limits for consumption.	Department of Agriculture, Ministry of Agriculture and Cooeratives		
	Ministry of Public Health		Used with Fresh food, cooked food, and processed food
Organic: Agricultural commodities which are produced without uses of GMO, synthetic fertilisers and chemical pesticides; farming carried out in the way that promotes biodiversity, and health of ecosystems and the environment; exercising care of humanity in production system. or Organic products (for those labelled Organic Thailand)	Organic Agriculture Certification Thailand (ACT)		Organic seal combining ACT logo and IFOAM accredited seal may be used after signing the organic seal user contract.
	Ministry of Agriculture and Cooperatives		To be used with organic products; encouraged to be used with the "Q" label
	Thai Organic Agriculture Foundation (TOAF) (example of a PGS certification)		
	Thai PGS Organic Plus (TPO+) is a collaborative network of the Earth Net Foundation		
	Northern Organic Standard Organization (example of a self-claim certification)		

Pesticide-free agricultural commodities. Such commodities are produced without the use of pesticides, and in many cases, also without the use of organic fertilisers and other synthetic chemicals. However, they are not necessarily organic, since the uses of GMO or non-organic seeds or materials are not prohibited or restricted, and the farming is not necessarily aimed at preserving or promoting biodiversity or trying to be environmentally safe. The produce of this category can be certified through self-claim or through Participatory Guarantee Systems (PGS), which are locally focused quality assurance systems that certify producers based on their active participation in a farmer union.[37]

Organic produce or organic food products. These are produced under the organic concept and organic standards or guidelines. Organic farmers can have a formal certification for the commodities or products through third-party certification, by which nationally-recognised organic-labels, such as ACT organic seal and "Organic Thailand," are granted.[38] However, they can have other types of certification granted by a self-claim organic producer association or by a PGS certification body.

11.3 Organic Food Manufacturing in Thailand

As in other countries, demands for organic food in Thailand has expanded its scope from fresh produce to organic processed food. The domestic demand is rising as the modern Thai lifestyle has shifted from home cooking to reliance on precooked or ready-to-eat food. From the point of view of the food producers, it is also natural to process some of the fresh produce in order to extend their shelf-life and reduce waste. International market demands also influence the production of organic food products in Thailand, for the major market for organic food and certified organic products is the export market.[39]

According to ACT organic standards, for a food product to be correctly labelled as "organic", using the seal "Organic Thailand," it is required that the product contain ≥95% (by weight) of organic ingredients in the total amount of all ingredients, excluding water and salt.[40] It is challenging for both small and big enterprises to produce completely organic food products or products that meet this requirement. Although a farmer or a manufacturer may have a supply of the main organic ingredient, still other minor ingredients have to come from organic sources. For example,

[37] Food and Agriculture Organization of the United Nations (FAO) (2018b) Participatory Guarantee System (PGS) for Sustainable Local Food Systems. Available at http://www.fao.org/3/I8288EN/i8288en.pdf

[38] National Bureau of Agricultural Commodity and Food Standards. (n.d.) Training document: Guidelines for international certification of organic products and use of organic seal. Available at: http://certify.dld.go.th/certify/images/project/organic/organic2562/pwR5/2.pdf. [article in Thai]

[39] Kongsom C and Panyakul V. (2016) Production and market of certified organic products in Thailand. International Journal of Economics and Management Engineering 10: 2723–2727.

[40] Organic Agriculture Certification Thailand. (2017a) ACT Organic Standards 2016 (Revision 2017) Available at: http://actorganic-cert.or.th/wp-content/uploads/2017/10/act_standards_2016_v4_revision24-02-17.pdf

in Northern Thailand, there is a rice cracker snack that is made of steamed glutinous rice soaked in watermelon syrup. The rice mixture is then shaped into round or square shape and sun-dried. The crackers are fried in oil, and can be topped with palm sugar icing. Just for this simple traditional Northern Thai snack, a food manufacturer will have to have the supply of glutinous rice, watermelon, sesame seeds, palm sugar, powdered brown cane sugar and vegetable oil, all being organic. Non-organic ingredients may be used, but not exceed 5%. Keeping a continuous supply of the organic ingredients makes it difficult for an organic food manufacturer to produce organic food that can constantly be supplied to the market, while maintaining the product quality. There is a need for strategies to address this complex problem of demand and supply of organic raw materials. There is a need to ensure that organic food products available in Thai market are authentic. The chapter proceeds now to outline the key challenges to the safety of organic food in Thailand.

11.4 Safety of Organic Produce and Organic Food Products in Thailand

11.4.1 Safety of Organic Produce

11.4.1.1 Chemical Safety

There are very limited studies about chemical and microbial contaminants in organic products worldwide, including Thailand, compared to conventional products. In general, organic commodities and organic food products are expected to be free from contamination of agrochemicals. The fact is, chemical residues were found in organic food produced in different areas of the world, including Thailand. Surveys on organic product safety in Thailand have presented evidence of chemical contamination of organic food, and this has alarmed the public and the responsible government agencies. An example is one of the most extensive surveys, conducted by Thailand Pesticide Alert Network (Thai-PAN) in 2016. It revealed a high incidence of chemical contamination in food products labelled with the quality and safety seal and the official organic seal. In this report, 25% of the supermarket products that were certified organic were found to contain chemical residues exceeding the recommended levels.[41]

If a properly coordinated approach is taken, prevention of chemical hazards in organic food can be achieved on the level of farming and food manufacturing. It is possible that some of the produce from organic farms is contaminated with pesticides and chemicals used in the neighbouring conventional farms, and such instances

[41] Bio Thai (2016) Thai PAN has published the results of their second round of testing on chemical pesticide residues for 2016. Available at: https://www.biothai.org/node/1427

have been recorded.[42] The materials and inputs used in organic farms have to be from organic sources as far as possible. In Thailand, this has some exception, for if all raw materials had to have organic origins, there would be insufficient supplies of some materials. An example of this is the permission to use rice straw from conventional rice fields to help protect soil surface and to control weeds.[43] A challenge remains in organic livestock production, in which organic feedstuff remains limited.

11.4.1.2 Microbiological Safety

Unlike chemical contamination of organic food which can be controlled by farmers, microbial contamination in organic food is unavoidable. Abstaining from using chemicals results in a greater presence of microbes in the environment, in air, soil, water, and on the produce. Microbes are abundant in plants, human, and animals, and can contaminate agricultural commodities via many routes, such as direct contact, ingestion, infection through feeds or wounds, and faecal-oral route. Some environmental microbial pathogens, such as endospore-forming *Bacillus cereus*, can be widely distributed in the environment and often found in various food ingredients and food products (Chitov et al. 2008).[44] Moreover, there can be other contamination sources during the post-harvest steps and food processing steps, such as processing equipment, food handlers, transport vehicles, packaging, etc.

In one of the studies of the author's research group, the possible sources of bacterial contamination in raw milk in a dairy farm were investigated. The research work was carried out at a collaborative dairy farm in Chai Prakan District, Chiang Mai. The farm operates under the ecological farming concept, making it as close to organic farming as it could possibly be. The water was supplied from a natural reservoir that collects water from mountain streams. Cows are free to graze, and fed with farm-grown Napier grass, straw, and feedstuff made of cassava and rice bran. From our survey of microbial quality of raw milk in two periods of time, we found that in one sampling set, all raw milk samples collected from milk containers (4 for each sampling period or each lot) had higher counts than the standard limit, which is set at 10^6 colony forming units of microorganisms per millilitre of milk. All raw milk samples from both lots were found to be contaminated with coliform bacteria and thermotolerant coliform bacteria or *Escherichia coli*. The environmental samples were collected during the same time periods to allow us to see whether water (for feeding and cleaning), milking containers, sieves used to filter raw milk, fresh

---

[42] Rama Healthy Farm Project. (n.d.) Lessons from Ranathibodi Healthy Farm. Medical Faculty, Ramathibodi Hospital, Mahidol University, Thailand. [article in Thai]

[43] Organic Agriculture Certification Thailand (2017c) ACT Organic Standards 2016 (Revision 2017); Appendix 1 List of approved inputs used in organic production, Part 1.1 Input used for soil fertilisers and soil conditioners. Available at: http://actorganic-cert.or.th/wp-content/uploads/2017/10/act_standards_2016_v4_revision24-02-17.pdf

[44] Chitov T, Dispan R, and Kasinrerk W. (2008) Incidence and diarrhegenic potential of *Bacillus cereus* in pasteurized milk and cereal products in Thailand. Journal of Food Safety 28: 467–481.

grass and straw, and the floor in the milking area were the possible sources of contamination into raw milk. Such equipment and environments contained coliforms, thermotolerant coliforms and *E. coli*, and were the possible sources of contamination of the organisms in raw milk.[45] The presence of coliforms indicates poor sanitation, which can be certainly improved through satisfactory sanitation measures.

International surveys show that it is not uncommon to have microbial pathogens spread around an organic farm or a food processing environment leading to contaminated in organic food. For example, Belgian researchers found *Bacillus cereus* group organisms in milk from organic dairy farms with the incidence of 81.3%.[46] There have been other examples. Research conducted in the U.S. also showed that 60% of the individual chickens from an organic free-range producer in the U.S. was positive for *Salmonella*, and the researchers advised consumers not to assume that free-range or organic conditions will affect the *Salmonella* status of the chicken.[47] A survey on vegetables by researchers in Korea also revealed food-borne bacterial pathogen contamination in organic vegetables, in which *Bacillus cereus* was detected in 11.1% of organic spinach samples, *Staphylococcus aureus* in 8% of organic sesame leaves, and *L. monocytogenes* in 6.4% of organic romaine lettuce and spinach samples.[48] Besides bacterial pathogens, foodborne viruses can commonly be transmitted to consumers through fresh produce.[49]

Although the nature of distribution of microbial contaminants is different from that of chemical contaminants, i.e. there will always be microbial contaminants in the organic produce, it does not mean that microbiological safety of organic produce is impossible to control. To help us understand how control measures can be applied, we can look at microbial hazards according to their natural habitat or sources in which they usually occur.[50]

1. **Environmental pathogens.** The pathogens in this group are ubiquitously distributed in the environment, such as in soil and water, and can occasionally cross-contaminate to the produce. Examples of these are *Bacillus cereus*, *Clostridium perfringens*, and *Pseudomonas aeruginosa*.

---

[45] Chitov T. (2018) Evaluation of dairy farm sanitation and microbiological quality of raw milk. Internal report of research investigation submitted to Science Faculty, Chiang Mai University.

[46] Coorevits A, De Jonghe V, Vandroemme J, Reekmans R, Heyrman J, Messens W, De Vos P, and Heyndrickx M. (2008) Comparative analysis of the diversity of aerobic spore-forming bacteria in raw milk from organic and conventional dairy farms. Systematic and Applied Microbiology 31: 126–40.

[47] Bailey JS and Cosby DE. (2005) *Salmonella* prevalence in free-range and certified organic chickens. Journal of Food Protection 68: 2451–2453.

[48] Tango CN, Choi N-J, Chung MS, and Oh DH. (2014) Bacteriological quality of vegetables from organic and conventional production in different areas of Korea. Journal of Food Protection 77: 1411–1417.

[49] Seymour IJ and Appleton H. (2001) Foodborne viruses and fresh produce. Journal of Applied Microbiology 91: 759–773.

[50] For further reference, see: Adams M, Moss MO, McClure P. (2016) Food Microbiology, 4th edition. RSC Publishing. UK.

2. **Foodborne pathogens that are part of the skin flora.** The best example of this group of foodborne pathogens is *Staphylococcus aureus*. This bacterium can colonise human and animal skin, and is carried over to food of animal-origin through direct contact or through food-handling personnel. *S. aureus* is also known to be able to infect the udder of dairy cattle, causing mastitis. It often occurred in milk from dairy cattle with mastitis, or milk from dairy cattle in farms where *S. aureus* mastitis is present.
3. **Foodborne pathogens that have reservoir in intestinal tract.** The foodborne pathogens of this type do exist in the environment, but they are usually found in high numbers in the gut of infected human and animals, or in human and animals that are carriers of the microorganisms. A large number of microbial cells can be released from the body of the host and the most common route of contamination in food is the faecal-oral route. Some of the microorganisms in this group are zoonotic pathogens, i.e. pathogens that can cause diseases in animals and the infections are spread between animals and people.[51] Among the most common microbial pathogens of this group are pathogenic bacteria such as *Campylobacter*, *Salmonella* and pathogenic *E. coli*; pathogenic protozoa such as *Giardia* and *Cryptosporidium*; and foodborne viruses such as Norwalk-like viruses and the Hepatitis A virus.
4. **Microbial toxins.** These can include bacterial toxins, mycotoxins (mould toxins), and algal toxins. Illnesses caused by microbial toxins are related to sufficiently high levels of toxins released into food, which are proportionally related to the number of microorganisms. The pre-formed toxins accumulate in agricultural produce and processed food, such as *Bacillus cereus*'s emetic toxin or *Staphylococcus aureus* enterotoxins in food, Aflatoxins in cereals, nuts, fruits and spices and algal-derived toxins that cause paralytic shellfish poisoning in aquaculture, for examples.

It is impossible to eliminate microbial contamination in organic commodities, however, prevention of contamination of organic produce from bacterial pathogens from all sources can be achieved by a better regulation of organic agricultural practice and sanitation/hygiene management in farming, pre-harvesting, post-harvesting, and food processing. For examples, in crop production, compost made using animal manures should be left for a sufficient time for microbial pathogens to cease, and one should not assume that manures from organically raised animals within organic farms are free of microbial pathogens. Rodents must be controlled in organic farms, not only because they can destroy crops, but also because they can carry microbial pathogens.[52] To control microbial toxins in fresh produce and food products, prevention of proliferation of microorganisms is an effective approach, which can be achieved by good post-harvest handling and proper storage of raw materials

[51] European Food Safety Authority (EFSA) (2019) Food-borne zoonotic diseases. Available at: https://www.efsa.europa.eu/en/topics/topic/food-borne-zoonotic-diseases

[52] Meerburg BG and Kijlstra A. (2007) Review role of rodents in transmission of *Salmonella* and *Campylobacter*. Journal of the Science of Food and Agriculture 87: 2774–2781.

and food products for the control of mycotoxins and bacterial toxins, and efficient monitoring of water quality for the control of algal toxins. At the moment, Thailand does not have clear regulations addressing prevention of microbial contamination in particular, but there are requirements in different guidelines or standards to assist such process. An example of this is seen in ACT organic standards, which requires that in organic farming, the use of fresh animal manures, or organic matter containing human or animal excrement, should be avoided, as this can result in cross-contamination of intestinal microbial pathogens on parts of the plants used for consumption.[53]

In regulating food product manufacturing, proper processing methods and conditions, such as temperature and time in pasteurisation must be prescribed and monitored to help control the safety of processed organic food products. Preventative measures through the application of internationally recognised protocols such as good manufacturing practice (GMP) and, if possible, hazard analysis critical control point (HACCP), should be encouraged throughout all stages of primary production, food processing, storage, and distribution. Systematic monitoring and control of food quality and safety by food producers are also key mechanisms for food safety. Because of the short shelf-life of most organic fresh produce, rapid analytical methods will be on demand and research and development works will definitely be required to support safety monitoring and control systems of organic food. Safeguards through the food production chain are still necessary for organic food, as well of food produced conventionally, for there is no definite conclusion of whether fresh produce and food products produced organically are of more advantage in terms of microbiological safety than those produced conventionally.[54]

11.4.2 The Ten Biggest Challenges to Organic Food Safety in Thailand

Regulating food production process will not be sufficient for ensuring the safety of organic products in Thailand. There are some factors that can potentially affect the safety of organic commodities, particularly. In Thailand, most organic farms that are operated in compliance with organic standards are those that have transitioned or are transitioning from conventional farming. Problems faced by farmers in different farm areas in Thailand, can be summarised, as follows[55]:

[53] Organic Agriculture Certification Thailand (ACT) (2017b) ACT Organic Standards 2016 (Revision 2017); Section 2.5 Soil, water, and fertilizers management. Available at: http://actorganic-cert.or.th/wp-content/uploads/2017/10/act_standards_2016_v4_revision24-02-17.pdf

[54] Lücke F-K. (2018) Microbiological safety of organic and conventional foods. Foodbalt-2017 conference paper; doi: https://doi.org/10.22616/foodbalt.2017.003

[55] See: Boonchai K. (2016) Management of agricultural and food systems that are safe from agrochemicals: lessons from modelled areas. A publication of National Health Commission Office

1. *Governmental Policies*

   Thai government policies such as securing sufficient domestic food supply and exports, and promoting production of single economic crops (e.g. sugar cane, cassava, rubber) have caused the deterioration of local natural resources and ecosystems. The organic farmlands have been decreased as a result of the increasing acreage of conventional farmland for the economic crops.
2. *Soil Quality*

   Poor quality soil, such as soil with insufficient nutrients and organic matter, acid soil, and saline soil, is an obstacle to organic agriculture in many areas of Thailand. Many organic farmers who start an organic farm in certain parts of Thailand are facing this problem and it takes a tremendous amount of effort, time and patience to improve soil quality.[56] Without serious commitment for organic agriculture, many farmers might have given up and returned to the use of chemical fertilisers for rapid (but non-sustainable) improvement of soil quality and increase of crop yield, thus causing recontamination of chemicals in soil and leaving residues in produce from the farms that were understood to be organic.
3. *Accumulation of Nondegradable Hazardous Chemicals in Soil*

   One main problem relating to the suitability of soil for organic farming in Thailand is the contamination of soil by agrochemicals. This can, and in most cases, occur during the use of land for conventional agriculture, and this takes a time period for the level of chemicals to decrease or be eliminated. It is possible that some heavy metals or chemicals in the class of persistent organic pollutants (POPs), which are resistant to environmental degradation through chemical, biological, and physical processes, will not degrade during that time, leaving traces of chemical residues in organic produce, both of plant and animal origins, even after the given "conversion" periods for transitioning farms are completed.
4. *Contamination of Chemicals from Neighbouring Conventional Farms*

   In Thailand, organic producers are still in the minority, and organic farms often are surrounded by conventional farms. Chemicals used in the neighbouring conventional farms can be transferred to organic farms through soil, water, and wind.
5. *Pollution of Surface and Ground Water*

   This problem is related to the problem above. As mentioned earlier, there is a problem to control the use of agrochemicals in conventional agriculture in Thailand. This also affects the organic agriculture system as a whole. Studies showed that water in many areas in Thailand were polluted with hazardous

[article in Thai]. These conclusions are also supported by personal interviews and observations of the author.

[56] Rama Healthy Farm Project. (n.d.) Lessons from Ranathibodi Healthy Farm. Medical Faculty, Ramathibodi Hospital, Mahidol University, Thailand. [article in Thai]

chemicals, especially heavy metals and nitrate.[57,58] Traces of the toxic elements and chemicals can diffuse into soil within organic farm areas, and in turn, cause contamination in organic produce.

6. *Management of Crops from Transitioning Farms*

   A farmer who is in transition to organic agriculture (who applies for organic certification) has to go through a conversion period, and during such period, must follow the organic standards set by Organic Agriculture Certification Thailand (ACT). The conversion periods vary, depending on the type of crop (annual or perennial) and the market (domestic or export). Before conversion periods are completed, this can be an economic burden to farmers as they have to sell their crop as non-organic while bearing the increasing cost of management of organic farms.

7. *Marketing Channels and Stability of the Organic Market*

   Marketing plays a significant role for sustainability of an organic production system. Marketing of organic produce in Thailand relies heavily on the strength of farmer groups or farmer associations. Marketing and safety of organic produce have a complex connection. Many organic farmers face the problem of distributing their produce, and eventually fall in the trap of the marketing mechanism, in which the price of the organic produce can no longer be set by the primary producers but by traders. In recent years, international traders have become more influential in setting the market price of organic produce, such as how prices of fruit orchard produce of the south-east areas of Bangkok are determined by Chinese market. Moreover, organic production does not always correlate positively with increased sale or farmers' income. The support from non-governmental sectors can change with time. Some leaders of the organic farmer association felt that the organic market still faces uncertainty and instability.

8. *Supply of Organic Commodities and Food Products*

   It is a general policy of domestic supermarkets and international trade agencies in the modern trade system to secure the supply of organic food for their regular customers. As a result, organic food producers, manufacturers and distributors in Thailand can be under great pressure to supply organic commodities or products in a certain amount for the big market demands. As a result, a manufacturer or a food supplier/distributor may combine organic food with food that are produced through conventional production. There is a case report of frozen durian, in which a manufacturer received the supplies of fresh durian from both organic orchards and conventional orchards. During the processing of fresh durian into the frozen durian product, the produce from organic

[57] Wongsasuluk P, Chotpantarat S, Siriwong W, and Robson M. (2013) Heavy metal contamination and human health risk assessment in drinking water from shallow groundwater wells in an agricultural area in Ubon Ratchathani province, Thailand. Environmental Geochemistry and Health 36: 169–182.

[58] Tirado R (2007) Nitrates in drinking water in the Philippines and Thailand. Greenpeace Research Laboratories Technical Note 10/2007.

orchards and conventional orchards were no longer divided.[59] Such situation can result in devaluing organic commodities, i.e. decreasing their prices, thus discouraging organic farmers to continue farming with strict organic standards.

9. *Authenticity and Mislabeling of Organic Products*

   Although there are regulation systems and mechanisms to control safety of organic food, like conventional food, it is unavoidable to have occasional problems with authenticity and mislabeling. These problems can happen intentionally or without any intention or knowledge on the part of the producers. According to ACT Organic Standards, a label of "Organic Product" with an ACT seal can be used only with unprocessed products in containers or packages in which all ingredients in the containers or packages are certified by ACT or, as we noted earlier, with processed products containing at least 95% (by weight, excluding water and salt) of their ingredients as organic. If the organic ingredients are not less than 70% but not up to 95%, its label can include claims of the product having been made with indicated percentage (examined or approved by ACT) of organic raw materials and should be placed near the ACT seal. Products with less than 70% of its agricultural ingredients as organic and in-conversion products cannot be labelled as organic products.[60] The fact is, once being processed in industrial scale, it is very difficult to trace authenticity of the products whether they are produced with certain amounts of organic ingredients as claimed. Efficient on-site production and on-shelf food product monitoring systems are required to reduce these problems. Efficient traceability systems are also needed. In Thailand, however, these systems are not strong or not thoroughly-established.

10. *Problems with Varieties of Food Products and Technical Problems in Food Product Manufacturing*

    The short shelf-life of most organic fresh produce calls for research and development of processed organic food. However, for farmers and small food enterprises, this is not their expertise. Therefore, organic food products are still limited to a few traditional products which have a narrow market. Moreover, food processing knowledge and technology are not reachable by, or made available to, small-scale food producers. The problem of food safety can come in through this situation because of the lack of knowledge in food processing. For example, an organic longan orchard owner who decides to manage the unsold fresh longan by processing it into dried longan might face the problem of a high mould count if the produce was not properly dried, or he may use sulfur dioxide in processing and rob the product of its organic, safe nature. For another example, an organic dairy farmer who wants to process ice cream from his organic

---

[59] Boonchai K. (2016) Management of agricultural and food systems that are safe from agrochemicals: lessons from modelled areas. A publication of National Health Commission Office [article in Thai].

[60] Organic Agriculture Certification Thailand (ACT) (2017c) ACT Organic Standards 2016 (Revision 2017); Section 7 Labelling and use of organic seal. Available at: http://actorganic-cert.or.th/wp-content/uploads/2017/10/act_standards_2016_v4_revision24-02-17.pdf

milk might face the problem of contamination of products by harmful bacteria such as *Listeria monocytogenes* or *Salmonella* because of the lack of sufficient knowledge of how to prevent their contamination or their growth during processing of ice cream. Therefore, mechanisms to address the lack of product varieties and processing knowledge and technology are also needed to support the safety of organic food.

11.4.3 Reducing Risks to Hazard Exposure and Enhancing Safety of Organic Food: Thai Traditional Knowledge and Modern Technology

Thailand has one advantage for organic farming: the heritage of Thai traditional knowledge of organic agriculture from the past, including expertise in pest and disease controls. The traditional knowledge and modern technology are very much needed for current organic agriculture in Thailand, which has a tropical climate that supports the growth and multiplication of pests and plant and animal- disease-causing microorganisms. Moreover, most organic farms are still situated among conventional farms; pests are, therefore, attracted to the organic farming areas, for the crops are free of chemical or synthetic pesticides. Although there are general ideas of organic farm management to assist production and ensure food safety, the natural remedies may take different forms depending on the local folk knowledge or heritage. This is why it is important for organic farmers to be connected and exchange their knowledge and experience. Natural remedies that have been used successfully in a case study organic farm included physical, biological, and (natural) chemical concepts. Some examples are[61]:

- using extract of fermented plant parts (root and leaf parts of herbs, fruits) to control pests and diseases;
- using herb extracts or essential oils to prevent insects;
- using *Trichoderma* which has antagonistic effect against fungi causing plant diseases;
- making a nursery in the middle of a pond to prevent rodent attack;
- using alternate plantation to prevent spreading of diseases; and
- planting insect repellent plants, such as basil or citronella grass, among the main crops.

Many of the substances used in traditional Thai agriculture are included in the list of approved inputs allowed for use in organic production, given in the ACT Organic standards, such as marigold plant, repellent plants like citronella grass, essential oil, neem, and botanical extracts. Some substances, although coming from

[61] Rama Healthy Farm Project. (n.d.) Lessons from Ranathibodi Healthy Farm. Medical Faculty, Ramathibodi Hospital, Mahidol University, Thailand. [article in Thai]

natural sources, have to be used with caution, as they might harm the plant or animals themselves or can be toxic to humans if used in large amounts. Farmers must also consider the organic standards they have to comply with, for they may have different requirements. For example, use of essential oils for insect control is allowed according to ACT and IFOAM standard, but not allowed according to European Union standards.[62]

There are significant differences in the regulatory approaches to organic food production between EU and Thailand. For example, in relation to microorganisms used for biocontrol, ACT and IFOAM organic standards allows the use of any non-GMO microorganisms, while EU standards allow certain species, or in some cases, certain sub-species of microorganisms (e.g. *Bacillus thuringiensis* supsp. *aizawai* or *B. thuringiensis* supsp. *kurstaki*, not just any *B. thuringiensis*).[63] This is because the EU food safety control programmes are based on precautionary principle, which only allows what has been approved as safe for humans and the environment to be used. However, the approved inputs in the list will have to be updated periodically, according to the scientific information. For example, there has been concern about the use of *Bacillus thuringiensis* (BT) to control insects in crop plants that are to be used as food because many strains of environmental BT harboured enterotoxin genes and the strains used as biopesticides were proved to express low level of enterotoxins, thus representing a food safety risk.[64]

Modern technology and methods have also been employed by organic farmers to enhance crop production and pest control efficiency. The use of data, machine learning, artificial intelligence, advanced sensor technology, and precision farming technology under the concept of modern agriculture, as examples, can be adopted in organic agriculture to offer an alternative, efficient way for safe food production with the least effects to the environment.

---

[62] Organic Agriculture Certification Thailand (2017b) ACT Organic Standards 2016 (Revision 2017); Appendix 1 List of approved inputs used in organic production. Available at: http://actorganic-cert.or.th/wp-content/uploads/2017/10/act_standards_2016_v4_revision24-02-17.pdf

[63] Organic Agriculture Certification Thailand (ACT) (2017a) ACT Organic Standards 2016 (Revision 2017); Appendix 1 List of Approved Inputs Used in Organic Production, Part 1.2 Products and methods used for pest, disease, weed control and growth regulators. Available at: http://actorganic-cert.or.th/wp-content/uploads/2017/10/act_standards_2016_v4_revision24-02-17.pdf

[64] Johler S, Kalbhenn EM, Heini N, Brodmann P, Gautsch S, Bağcioğlu M, Contzen M, Stephan M, and Ehling-Schulz M. (2018) Enterotoxin production of *Bacillus thuringiensis* isolates from biopesticides, foods, and outbreaks. Frontiers in Microbiology 9: 1915; doi: https://doi.org/10.3389/fmicb.2018.01915

11.5 Conclusion

In Thailand, the advantages of organic production are not clear in the minds of consumers. An average consumer has little concern about organic food or safe food. However, there are significant numbers of consumers who are well aware of the danger of unsafe food and are positive about organic production. They are mostly educated consumers and organic producers themselves, and the rapid rise in demand for organic food domestically and worldwide have proved the value of organic production. It is crucial for organic farmers to be committed to authentic organic production in order to ensure the safety of organic food.

Ensuring the safety of organic food is a complex process that requires strong supports from all sectors: governmental and non-governmental, individuals and networks, non-profit organisations and business, to help organic farmers achieve this goal. Organic agriculture in Thailand is on the rise, but we must make sure that control mechanisms that can sustain organic production and safeguard organic food are not lagging behind. In the Thai environmental, social, political, and economic contexts, this is not a process that can naturally occur, but it needs much effort, cooperation, and persistence to firmly establish thorough organic food safety control systems. Some regulatory systems are already available, and they are like tools that need to be used skillfully or sharpened. There are many weaknesses in organic production in the Thai context to fight with, such as the widespread and heavy contamination of agrochemicals in the soil and water, the unstable market, and the enormous amount and variety of pests and diseases. However, there are also advantages to look up to, such as the strength of non-governmental organisations and networks, the heritage of Traditional Thai agricultural culture, and the supportive movement in the government sector in recent years. Considering the present situations, we are still a long way from reaching the goal. The way ahead is long and full of obstacles, but we are on the right path.

References

Adams M, Moss MO, McClure P (2016) Food microbiology, 4th edn. RSC Publishing, Cambridge

Bailey JS, Cosby DE (2005) *Salmonella* prevalence in free-range and certified organic chickens. J Food Prot 68:2451–2453

Bio Thai (2016) Thai PAN has published the results of their second round of testing on chemical pesticide residues for 2016. Available at: https://www.biothai.org/node/1427

Boonchai K (2016) Management of agricultural and food systems that are safe from agrochemicals: lessons from modelled areas. A publication of National Health Commission Office [article in Thai]

Chitov T (2018) Evaluation of dairy farm sanitation and microbiological quality of raw milk. Internal report of research investigation submitted to Science Faculty, Chiang Mai University

Chitov T, Dispan R, Kasinrerk W (2008) Incidence and diarrhegenic potential of *Bacillus cereus* in pasteurized milk and cereal products in Thailand. J Food Saf 28:467–481

Coorevits A, De Jonghe V, Vandroemme J, Reekmans R, Heyrman J, Messens W, De Vos P, Heyndrickx M (2008) Comparative analysis of the diversity of aerobic spore-forming bacteria in raw milk from organic and conventional dairy farms. Syst Appl Microbiol 31:126–140

Destoumieux-Garzón D, Mavingui P, Boetsch G, Boissier J, Darriet F, Duboz P, Fritsch C, Giraudoux P, Le Roux F, Morand S, Paillard C, Pontier D, Sueur C, Voituron Y (2018) The one health concept: 10 years old and a long road ahead. Front Vet Sci 5:14. https://doi.org/10.3389/fvets.2018.00014

European Food Safety Authority (EFSA) (2019) Food-borne zoonotic diseases. Available at: https://www.efsa.europa.eu/en/topics/topic/food-borne-zoonotic-diseases

Food and Agriculture Organization of the United Nations (FAO) (2018a) Country fact sheet on food and agriculture policy trends: Thailand. Available at: http://www.fao.org/3/I8683EN/i8683en.pdf

Food and Agriculture Organization of the United Nations (FAO) (2018b) Participatory Guarantee System (PGS) for Sustainable Local Food Systems. Available at http://www.fao.org/3/I8288EN/i8288en.pdf

Food and Agriculture Organization of the United Nations (FAO) (n.d.) AGP-Agriculture and soil biodiversity. Available at: http://www.fao.org/agriculture/crops/thematic-sitemap/theme/spi/soil-biodiversity/agriculture-and-soil-biodiversity/en/

Food and Drug Administration. (n.d.) Unofficial translation of Food Act B.E. 2522. Available at: http://food.fda.moph.go.th/law/data/act/E_FoodAct2522.pdf

International Atomic Energy Agency (IAEA) (2019) Greenhouse gas reduction. Available at: https://www.iaea.org/topics/greenhouse-gas-reduction

Jaipieam S, Visuthismajarn P, Sutheravut P, Siriwong W, Thoumsang S, Borjan M, Robson M (2009) Organophosphate pesticide residues in drinking water from artesian wells and health risk assessment of agricultural communities, Thailand. Hum Ecol Risk Assess 15:1304–1316

Johler S, Kalbhenn EM, Heini N, Brodmann P, Gautsch S, Bağcioğlu M, Contzen M, Stephan M, Ehling-Schulz M (2018) Enterotoxin production of *Bacillus thuringiensis* isolates from biopesticides, foods, and outbreaks. Front Microbiol 9:1915. https://doi.org/10.3389/fmicb.2018.01915

Junpen A, Jirataya Pansuk J, Kamnoet O, Cheewaphongphan P, Garivait S (2018) Emission of air pollutants from rice residue open burning in Thailand, 2018. Atmos 9:449. https://doi.org/10.3390/atmos9110449

Kongsom C, Panyakul V (2016) Production and market of certified organic products in Thailand. Int J Econ Manag Eng 10:2723–2727

Lücke F-K (2018) Microbiological safety of organic and conventional foods. Foodbalt-2017 conference paper; https://doi.org/10.22616/foodbalt.2017.003

Meerburg BG, Kijlstra A (2007) Review role of rodents in transmission of *Salmonella* and *Campylobacter*. J Sci Food Agric 87:2774–2781

Ministry of Agriculture and Cooperatives (2016) Vision and mission page. Available at: https://eng.moac.go.th/ewt_news.php?nid=101

Naewbanij JO (2005) Food safety overview in Thailand. E-document of the presentation at SELAMAT EU-Asian Food Safety Platform Lisbon, Portugal, April 11, 2005. Available at: https://selamat.net/upload_mm/7/5/3/6be94190-ef9f-4753-9c80-126732c76da7_p11.pdf

Nandwani D, Nwosisi S (2016) Global trends in organic agriculture. In: Nandwani D (ed) Organic farming for sustainable agriculture. Springer, Cham, p 22

National Bureau of Agricultural Commodity and Food Standards (2018) Manual of use and display of standard certification under the Agricultural Standard Act B.E. 2551 (2008). [article in Thai]

National Bureau of Agricultural Commodity and Food Standards (n.d.) Training document: Guidelines for international certification of organic products and use of organic seal. Available at: http://certify.dld.go.th/certify/images/project/organic/organic2562/pwR5/2.pdf. [article in Thai]

National Bureau of Agricultural Commodity and Food Standards, Ministry of Agriculture and Cooperatives (2003) Thai Agricultural Standard TAS 9000–2003: Organic agriculture part 1:

the production, processing, labelling and marketing of organic agriculture. Available at: https://www.acfs.go.th/standard/download/eng/TAS-9000.pdf
National Bureau of Agricultural Commodity and Food Standards, Ministry of Agriculture and Cooperatives (2013) Thai Agricultural Standard TAS 9001–2013: good agricultural practices for food crop
Nicolopoulou-Stamati P, Maipas S, Kotampasi C, Stamatis P, Hens L (2016) Chemical pesticides and human health: the urgent need for a new concept in agriculture. Front Public Health 4:148–155
Office of Agriculture Regulation (OAR), Department of Agriculture (2018) Summary of imported hazardous agrochemicals 2018 Available at: http://m.doa.go.th/ard/
Office of the Council of State (2008) Unofficial translation of Hazardous Substance Act, B.E. 2535 (1992) https://www.jetro.go.jp/thailand/e_activity/pdf/hazsubact2535.pdf
Office of the Council of State of Thailand's Law for ASEAN project (n.d.) Agricultural Commodity Standard Act, B.E. 2551 (2008) Initial version. Available at: http://web.krisdika.go.th/data/outsitedata/outsite21/file/AGRICULTURAL_COMMODITY_STANDARDS_ACT,_B.E._2551_(2008).pdf
Organic Agriculture Certification Thailand (2017a) ACT Organic Standards 2016 (Revision 2017) Available at: http://actorganic-cert.or.th/wp-content/uploads/2017/10/act_standards_2016_v4_revision24-02-17.pdf
Organic Agriculture Certification Thailand (2017b) ACT Organic Standards 2016 (Revision 2017); Appendix 1 List of approved inputs used in organic production. Available at: http://actorganic-cert.or.th/wp-content/uploads/2017/10/act_standards_2016_v4_revision24-02-17.pdf
Organic Agriculture Certification Thailand (2017c) ACT Organic Standards 2016 (Revision 2017); Appendix 1 List of approved inputs used in organic production, Part 1.1 Input used for soil fertilisers and soil conditioners. Available at: http://actorganic-cert.or.th/wp-content/uploads/2017/10/act_standards_2016_v4_revision24-02-17.pdf
Organic Agriculture Certification Thailand (2019) Bout ACT. Available at: http://actorganic-cert.or.th/about/
Organic Agriculture Certification Thailand (ACT) (2017a) ACT Organic Standards 2016 (Revision 2017); Appendix 1 List of Approved Inputs Used in Organic Production, Part 1.2 Products and methods used for pest, disease, weed control and growth regulators. Available at: http://actorganic-cert.or.th/wp-content/uploads/2017/10/act_standards_2016_v4_revision24-02-17.pdf
Organic Agriculture Certification Thailand (ACT) (2017b) ACT Organic Standards 2016 (Revision 2017); Section 2.5 Soil, water, and fertilizers management. Available at: http://actorganic-cert.or.th/wp-content/uploads/2017/10/act_standards_2016_v4_revision24-02-17.pdf
Organic Agriculture Certification Thailand (ACT) (2017c) ACT Organic Standards 2016 (Revision 2017); Section 7 Labelling and use of organic seal. Available at: http://actorganic-cert.or.th/wp-content/uploads/2017/10/act_standards_2016_v4_revision24-02-17.pdf
Panuwet P, Siriwong W, Prapamontol T, Ryan PB, Fiedler N, Robson MG, Barr DB (2012) Agricultural pesticide Management in Thailand: situation and population health risk. Environ Sci Policy 17:72–81
Pracharuengwit P, Chiaravutthi Y (2015) Consumer willingness to pay for organic food in Thailand: evidence from the random *n*th-price auction experiment. Bus Adm J 146:52–70
Rama Healthy Farm Project (n.d.) Lessons from Ranathibodi Healthy Farm. Medical Faculty, Ramathibodi Hospital, Mahidol University, Thailand. [article in Thai]
Roitner-Schobesberger B, Darnhofer I, Somsook S, Vogl CR (2008) Consumer perceptions of organic foods in Bangkok, Thailand. Food Policy 33:112–121
Sangkumchalianga P, Huang W-C (2012) Consumers' perceptions and attitudes of organic food products in Northern Thailand. Int Food Agribus Manag Rev 15:87–102
Seymour IJ, Appleton H (2001) Foodborne viruses and fresh produce. J Appl Microbiol 91:759–773
Shelef O, Weisberg PJ, Provenza FD (2017) The value of native plants and local production in an era of global agriculture. Front Plant Sci 8:2069. https://doi.org/10.3389/fpls.2017.02069

Tango CN, Choi N-J, Chung MS, Oh DH (2014) Bacteriological quality of vegetables from organic and conventional production in different areas of Korea. J Food Prot 77:1411–1417
Thai Organic Trade Association (2011) Overview of organic agriculture in Thailand. Available at: http://www.thaiorganictrade.com/en/article/442
Thailand International Development Cooperation Agency (n.d.) Thailand's best practices and lessons learned in development, Volume 1. Available at: https://www.undp.org/content/dam/thailand/docs/TICAUNDPbpVol1.pdf
Thailand National Food Committee (TNFC) (2012) Strategic framework for food management in Thailand (2012–2016). Available at: https://www.foodsafety.moph.go.th/document/Info_general/Strategic_Framework.pdf
Tirado R (2007) Nitrates in drinking water in the Philippines and Thailand. Greenpeace Research Laboratories Technical Note 10/2007
Tirado R, Englande AJ, Promakasikorn L, Novotny V (2008) Use of agrochemicals in Thailand and its consequences for the environment. Greenpeace Research Laboratories Technical Note 03/2008
Wanwimolruk S, Phopin K, Boonpangrak S, Prachayasittikul V (2016) Food safety in Thailand 4: comparison of pesticide residues found in three commonly consumed vegetables purchased from local markets and supermarkets in Thailand. Peer J 4:2432
Win HE (2017) Organic agriculture in Thailand. FFTC Agricultural Policy Article. Available at: http://ap.fftc.agnet.org/ap_db.php?id=734&print=1
Wongsasuluk P, Chotpantarat S, Siriwong W, Robson M (2013) Heavy metal contamination and human health risk assessment in drinking water from shallow groundwater wells in an agricultural area in Ubon Ratchathani province, Thailand. Environ Geochem Health 36:169–182

Thararat Chitov has had teaching and research experience in the fields of food microbiology and food safety for more than 20 years. Her passion in food safety began when she studied at the University of Strathclyde, Glasgow, United Kingdom, and continued to grow during her years of academic service in the Department of Food Technology, Maejo University, and in the Department of Biology, Chiang Mai University, Thailand, where she currently works. Some of her other experiences include science education, consultancy services for food industry and analytical laboratory, working with dairy farms and mixed crop-livestock farms to assist transitioning from non-organic to organic farming and, together with her family, running the family's own organic pomelo orchard.

Chapter 12
Perception of Challenges in Opportunities for Organic Food Research and Development in Vietnam

Van Kien Nguyen

Abstract Since the inception of the green revolution during the 1970s, Vietnamese farmers converted their natural ways of farming into conventional agriculture using short-term rice and vegetables hybrids and intensive inputs of agrochemicals. Unfortunately, the Vietnamese society nowadays has a considerable concern about the side-effect of conventional farming on human health and the natural environment. There is a strong demand for safe food or organic food in urban areas as well as the rural communities. In response to this issue, the government has initiated the organic agriculture strategy for Vietnam in 2017 while growers and processors continue to practice organic agriculture for domestic consumption and exports over the last decade. Drawing on evidence from 36 in-depth interviews (qualitative research) with government officials, researchers, NGOs, media, processors and organic growers in Vietnam, this chapter aims to shed light on the organics industry. It places a spotlight on the perceptions of stakeholders with particular reference to (1) scope of adoption of organic agriculture, especially organic rice farming, (2) awareness of domestic and international standards and markets for organic food, (3) perception of government policies for the organic agriculture movement in Vietnam, and (4) the perception of current research activities on organic farming and practices inside the country. There is also a review of the literature and a profile of current research activities in key agricultural research institutions and universities in Vietnam is given.

Keywords Organic agriculture · Organic standards and certification · Perception · Stakeholders · Vietnam

---

V. K. Nguyen (✉)
Fenner School of Environment & Society, The Australian National University, Canberra, Australia

Research Center for Rural Development, An Giang University, Vietnam National University, HCM City, Long Xuyên, Vietnam

Mekong Organics Pty Ltd, Canberra ACT, Australia
e-mail: nv.kien@anu.edu.au

GOH Bee Chen, R. Price (eds.), *Regulatory Issues in Organic Food Safety in the Asia Pacific*, https://doi.org/10.1007/978-981-15-3580-2_12

12.1 Background of Organic Agriculture Research and Development in Vietnam

Preventing the growing and distribution of unsafe food is a top priority of Vietnam's government. Most consumers and buyers in Vietnam show strong concerns about safe vegetable quality (Wang et al. 2012). Some 88.5% of Hanoi residents are worried about the quality of vegetables due to the widespread use of agrochemical inputs (Wang et al. 2012).

The Vietnamese government recognized that unsafe food is a root cause of global health problems and negative impacts on socio-economic development (Minh 19/04/2018). The government has a special concern about food safety for the nation and export.

Promoting organic agriculture is a key option to improve food safety nutrition and health. According to IFOAM (2015), organic agriculture is defined as "a production system that sustains the health of soils, ecosystems and people. It relies on ecological processes, biodiversity and cycles adapted to local conditions, rather than the use of inputs with adverse effects. Organic Agriculture combines tradition, innovation and science to benefit the shared environment and promote fair relationships and good quality of life for all involved". From this perspective, organic agriculture is important to support not only for the health of soils but also the health of the ecosystems and people.

Vietnam's organic agriculture can be classified into three forms: traditional organic farmers, reformed organic farmers and certified organic farmers. Many farmers developed small scale organic farms at the individual levels without organic certification and inspection. Organic certifications in Vietnam are grouped into three types: the international certification standards, PGS (participatory guarantee system) initiated by IFOAM International, and the Vietnam National Organic Certification Standard 2017 (Ministry of Agriculture and Rural Development 2019). However, International Certifications and Participatory Guarantee System (PGS) are active in Vietnam. These standards include the USDA, JAS, and EU.

Vietnam's national organic standard was issued in 2017 but is not active yet. According to FiBL and IFOAM (2017), Vietnam has 76,666 ha of organic agriculture land, which was much lower than some Southeast Asian countries such as Indonesia and the Philippines (Table 12.1). Back in 2013, however, the certified organic area in Vietnam was 31,700 ha including 22 ha of PGS (Participatory Guarantee System), among these areas, 23,400 ha of agricultural land, 7000 ha of

Table 12.1 Organic agriculture land by country, 2012–2015

Countries	2012 ha	2013 ha	2014 ha	2015 ha	1 year growth	10 years growth
Indonesia	88,247	65,688	113,638	130,384	+16,746	+89,965
Philippines	80,974	86,155	110,084	234,642	+124,558	+228,951
Vietnam	36,285	37,490	43,007	76,666	+33, 659	+54,799
Thailand	32,577	33,840	37,684	45,587	+7903	+23,037
Cambodia	9055	9889	9889	12,058	+2169	+10,607

Sources: FiBL and IFOAM (2017)

aquaculture, and 1300 ha for wild food collection were reported (Huong 2017). In 2015, Vietnam had €5 million in locally retailed organic produce, and exported € 817 million Euro (USD 5.5 Million and USD 900, 000, 000 respectively) (FiBL and IFOAM 2017).

However, the organic agriculture movement in Vietnam is in the early stage and requires extensive research and investment for sustainable development. Minister of Agriculture & Rural Development Nguyen Xuan Cuong said at the first national Congress on organic food in December 2017 that organic farming in Vietnam is still underdeveloped (Dung 2017). At this event, Prime Minister Nguyen Xuan Phuc asserted that the development of organic agriculture is especially important and is more than a fashion (Vietnam Organic Agriculture Association 27 Dec 2017). The Congress has raised several key questions: (1) what is the key market for organic agriculture? (2) who are the growers? (3) what is the optimum scale for organic agriculture in Vietnam? (4) what crops, or animals should be key organic products for Vietnam? and (5) what standards are appropriate for Vietnam? All such questions point toward development of policies for promoting this relatively new agricultural industry. The Congress also pointed out that Vietnam has particular potential for promoting organic vegetables, mushroom, fruit, tea, herbs, medicinal plants, aquaculture (shrimp), honey, coffee and paper (Vietnam Organic Agriculture Association 27 Dec 2017).

The organic agriculture policy discourse has grown over recent years. At the 2019 workshop in Ho Chi Minh City, Ministry of Agriculture & Rural Development asserted that Vietnam faces challenges to develop organic farming including (1) lack of a special policy framework for organic agriculture development, (2) incomplete organic standards, (3) few enterprises having international organic certified certificates, and (4) small farm size (Bo Nong Nghiep and Phat Trien Nong Thon (MARD) 2019).

To highlight the state of knowledge about organic agriculture and food production in Vietnam, this chapter provides insight into the perception of organic agriculture research and development from different stakeholders in Vietnam. In particular, this chapter explores four issues: (1) the perception of stakeholders about adoption of organic agriculture in Vietnam, (2) the perception of international and domestic organic agriculture in its standards, (3) perception of its organic agriculture policies and (4) the perception of current research activities on organic farming and practices inside the country.

12.2 Review of Literature about the Perception of Organic Agriculture

There is a growing body of literature on the perception of organic agriculture in recent years. Largely, perception research on organic agriculture has been explored in developed countries (Gamboni and Moscatelli 2015; Wägeli and Hamm 2016; Janssen and Hamm 2011). In Southeast Asia, there are only a few examples of research on perception of challenges and opportunities of organic farming, most notably from Indonesia and the Philippines (Jouzi et al. 2017; Wägeli and Hamm 2016; My et al. 2017).

12.2.1 Opportunities of Organic Farming Globally

Researchers found that organic agriculture has several advantages. Jouzi et al. (2017) showed the most significant benefits are higher resilience to environmental changes, as well as improving farmer income, social capacity and employment opportunities. In France, dairy farmers perceived that organic farming is less risky and provides a stable price (Bouttes et al. 2019). Total water, energy and greenhouse gases can be reduced in the organic food production system (Wood et al. 2006). A range of contradictory beliefs and practices have derived from the discursive conflict between conventional and organic food industries over environmental, health and safety claims (Lockie et al. 2002). A study by Mergenthaler et al. (2009) showed that consumers metropolitan areas of Vietnam are willing to pay an average price premium of 60% for chemical-free Chinese mustard. So, the demand for organic food at higher prices suggests that Vietnamese consumers are increasingly affluent and concerned about safe food for health. The price of imported organic food is 5 times higher than the conventional food in the Asian market (Mutiara and Satoshi 2017).

12.2.2 Challenges of Organic Farming Globally

However, organic farming in Southeast Asia must overcome significant challenges. Costs of international organic certification are high for farmers in developing countries (Mutiara and Satoshi 2017). Although consumers want to buy an organic product with a reasonable price, to receive an organic label certification requires high cost which leads to a high selling price (Mutiara and Satoshi 2017). Organic food remains a niche product, consumed by a minority of consumers (Lyons 1999). However, by strengthening the communication of organic food quality consumers with strong environmental values can help to increase the price for organic food (Loebnitz and Aschemann-Witzel 2016). Pham and Shively (2019) show that the profit from organic vegetables was lower than conventionally grown vegetables. It has also been contended that here is no significant difference between social and environmental benefits of conventional and organic systems (Baumgartner and Nguyen 2017). Moreover, farmers have difficulties in soil management (Jouzi et al. 2017).

Awareness of organic certification standards is low in both developed and developing countries. My et al. (2017) found that consumers had low awareness of food quality-related standards such as organic and Global GAP, and Hazard Analysis and Critical Control Points (HACCP) in Vietnam. In Italy and the UK, the great majority of participants were not aware of any differences between the schemes that were discussed. In the Czech Republic, Denmark and Germany, several participants preferred a particular organic certification scheme over others (Janssen and Hamm 2011).

Quality control of organic food is one of the most concerns by consumers when purchasing food in Vietnam. Wang et al. (2012) concluded that group leadership and promoting the farmers' sense of responsibility were the main ways for current Vietnamese farmer organizations to ensure production of a high-quality product (Wang et al. 2012).

Yield is a critical issue in adopting organic agriculture. The organic yields of individual crops are on average of 80% of the conventional yields (Ponti et al. 2012; Jouzi et al. 2017).

The organic agriculture industry in Southeast Asia faces a challenge for both non-perishable and perishable organic exported products. In the case of Indonesia, non-perishable exported products (coffee, tea, rice, etc.) faced organic certification issues while perishable organic products (raw horticultural products) are over-supplied (David and Ardiansyah 2017).

Economic return from organic rice is not attractive to farmers in some countries. The economic return from organic rice farming is lower than conventional rice in the experience of Bhutan. If organic rice met the premium price, then the Benefit-Cost Ratio (BCR) of an organic system may become similar to or higher than a conventional system in Bhutan (Tashi and Wangchuk 2016).

Organic rice farmers faced many challenges in maintaining sustainability. Heryadi et al. (2018) found over 85% of organic rice farmers converted to conventional rice farming in Indonesia because of economic, technical, institutional barriers, certification, and organic rice requirements. Additionally, the Pilipino rice farmers primarily received support from NGOs and private institutions in the intensification of organic rice production rather than the governments (Digal and Placencia 2019; Piadozo et al. 2014). Rice farmers have low to medium level of awareness of organic farming practices and markets in the Philippines (Piadozo et al. 2014). Bangladeshi farmers and consumers were aware of the hazards of chemical compounds but had little knowledge about organic rice (Hossain et al. 2007). Vietnamese rice farmers were mostly concerned about agrochemical use in extensive rice-shrimp systems to prevent harm to shrimps (Braun et al. 2019).

In adopting organic farming practice in Vietnam, Pham and Shively (2019) suggested that farmers need to know the economic benefits and challenges of organic farming. They further suggested that farmers need to develop a marketing plan and work together to promote sustainable organic farming in Vietnam (Pham and Shively 2019).

12.3 Qualitative Research Approach

Having conducted a brief literature review on key perceptions, a qualitative method is to be adopted. In-depth interviews with experts and the key informants were employed. In-depth interviews are conducted by face-to-face encounters between researchers and informants directed toward understanding informants' perspectives on their lives, experiences, or situations as expressed in their own words (Creswell 1994).

Purposive sampling assists a researcher to select information-rich cases for an in-depth understanding of the topic (Liamputtong 2009:11–12). This method was applied for interviewing specific stakeholders in Vietnam. In this research, I interviewed NGO practitioners, government agencies, private sectors (processors, retailers, wholesalers and small enterprises), researchers, organic growers, media and international and domestic organic certification bodies who are working in the fields of organic agriculture farming, processing, marketing and research. This paper also employed a snowball sampling method to approach interviewees (Creswell 1994). A total 36 in-depth interviews were undertaken using a semi-structured questionnaire through skype and emails during April and October 2019 (Table 12.2).

Table 12.2 List of interviewees

Participant Code	Types of stakeholders	Sample Size	Province (name of the stakeholder)
F	Growers/small enterprises	10	1. An Giang: Ech Op Vegetable Farm
			2. Binh Phuoc: Viet Ha Organic Farm
			3. Dong Thap: Tam Viet Ecological Farm
			4. Can Tho: Anh Ba Viet Nam Social Enterprise
			5. Soc Trang: Ngoi Sao Nho Farm
			6. Hai Phong: Gao Ruoi Farm
			7. Dong Thap: Co May Rice Company
			8. An Giang: Tan Vuong Rice Company
			9. Kien Giang: Organic Rice Shrimp
			10. Dong Thap: Tam Viet Ecological Farm
P	Processors	2	1. HCM City: Ong Tho Rice Company
			2. Binh Phuoc: Viet Ha Farm/Processor
S	Retailors/wholesalers	3	1. HCM City: Ong Tho Rice Company
			2. Can Tho: Anh Ba Viet Nam Social Enterprise
			3. An Giang: Ech Op Vegetable Farm
R	Researchers	11	1. An Giang: An Giang University (4 cases)
			2. Can Tho: Can Tho University (2 cases)
			3. Kien Giang: Kien Giang University (1 case)
			4. HCM City: HCM National University-University of Social Sciences and Humanity (1 case)
			5. Hanoi: Institute Policy Strategy for Agriculture & Rural Development (IPSARD)
			6. Can Tho: Tay Do University (1 case)
			7. Hanoi: Vietnam Agricultural Academy (1 case)

(continued)

Table 12.2 (continued)

Participant Code	Types of stakeholders	Sample Size	Province (name of the stakeholder)
A	Associations/NGOs	4	1. Hanoi: Vietnam Organic Agriculture Association
			2. Can Tho: Vietnam Union of Science and Technology Associations
			3. Can Tho: Mekong Youth Farm
			4. Ha Tinh: HEPA-Eco-farming School
G	Government staffs	3	1. An Giang: Department of Agriculture & Rural Development
			2. Dong Thap: Department of Agriculture & Rural Development
			3. Kien Giang: Department of Sciences & Technology
C	International Organic Certification Body	1	1. HCM city: Control Union
C	Domestic Organic Certification Body	1	1. Hanoi: PGS Vietnam
M	Social Media	1	1. Can Tho: The Gioi Hoi Nhap newspaper
Total		36	

Note: Questionnaire was sent and received via personal email

12.4 Results

12.4.1 Benefits of Organic Farming from 'Stakeholders' Perspectives in Vietnam

This research was advanced through expert interviews about the adoption of organic farming in Vietnam from leaders of organic agriculture associations, senior researchers at universities and research institutions, private business sector, NGOs, and organic growers across Vietnam. The results indicated that most participants address the benefits of organic agriculture for many reasons: promoting food safety, changing the habit of current farming practices which overuse agri-chemicals, improving human health, emerging organic food demand, incentives for investment from the business sector and support from the Vietnamese government.

12.4.1.1 Organic Farming: Environmental and Health Issues

According to an organic food researcher in Southern Vietnam (in a personal communication in July 2019), the application of organic agriculture in Vietnam is very timely important and urgent. Current conventional farming systems destroy the ecosystem, environment, and human health. Additionally, the low awareness of food safety using preserved chemicals to store and process food that affects human health. To promote organic farming, a policy is very important to sustain the organic agriculture in the longer term.

In a dialogue with a NGO practitioner in Hanoi (personal communication in May 2019) it was revealed that adoption of organic farming is a selection process of soil, water, and human resources. The development of organic farming in Vietnam does not necessarily mean that everyone or everywhere will adopt organic agriculture. To produce organic food, the choice of suitable land area, assessment of producing conditions, current and future risks, growers must learn about organic farming practices, organic certification standards and processes. At the moment, most Vietnamese people understood simply that organic farming means not using chemical fertilizers or pesticides. According to my interviewee, most growers and consumers did not fully understand the process of organic farming.

In Vietnam, my interviewee was confident that many regions can be converted into organic agriculture for a diversity of crops, animals, and aquaculture. Depending on the ecological conditions, different organic crops can be applied in different areas. Organic rice can be grown in the Mekong Delta, central highlands and in mountainous areas. However, small scale farming size is a critical challenge for developing organic farming in Vietnam. In particular, water management, irrigation, and pest management are most challenging because of small scale farm size which is closed to conventional farms. With 60% of small-scale farmers in Vietnam, the challenge of losing crops in the first 2 years discourages farmers to convert to organic.

A practitioner of ecological farming in Northern Vietnam (in a personal communication in July 2019) contended that adoption of organic agriculture amongst smallholders (i.e. direct farmers) remain low. Initiatives around this stay in small, fragmented linking, supported by local NGOs, or other small-scaled actions. Initiatives supported by bigger donors may see the benefits of exporting these organic produces back to their market(s), so the costs involved are too high for any smallholders (direct farmers) to apply.

According to my interviewee, adoption of organic agriculture can increase if there were clear, motivated and passionate government policies and supportive schemes (along with labelling organic agriculture in simpler terms for the local market). There will be affordable organic products for Vietnamese food consumers i.e. farmers first and then out to retail customers/consumers (regionally and internationally) in the near future. The competitive trend of organic agriculture amongst initiatives in the SEA countries/region may be a factor to engage public and youths and other social actors making these transformations along with the social media supports.

After an interaction with an international organic certifier HCM City (personal communication in July 2019) it was clear that organic consultants play a key role in linking Vietnam organic agriculture and global trends. Organic farming has just developed in Vietnam recently, in the last 6–7 years. Many people knew about organic farming, but only in very general terms. The question remains: what is effective organic farming? How to do it? What is the market for organic food? These are the critical questions for the majority of farmers in Vietnam.

Additionally, due to the fragile organic agriculture development, most people tend to think that the adoption of organic farming is a luxurious exercise. Culture of Vietnamese growers prefers to go fastly and successfully, rather than slowly. Therefore, there is an urgent need to have the right organic consultants to advise farmers of the major risks to avoid.

A government official in An Giang province (personal communication in July 2019) contended that adoption of organic farming in the Mekong Delta has several challenges due to unclear government policies. The organic agriculture movement has increased in quantity but not quality. The market does not distinguish between organic and non-organic food, so it is difficult to encourage farmers to do so.

A researcher in HCM City (personal communication in July 2019) emphasised that, application of organic farming is essential. However, the high production costs of organic farming practice may not be applicable for scaling up the processes, but it can be applied for a niche market in the urban area. The organic farming practice is still *ad hoc* and at a small scale. The government can promote appropriate land use planning for organic farming and enforce equity of access to organic food.

Besides technical issues, organic farming development requires well trained human resources. Farmers and farming services extension workers must be trained about organic farming processes. To be an organic farmer, she or he must know about the certification process, the market, and have other business skills. However, little training was provided by the formal farm extension service program across the country. Organic growers primarily received training supports from NGOs and private sector actors or organic consultants. Additionally, farmers learnt about the conventional farming practice that uses chemicals for many years. To promote organic farming, it may take quite a long time.

Rice is the most common crop in Vietnam. Vietnam is the top five world rice exporter since the 2000s (Demont and Rutsaert 2017). However, an organic rice grower in Northern Vietnam's Hai Phong City (personal communication in September 2019) confirmed that the country only focuses on conventional rice for food security and export. Organic rice is underdeveloped and not documented in Vietnam. Shifting from conventional to organic has emerged recently (Hay et al. 2018). In the Mekong Delta, organic rice was grown by private companies and agricultural cooperatives, and individual farmers (Table 12.3).

An ecological rice grower, in Dong Thap province (personal communication in July 2019), expressed a view that rice farmers in Dong Thap province of the Mekong Delta (Mr Vo Van Tieng's hometown[1]) overused pesticides, herbicides and chemical fertilizers that caused environmental pollution. This created a negative effect on the environment, including pest attacks and diseases, damage to the ecosystem and crop

---

[1] Mr Vo is a young farmer who recognized the problems of using chemicals in his rice paddy fields and surrounding rice-based landscape. In 2014, he recovered 2 ha of ecological rice which was used to grow three crops of conventional rice per year in Thuong Thuong Tien commune of Hong Ngu District of Dong Thap province (Mekong Delta). After this success with ecological rice, he decided to rent 40 ha of neighbouring farmland to grow ecological rice in 2016. In 2018, he extended to 80 ha of the ecological rice model in another area in Long An Province (Mekong Delta).

Table 12.3 Organic rice farming in the Mekong Delta, Vietnam

Names of companies/ growers	Cultivated area (ha)/households	Certification/Standards	Location
Tan Vuong Company	250 ha	USDA	Ca Mau province
Vien Phu Organic Rice	300 ha	USDA and EU	Ca Mau province
Co May Croup	50 ha	USDA	Ca Mau province
Dai Duong Xanh Company	80 ha	USDA	Ca Mau province
Gao Ong Tho	170 ha (2019)	Local procedure using organic procedure (not certify)	Ca Mau province
	136 ha (2018)	She sold all rice until May 2019 – if rice is stored longer, quality goes down.	
		Lack facility to store round year – lack investment	
Hoa Nang Company	50 ha	USDA	Ben Tre Province
Tan Vuong Company	10 (5 households)	USDA	Kien Giang province
Trung An Company	100 (Company farm)	In progress of international certification	Kien Giang province
Lua Tom Huu Company	37 ha (2018), 45 households/farmers	Local produce – using organic farming procedure (not certify)	Kien Giang province
	34 ha (2019)		
Ho Quang	9.3 ha	USDA	Soc Trang province
Ecotiger Ltd., Company	50 ha	USDA, EU, JAS	Tra Vinh province
Tan Dat Rice Cooperative	60 ha	USDA	Vinh Long province
Tam Viet Ecological Rice	80 ha	Not certify	Long An province
Former Tam Viet Ecological Rice	40 ha	Not certify	Hong Ngu District, Dong Thap province

Source: Synthesised from various sources (in-depth interviews) – data was collected in July 2019

losses. The health of consumers and farmers have been negatively affected. Additionally, the Mekong delta is affected by climate change, droughts, and sea-level rise. Agriculture with a diversity of ecosystem can help to cope with these challenges. Ecological agriculture will supply safe food not only for Vietnam but also the world. The adoption of ecological farming will also create jobs for local young people and small-scale farmers. This ecological rice grower also took the view that consumers are not only unclear about the organic standards, but they are not aware of the impacts of pesticides residuals on their health.

An organic rice farmer in Hai Phong province (personal communication on July 2019) attested that farmers should not only know about organic rice farming but also understand the aims and values of the organic rice-based system. The income from milled rice itself does not encourage rice farmers to grow organically. It takes time and resources to prepare soil through organic fertilizer, but it has a low yield, when the ecosystem is not stable. Farmers need to explore the range of embedded values of the organic rice farming system such as biodiversity value and the price premium of organic rice. If there is no price premium for organic rice, farmers will sell it as conventional rice. However, the organic rice farming system has a diversity of values that can be argued as an improvement to livelihood.

An organic rice wholesaler in HCM City (personal communication in September 2019) ventured an opinion that there was misunderstanding and mistrust about organic standards. Many people misunderstood about organic certifications in particular. They understood organic agriculture as safe farming without using chemicals. They perceive that VietGAP (Vietnam Good Agriculture Practice) or Global GAP (Good Agriculture Practices) as organic agriculture. For example, a rice farmer group in Tam Binh District of Vinh Long province, did not distinguish between GAP and international organic standards. They applied VietGAP standard on rice in 2017 and believe that it is organic rice.

My interviewee further asserted that Organic rice production has more advantages over fruit production because the on-farm expenditure on rice is lower than fruits and vegetables. The rice-shrimp system in the Mekong Delta has very good potential because farmers are less likely to use chemicals on shrimp and rice.

In a focus group discussion between myself and four other researchers in September 2018 it emerged that there is a slight difference in the perception of an international organic and domestic standard by laypeople. Only a few businesses and wholesalers explored this difference. Consumers have little concern about the organic standard. Affluent Vietnamese consumers have a consuming culture of imported products, and they are willing to pay for imported organic food from other countries.

In a personal communication in September 2019 between myself and two other researchers, there were trust issues expressed concerning organic rice in the Ho Chi Minh (HCM) market. Consumers do not trust local organic certification, but international certification fees are high. Farmers and private sector participants are not able to apply. Some private rice companies attempted to get the international certification, but there are few export markets for organic rice, so they keep rice for 2 years in the storage. A small proportion of consumers are interested in the international standard because it is held with high trust by consumers. Consumers generally have little trust in the domestic market (organic vegetable social enterprises).

In a personal communication, a government official in An Giang province took the view that growers who exported food, look for international certifiers such as the EU to certify their products, and they may know about the standard. However, the majority of small-scale farmers do not know or recognize that organic vegetables/products must be grown based on the standard, including approval, evaluation and monitoring. That is why many small-scale farmers bring their uncertified products to market as organic products. This affects the trust of the organic market in Vietnam.

An organic certifier in HCM City (personal communication in July 2019) claimed that awareness of domestic and international standards for organic food and the market for organic food has been increased. The international standard is tough at first, but Vietnam needs a good system, including systematic monitoring and enforcement of organic agriculture and organic industry (linking cross-sectors and multi-agents approach, as well as one door information sharing). This will go a long way toward better management of the standard.

Another perception is that consumers do not need certification, they need real, safe food. This was the view put by an ecological agriculture practitioner in Northern Vietnam in a personal communication in July 2019. Some international organic certification organizations can be trusted. However, many VietGAP certifications could not be trusted. Most companies do not have money to request or repeat the inspections. They used the expired certification to trade products. They cheated consumers. For example, only 15 farmsteads were certified but the products were collected from 30 farmsteads.

A media reporterin Can Tho City (in a personal communication in July 2019) found that in the Vietnamese market, there is no precise definition of an organic standard for rice and other products. Many consumers are confused with the organic labels or market. Most consumers buy rice products because they trust it, or they trust influential people or professional people who introduced them to the products. A small enterprise is not able to certify to an international standard while consumers trust to buy rice because of its flavour and brand name. The international certified organic products made the production costs high and this discourages medium or low-income clients from buying certified organic rice.

A safe vegetable grower in Long Xuyen city, An Giang, believes there to be a high demand for organic food in the urban areas (personal communication in July 2019). However, consumers did not understand fully about the organic products, even in the professional groups also are not aware of organic food. There are many labels about organic foods in HCM city market. Some products are international certified, few products are domestic, but the others are organic self-labelled. There is a mix of real and false organic products in the market which made consumers confused. Consumers tend to trust the brand name of the products rather than the standard of organic certification.

A large-scale organic grower in Binh Phuoc province (personal communication in July 2019) had concerns about the international standard rather than the local or domestic standard. The domestic standard is often easy to be certified, low in quality and trust. Surprising that growers want to certify international standard certification for the domestic market.

An ecological rice grower in Dong Thap province found that many farmers want to purchase international certified organic food for several reasons (personal communication in July 2019). They have little trust about the quality of local certified products. For example, most people perceived that VietGAP certified vegetables are similar to the uncertified vegetables in the local marketplaces. This grower had experienced with the weak enforcement and monitoring of the VietGAP certification process, leading to low quality and low trust.

An international certifier in HCM City (personal communication in July 2019) stated that the National Vietnam organic standard is not accredited by other international standards or approved by importing countries. The accreditation is particularly important because it states the acceptance of the exported market. If the accreditation is not approved, the trade will be severely affected.

12.5 Prospects of Organic Agriculture Policy and Research in Vietnam

Vietnam's agricultural policy although becoming more specific, can use further improvement. An organic practitioner in Northern Vietnam maintained that since the first organic agriculture project was introduced in 2005, policies for organic farming had been neglected in Vietnam (personal communication in July 2019). After 7 years of project implementation, only 20 farmer groups followed the project to produce organic vegetables in the market. Consumers tend to know more about organic agriculture since then.

Recently, the Vietnam government issued legal framework and policies to promote organic agriculture. In Vietnam, the organic standard (TCVN 110412017) was issued on 29 December 2017 by the Ministry of Science & Technology which included (1) TCVN 11041-1: 2017 – organic agriculture – part one: general requirement for production, processing and labelling for organic products; (2) TCVN 11041-2:2017–organic agriculture–part 2: organic crops; (3) TCVN 11041-3:2017-organic agriculture–part 3: organic livestock; (4) TCVN 11041-4:2017- organic agriculture–part 4: requirement for evaluation and certification of organic producing and processing system organizations (Ministry of Agriculture and Rural Development 2019).

On 26 December 2018 the Ministry of Science and Technology issued four organic standards for (1) TCVN 11041-5:2018 organic agriculture – part 5: Organic Rice; (2) TCVN 11041-6:2018 organic agriculture – part 6: Organic tea; (3) TCVN 11041-7:2018 organic agriculture – part 7: Organic milk; (4) TCVN 11041-8:2018 organic agriculture–part 8: Organic shrimp. However, there is no organic standard for aquaculture, medicinal plants, vegetables, fruits, coffee, and paper.

Ministry of Agriculture and Rural Development developed a project to promote organic agriculture from 2019 to 2025 and the vision to 2030 (Ministry of Agriculture and Rural Development 2019). Some policies encourage investment, but not provide details for small scale farmers and small enterprises to access the policies. There is no guideline to implement the national standard according to a member of a Vietnam organic agriculture association in a personal communication in July 2019. None of the Vietnamese organic standards are applied in practice. The current Vietnamese standard is not accredited by other international and regional standards.

In the view of an international certifier in HCM City (personal communication in July 2019), it is difficult to access fund from the government due to lack of information about support methods, complicated admin process that discourages growers. Big companies can access loan from government's policies, but small companies and growers could not access this fund.

The Vietnamese government issued policies to support organic agriculture. A government staff member conceded that it is still very general and lacked a detailed process for farmers to access (personal communication in July 2019).

In my conference with an organic rice grower in Dong Thap and retailer in HCM City in July 2019, a few key themes emerged. There are policies to support farmers and enterprises to implement organic farming. However, the budget of the policies does not go to the right people for several reasons. First, there are polices but it lacks the details and clear direction for farmers to access. Secondly, polices were stuck at the level of local authorities who supported their relatives who, themselves, were not interested in implementing organic farming. A new regulation (107/2018) reduced the bottleneck for exported rice companies, such as organic and nutritional rice. There is no rule or regulation to manage organic trademark (unregulated market). This leads to self-regulation of the organic products standard leading to confused consumers. This affected the organic market. There is a need for a mechanism to manage it.

12.5.1 Perception of Organic Agriculture Research in Vietnam

Research on organic agriculture has not been developed yet in Vietnam although organic agriculture policies were issued recently.

Four Deans of Faculty of Agriculture in four public universities and one private university, Research Institute (IPSAR) under Ministry of Agriculture and Rural Development (MARD), Vietnam academy of agriculture, and one international organic certification in the South of Vietnam, researchers, farmers, and private sectors were interviewed. Most of them said that there are many opportunities for organic agriculture development, but there is little research in this area. Can Tho and Kien Giang University have not apparently undertaken any projects to research on organic farming. IPSAR is a think tank of MARD whose responsibilities are undertaking policy research for agriculture. However, this research institute has few research projects and activities on organic agriculture. Some researchers studied organic agriculture (crops), but no policy research on organic agriculture has been explored. In the future, the IPSAR will undertake policy research to support organic agriculture in Vietnam.

Can Tho University is a focal point of agricultural research in the South of Vietnam; however, the faculty of agriculture has reported that they have not invested in this research area. Kien Giang and An Giang University are provincial universities that have a faculty of agriculture, but very little research has been done about organic farming. Tay Do University is a private educational institution that did not research organic agriculture.

An interesting point raised by an international organic certification body is about the lack of research for alternative inputs to replace chemicals. Vietnam does not have research to replace chemicals by bioproducts, pest management, alternative products, or to support farmers to convert to organic farming. So far, there are few research projects on this topic.

12.5.1.1 Drivers and Lessons of Promoting Organic Agriculture in Vietnam

Private small enterprises are the key actor in investing in organic farming in the South of Vietnam. They have a different perception of the adoption of organic agriculture. A rice company in An Giang province believes investment in organic rice is the marketing philosophy of the company. Other organic rice-shrimp company invested in organic farming because of increased values for farmers, the company, and the environment. One organic rice wholesaler believes three drivers of adopting organic farming: (1) increase the competitiveness of the company, (2) improve income for farmers, and (3) create safer food for consumers. Another organic rice grower has three reasons to desire recovery of ecological rice production: (1) traditional experience of ecological farming, (2) overuse of chemical leading to pollution, and (3) preservation of traditional farming practices

12.5.1.2 Lessons from Developing Organic Farming

One rice company collaborated with farmers to grow 50 ha of organic rice on a shrimp farm in Ca Mau province in 2018. However, just over 50% (27 ha) was certified in 2018 because many farmers did not comply with international organic certification procedures.

A private rice company invested in their rice-producing area where she can cooperate with farmers to produce organic rice (not certify). In the 2018 crop, she contracted 136 ha without international certification. She has a network of organic consumers in HCM city where she trusts that all of her organic rice was sold easily.

A private wholesaler in HCM City believes that consumers do not trust the product if it is not organically certified. However, some people still do not trust the products even it is certified.

12.6 Conclusion

Adoption of organic farming is possible in Vietnam. However, some people perceived that small-scale farm size has disadvantages in converting from conventional to organic farming. It is the fact that it is difficult to manage organic farms and conventional farms next to each other. Stakeholders awareness of organic farming is low. This is relevant to previous studies in Southeast Asia and Europe (My et al. 2017; Mutiara and Satoshi 2017; Hossain et al. 2007; Janssen and Hamm 2011). Rice is the most common crop in Asia. Conventional rice farming requires using too many chemicals that affect the natural environment and fish (Nguyen et al. 2018). Converting to organic rice farming has been done in Southeast Asia (the Philippines and Indonesia) since the 2000s (Piadozo et al. 2014; Digal and Placencia 2019; Heryadi et al. 2018). Private sector actors and NGOs play a key role in supporting organic farming in Vietnam while the role of government extension services is limited. This finding is relevant to perception studies in organic rice farming in the Philippines and Indonesia (Piadozo et al. 2014; Heryadi et al. 2018). Trust in national and domestic standards was low but high in international standards. Government policies on organic farming were initiated by the Vietnamese government, however, it is difficult for small enterprises and farmers to access these resources. Limited research was undertaken to study alternative farming inputs (organic fertilizer and pesticides), pest management, organic crops and animals in research organizations and universities. It is important to support small scale growers and enterprises to convert organic farming as it will bring social, environmental and economic benefits.

Acknowledgments I would like to thank participants who responded the questions. In particular, I wish to thank Vietnam PGS team leader (Mrs Tu Tuyet Nhung) in Vietnam, government research organization (IPSAR), universities (HCM National University, University of Social Science and Humanity, Vietnam Agriculture Academy, An Giang, Kien Giang,Can Tho, and Tay Do), private sector organisations (Ong Tho Rice, Tan Vuong Company, Co May Group, Ech Op Farm, Tam Viet Farm, and Viet Ha Organic Agriculture Company), NGOs (HEPA Eco-Farming School, Anh Ba Việt Nam and Y-farm Mekong), international organic certified organizations in Vietnam, growers, associations, and media who responded to the in-depth interviews for this research. I wish to acknowledge the contribution of SEARCA (The Southeast Asian Regional Center for Graduate Study and Research in Agriculture) in funding this research.

References

Baumgartner U, Nguyen TH (2017) Organic certification for shrimp value chains in Ca Mau, Vietnam: a means for improvement or an end in itself? Environ Dev Sustain 19:987–1002

Bo Nong Nghiep & Phat Trien Nong Thon (MARD) (2019) Báo cáo tóm tắt đề án phát triển nông nghiệp hữu cơ Việt Nam giai đoạn 2020–2030 (Draft summary report: organic agriculture development plan for 2020–2030). Bo Nong Nghiep & Phat Trien Nong Thon, Hanoi

Bouttes M, Darnhofer I, Martin G (2019) Converting to organic farming as a way to enhance adaptive capacity. Org Agric 9(2):235–247. https://doi.org/10.1007/s13165-018-0225-y

Braun G, Braun M, Kruse J, Amelung W, Renaud FG, Khoi CM et al (2019) Pesticides and antibiotics in permanent rice, alternating rice-shrimp and permanent shrimp systems of the coastal Mekong Delta, Vietnam. Environ Int 127:442–451
Creswell JW (1994) Research design: qualitative and quantitative approaches. Sage, London
David W, Ardiansyah (2017) Organic agriculture in Indonesia: challenges and opportunities. Org Agric 7(3):329–338. https://doi.org/10.1007/s13165-016-0160-8
Demont M, Rutsaert P (2017) Restructuring the Vietnamese rice sector: towards increasing sustainability. Sustainability 9(2):1–15. https://doi.org/10.3390/su9020325
Digal LN, Placencia SGP (2019) Factors affecting the adoption of organic rice farming: the case of farmers in M'lang, North Cotabato, Philippines. Org Agric 9(2):199–210. https://doi.org/10.1007/s13165-018-0222-1
Dung T (2017, Dec 19) Vietnam seeks to develop sustainable organic agriculture. Vietnam Economy
FiBL, IFOAM (2017) The world of organic agriculture statistics and emerging trends 2017. FiBL & IFOAM
Gamboni M, Moscatelli S (2015) Organic agriculture in Italy: challenges and perspectives. Org Agric 5(3):165–177. https://doi.org/10.1007/s13165-015-0098-2
Hay DV, Thanh NC, Kha LQ, An NV, Manh NV, Anh TT et al (2018) Model of public–private partnership in organic rice production in rice–shrimp cultivation system in Cuu long delta, Vietnam. MOJ Ecol Environ Sci 3(2):94–102
Heryadi DY, Noor TI, Deliana Y, Hamdani JS (2018) Why organic rice farmers switch back to conventional farming? J Econ Sustain Dev 9(9):179–185
Hossain ST, Sugimoto H, Ueno H, Huque SMR (2007) Adoption of organic rice for sustainable development in Bangladesh. J Org Syst 2(2):27–37
Huong NT (2017) The fact of organic farming in Vietnam. Department of Medicinal and Aromatic Plants, Faculty of Horticultural Sciences, Szent Itsvan University, Budapest
IFOAM (2015). Definition of organic agriculture. https://www.ifoam.bio/en/organic-landmarks/definition-organic-agriculture2018
Janssen M, Hamm U (2011) Consumer perception of different organic certification schemes in five European countries. [journal article]. Org Agric 1(1):31–43. https://doi.org/10.1007/s13165-010-0003-y
Jouzi Z, Azadi H, Taheri F, Zarafshani K, Gebrehiwot K, Passel SV et al (2017) Organic farming and small-scale farmers: main opportunities and challenges. Ecol Econ 132:144–154
Liamputtong P (2009) Theoretical frameworks and sampling in qualitative reesearch. In: Liamputtong P (ed) Qualitative research methods. Oxford University Press, South Melbourne
Lockie S, Lyons K, Lawrence G, Mummery K (2002) Eating 'Green': Motivations behind organic food consumption in Australia. Sociol Rural 42(1):23–40
Loebnitz N, Aschemann-Witzel J (2016) Communicating organic food quality in China: consumer perceptions of organic products and the effect of environmental value priming. Food Qual Prefer 50:102–108
Lyons K (1999) Corporate environmentalism and organic agriculture in Australia: the case of Uncle Tobys. Rural Sociol 61(2):251–266
Mergenthaler M, Weinberger K, Qaim M (2009) Consumer valuation of food quality and food safety attributes in Vietnam. Appl Econ Perspect Policy 31(2):266–283
Ministry of Agriculture & Rural Development (2019) Đề án phát triển nông nghiệp Việt Nam Giai Đoạn 2020–2030 (Vietnam organic agriculture development plan 2020–2030). Ministry of Agriculture & Rural Development, Hanoi
Mutiara VI, Satoshi A (2017) The challenges in organic agriculture products market in Southest Asia. Rev Agric Sci 5:36–44
My NHD, Rutsaert P, Loo EJV, Verbeke W (2017) Consumers' familiarity with and attitudes towards food quality certifications for rice and vegetables in Vietnam. Food Control 82:74–82
Nguyen VK, Dumaresq D, Pittock J (2018) Impacts of rice intensification on rural households in the Mekong Delta: emerging relationships between agricultural production, wild food supply and food consumption. Food Sec:1–15. https://doi.org/10.1007/s12571-018-0848-6

Pham L, Shively G (2019) Profitability of organic vegetable production in Northwest Vietnam: evidence from Tan Lac District, Hoa Binh Province. Org Agric 9:211–223

Piadozo MES, Lantican FA, Pabuayon IM, Quicoy AR, Suyat AM, Maghirang PKB (2014) Rice farmers' concept and awareness of organic agriculture: implications for sustainability of Philippine organic agriculture program. J ISSAAS 20(2):142–156

Ponti T d, Rijk B, van Ittersum MK (2012) The crop yield gap between organic and conventional agriculture. Agric Syst 118:1–9

Tashi S, Wangchuk K (2016) Organic vs. conventional rice production: comparative assessment under farmers' condition in Bhutan. Org Agric 6(4):255–265. https://doi.org/10.1007/s13165-015-0132-4

Vietnam Organic Agriculture Association (2017) The solution for organic agriculture development in Vietnam. In: The first national forum on organic agriculture. Agriculture Publisher (Nha Xuat Ban Nong Nghiep), Ho Chi Minh City

Wägeli S, Hamm U (2016) Consumers' perception and expectations of local organic food supply chains. [journal article]. Org Agric 6(3):215–224. https://doi.org/10.1007/s13165-015-0130-6

Wang H, Moustier P, Loc NTT, Tho PTH (2012) Quality control of safe vegetables by collective action in Hanoi, Vietnam. Proc Econ Finance 2:344–352

Wood R, Lenzen M, Dey C, Lundie S (2006) A comparative study of some environmental impacts of conventional and organic farming in Australia. Agric Syst 89(2–3):324–348

Van Kien Nguyen grew up in a small-scale rice-based farming family in Sóc Trăng Province of Mekong Delta, Vietnam. He is a current honorary senior lecturer in Agroecology at Fenner School of Environment and Society, The Australian National University, as well as the Founder of the company Mekong Organics Pty Ltd. He received his BSc in Land Management (2000) from Can Tho University, his master's degree in Environmental Management and Development (2006), his PhD in Sociology (2013) from The Australian National University (ANU) and his Postdoctoral Fellowship (2014–2016) from Luc Hoffmann Institute (LHI), ANU Fenner School of Environment and Society. He has been leading the Research Center for Rural Development in An Giang University, Vietnam, for 5 years (2013–2018). His research focuses on agroecology, sustainable agriculture and livelihoods, food security and water governance and climate change adaptation of rural households in the Mekong Region. Since 2017, he has been involved in linking Australia and Southeast Asia in promoting organic farming.

Chapter 13
Food Safety Law in the United States: Risk Management in the Organic Food Supply Chain

Anastasia Telesetsky

Abstract Organic food products both processed and unprocessed are not immune from system-wide food safety concerns. In the United States, consumer demand and concern over litigation exposure from companies led to the passage of the Food Safety Modernization Act in 2011. This chapter highlights the development and implementation of this law which applies both to organic and conventional products in the United States' food supply chain. This chapter also offers a brief commentary on challenges to implementation of organic food safety standards.

Keywords United States food safety · Organic · Food Safety Modernization Act

On the United States dollar, the words "In God we Trust" are printed. But each day that Americans shop at supermarkets and dine at restaurants, they also trust in the Food and Drug Administration to certify that the food they consume, both organic and conventional, is free from contamination. This trust is put in jeopardy when Americans are exposed to foodborne pathogens. In 2011, researchers calculated that 9.4 million people in the United States experienced some foodborne illness with 56,000 of these individuals requiring hospitalization.[1] These alarming food safety statistics apply to both organic and conventionally produced products. Based on reviewing the National Outbreak Reporting System, researchers have identified at least 15 outbreaks of foodborne illness related to an assortment of organic products.[2]

[1] Elaine Scallan et al., (January 2011) "Foodborne Illness Acquired in the United States-Unspecified Pathogens," *Emerging Infectious Diseases* 17(1): 7–15.

[2] Reid Harvey, Christine M. Zakhour and L. Hannah Gould, (2016) "Foodborne Disease Outbreaks Associated with Organic Foods in the United States." *Journal of Food Protection* Vol. 79, No. 11, 1953–1958. (Finding foodborne contaminants associated with eggs, unpasteurized milk, nuts/seeds, and multi-ingredient products).

A. Telesetsky (✉)
University of Idaho, College of Law, Moscow, ID, USA
e-mail: atelesetsky@uidaho.edu

GOH Bee Chen, R. Price (eds.), *Regulatory Issues in Organic Food Safety in the Asia Pacific*, https://doi.org/10.1007/978-981-15-3580-2_13

The purpose of this chapter is to highlight some organic food safety issues in the U.S. and describe how its laws and regulations, especially the Food Safety Modernization Act, address food system safety issues.

13.1 Food Safety Challenges in the United States Organic Industry

The globally booming organic food industry in the United States is a response to a spectrum of consumer concerns including threats to health associated with monoculture production as well as ethical concerns for food producers and ecosystems. When polled, organic food consumers have indicated that food safety is of paramount concern. That is why they chose organically produced food that has not been, for example, genetically engineered, bathed in chemical pesticides, or fertilized with treated sewage sludge.[3] Some organic food products may indeed be safer for overall human health. Organic grains grown with organic fertilizers appear to have lower levels of absorption of the heavy metal cadmium from soil than their conventional counterparts.[4] Organic fruits and vegetables have lower levels of pesticide residue and organic meats have lower levels of antibiotic resistant bacteria.[5]

Other organic products, however, face many of the same food safety challenges as conventionally produced food products, with some organic farming practices creating additional exposure to food contamination where practices are not properly implemented. For example, composted manure from livestock is often used as a soil amendment on U.S. organic farms. If the manure has not been properly composted at high enough temperatures to kill pathogens,[6] then consumers may be exposed to high levels of *Escherichia coli* (*E. coli*), shigella and salmonella in food that comes

[3] Gemma Harper and Aikaterini Makatouni, (2002) "Consumer Perception of Organic Food Production and Farm Animal Welfare", *British Food Journal* 104 Issue: 3/4/5, 287–299, https://doi.org/10.1108/00070700210425723; *See also* 7 U.S.C.A. 6508 (Organic Foods Production Act)

[4] Roman Grüter et al., (June 2019), "Long-term Organic Matter Application Reduces Cadmium but not Zinc Concentrations in Wheat", Sci *Total Environ.* 669: 608–620, https://doi.org/10.1016/j.scitotenv.2019.03.112

[5] Axel Mie et al. (October 2017) "Human health implications of organic food and organic agriculture: a comprehensive review" *Environ. Health* 16, 111–133, https://doi.org/https://doi.org/10.1186/s12940-017-0315-4

[6] Under USDA regulations, producers are restricted by regulation in what compost may be applied but researchers indicate that the regulations may not be sufficiently protective of human health. For example, organic producers are prohibited from applying raw non-composted manure to crops if harvesting will take place before 90 or 120 days have elapsed, depending on whether manure has been indirectly or directly applied to crops. 7 CFR 205.203(c) (1)(ii) and (iii). However, researchers note that some strains of *E. Coli* can persist for over 150–190 days. Mahbub Islam et al. (2005) "Survival of Escherichia coli O157:H7 in Soil and on Carrots and Onions Grown in Fields Treated with Contaminated Manure Composts or Irrigation Water" *Microbiology* 22, 63–70 (Proposing a 210 day period before harvesting after raw manure has been applied).

into contact with the amended soil.[7] Some of the pathogens including *E. coli 0157* persist in the soil for many weeks after compost has been applied and may also be found in groundwater used for irrigation or for processing fruits and vegetables for market. As organic products may require processing, they face many of the same safety risks associated with food products produced by industrial farming.[8]

Certain organic products are at the center of a food safety controversy that does not impact conventional products. Some consumers drinking unpasteurized milk as part of an organic diet have been unwittingly exposed to Brucella, Campylobacter, Cryptosporidium, *E. coli*, Listeria, and Salmonella, where farms have engaged in lax production and handling of milk products. In 2019, contaminated organic raw milk crossed state lines potentially in violation of state food safety laws.[9] Reflecting collision between regulation and religious food culture, have been increasing concerns about certain individuals in traditionally organic food producing communities, including the Amish and Mennonite communities, intentionally refusing to participate in required national food safety inspections when shipping products across State borders.[10]

13.2 Food Safety Regulatory Framework

The food safety regulatory framework in the U.S. is a patchwork of laws and normative practices that has evolved more reactively than with method. The network of laws and norms binds together public and private actors in a complex set of public regulation and private ordering. It is beyond the scope of this chapter to cover the safety framework that crosses a number of federal and state regulatory agencies, trade organizations, and industry bodies.[11] Accordingly, this section will highlight in a linear fashion some of the major developments that control both organic and non-organic producers selling into the U.S. market.

---

[7] E.B. Solomon, S. Yaron, and K. R. Matthews (2002) "Transmission of Escherichia coli O157:H7 from contaminated manure and irrigation water to lettuce plant tissue and its subsequent internalization". *Appl. Environ. Microbiol.* 68:397–400.

[8] See e.g. United States Department of Agriculture, News Release, Perdue Foods LLC Recalls Perdue Simply Smart Organics Poultry Products Due to Possible Foreign Matter Contamination (May 31, 2019) (Containing shards of bones); U.S. Food and Drug administration, Wakefern Food Corp. Voluntarily Recalls Wholesome Pantry Organic Nut Butters Because of Possible Health Risk (March 28, 2019) (Company identified positive listeria tests at their supplier and recalled products).

[9] Center for Disease Control, CDC Food Safety Alert: Brucellosis exposures from raw milk, Media Statement, (February 11, 2019).

[10] Dan Flynn, Amos Miller's Days of Ignoring Food Safety laws may be coming to an End (April 19, 2019) foodsafetynews.com (Describing how the US Department of Justice is seeking compliance the Miller's Organic Farm's compliance with basic food safety regulations including record-keeping and testing).

[11] Just within the federal government, there are 15 agencies across various federal government departments implementing 35 food safety related laws.

The major federal actors in food safety in the United States are the Food and Drug Administration (or FDA, based in the Department of Health and Human Services [DHHS]), the Center for Disease Control (also based in DHHS), and the United States Department of Agriculture. Other organizations contribute to various aspects of food safety including the Federal Trade Commission which regulates advertising, the National Marine Fisheries Service which inspects seafood, and the Environmental Protection Agency which promulgates regulations on pesticides. There have been proposals to merge many of the food safety activities of the federal government into a Consumer Safety Agency but that proposal has not received much traction. The number of government actors in food safety inspection and regulation has led to confusion. The most often quoted example of fragmentation in the U.S. oversight system is the USDA's regulation of egg products and the FDA's jurisdiction over whole eggs. In 2018, the executive branch proposed consolidating the food safety efforts of the FDA and the USDA into a single agency within the USDA.[12]

In 1906, with concerns of rampant cases of adulterated food, Congress passed *the Pure Food and Drug Act* that prohibited the sale of any adulterated food in interstate trade and the *Federal Meat Inspection Act*. Congress assigned implementation responsibilities for these two acts to USDA which organized a Bureau of Animal Industry and Bureau of Chemistry to identify contaminated food. During the New Deal Administration, Franklin Roosevelt championed the *Federal Food, Drug, and Cosmetic Act* of 1938 (FDCA). That led to the creation of the Food and Drug Administration (formerly USDA's Bureau of Chemistry). This new administration was moved out of the supervision of the USDA and into an agency that later became the DHHS. Today, the USDA's Federal Safety and Inspection Service (FSIS) is responsible for all meat and poultry inspections. This includes a review of whether a product can be labeled as "certified organic by [a recognized given entity]" under the *Organic Foods Production Act*.[13] The FDA is responsible for overseeing food production of almost all foods that are not meat or seafood.

Food safety has been a moving regulatory target for over a century in the United States. Some of the responses have been litigation-driven. Beginning in around 1975, the federal government has criminally prosecuted individuals who have sold contaminated food under the FDCA but this strategy has only been applied to the more egregious cases of dangerous food safety practices. Regulators and industries have instead focused their attention on improving food processing practices.

With outbreaks of contamination associated with both organic and conventionally produced products, regulators and companies have sought to enhance food

[12] Developing Government Solutions in the twenty-first Century, https://www.performance.gov/GovReform/Reform-and-Reorg-Plan-Final.pdf (2018)

[13] The purpose of the Organic Foods Production Act is to establish national standards for products that are being produced and retailed as organic foods. The Act does not specifically address consumer food safety regulation. The focus is instead on production. Organic production is defined as "A production system that is managed in accordance with the Act and regulations in this part to respond to site-specific conditions by integrating cultural, biological, and mechanical practices that foster cycling of resources, promote ecological balance, and conserve biodiversity." 7 CFR 205.2

safety across the supply chains. In response to outbreaks that have economically threatened whole industries such as the packaged greens industry, companies which are exposed to risk have voluntarily adopted more rigorous food safety practices. For example, in 2007, a food safety working group composed of both public and private stakeholders developed the California Leafy Greens Products Handler Marketing Agreement in 2007 in hopes of avoiding the costly mistakes of an *E. coli* outbreak 2006 on four California farms that led to millions of dollars of damages across the industry. The Agreement created good agricultural practice metrics requiring handlers of products to only purchase from farms who test against a set of risk-based thresholds for pathogens in irrigation water and compost; they must also comply with other safety related practices such as not picking crops where there is the possibility of contamination from wild animal droppings.[14] Notably, this self-regulating model only applied to "handlers" and not to growers or retailers. Proponents of the agreement worked to create a national version, but this provoked a vocal response from small and medium sized farms particularly for organic farms that supplied both animal products and produce.

13.2.1 Recent Regulatory Developments

While certain industries were adopting "good agricultural practices",[15] consumer groups have been frustrated by what they perceived to be a laxness in the general food safety system and pushed for the adoption of a more national approach. The debate around food safety that began in earnest in Congress in 2008 culminated in bipartisan support for *Food Safety Modernization Act* (FSMA) that President Obama signed into law on January 4, 2011.[16] This law required agencies to identify science-based minimum standards for hazard identifications in order to develop a preventive approach to food safety risks rather than an ongoing reactive approach. Owners, operators, and agents of regulated facilities are required to develop "hazard analysis and risk-based preventive controls."[17] Title II strengthened FDA's mandate to recall food items. Title II also provided for more inspections that would improve industry compliance with safety standards as well as strengthening FDA's mandate to recall food items. Title III required most food importers to verify that their suppliers comply with US food safety laws, granted the FDA the power to certify high-risk foods coming from foreign producers, and gives it power to inspect foreign facilities.

---

[14] The most recent version of the Food Safety Guidelines (2019) adopted by the California Leafy Greens Products Handler Marketing Agreement Advisory Board is available at https://lgma.ca.gov/wp-content/uploads/2019/06/190419-CA-LGMA-Metrics-with-Decision-Trees_2.pdf

[15] "Good agricultural practice" is a term of art used in food safety law in the United States to refer to both best industry practice and FDA guidance.

[16] Pub. L. No. 111-353, 124 Stat. 3885 (2011).

[17] Id. at Title I

Notably, the FSMA only applies to relatively large-scale producers and growers of food products. When it was still being debated, Senator Tester introduced amendments into the final drafts of the law providing that the Act would not impose excessive regulatory burdens on small, local farmers. The amendment exempted the smallest farmers. This was considered an important win for family and small-scale organic farms likely to be selling into community sustained agriculture projects or farmer's markets. State and local regulations regulate these small-scale producers. This outcome where large organic producers are regulated under the federal Act, while small producers are exempt, has generated disagreement among food safety proponents about the efficacy of the Act in preventing future food contamination.[18]

Under the FSMA, the FDA was responsible for promulgating implementation rules for the Act.[19] The "Standards for the Growing, Harvesting, Packing, and Holding of Produce for Human Consumption" also known as the *Produce Safety Rule* was a long awaited regulation finally coming into force in 2015. The drafters decided to require the same safety standards across the entire produce industry instead of focusing on particular known high-risk commodities such as melons, tomatoes, and greens.

The content of the rule built on the efforts of groups such as California Leafy Greens Product Handlers by providing specific metrics to govern as "good agricultural practices" in a number of areas including water quality, animal proximity, compost, worker hygiene, and maintenance of buildings and equipment. The final revised rule paralleled the organic regulations requiring 120 days waiting period between manure application and harvest if being applied directly and 90 days if being applied indirectly. The organic lobbying industry had advocated strongly for this even though some soil pathologists were concerned that there was a higher risk for certain types of produce contamination.

The rule is not intended to apply to the smallest producers who earn less than $25,000 per year (as averaged over 3 years). It would not impact the smallest organic producers such as individuals who sell for a short growing season at a farmer's market. For other producers, there are timelines for compliance with very small farms ($250,000 sales of produce or less per year as averaged over 3 years) having 4 years to comply until 2020, small farms ($500,000 sales of produce or less per year as

[18] Representative Joe Pitts made the following remark at the time of the debates that captured the sentiment of Congressional members who did not want a broad exception for small-scale farmers: "[W]hile we do not want to overly burden small facilities and small farms, we've learned in our committee hearings that food-borne pathogens don't care if you're a big facility or a small facility, a big farm or a small farm. They affect everyone." 56 CONG.REC. H8885 (Dec. 21, 2010) (statement of Rep. Joe Pitts)

[19] The foundational rules required to be implemented under the act include: (1) Accredited Third-Party Certification; (2) Current Good Manufacturing Practice and Hazard Analysis and Risk-Based Preventive Controls for Human Food; (3) Current Good Manufacturing Practice and Hazard Analysis and Risk-Based Preventive Controls for Food for Animals; (4)Foreign Supplier Verification Programs; (5) Mitigation Strategies to Protect Food Against Intentional Adulteration; (6) Sanitary Transportation of Human and Animal Food; and (7) Standards for the Growing, Harvesting, Packing, and Holding of Produce for Human Consumption

averaged over 3 years) had 3 years until to comply, and most other farms had 2 years to comply. For farms that sell directly to individuals or retailers within a state (or within 275 miles), producers do not have to comply with the farm metrics described above but are still subject to compliance and enforcement under the Produce Rule for recordkeeping and labelling the food at point of sale. By requiring farms to follow various metrics, the intent of the *Produce Safety Rule* is to create a food safety culture that is prevention-based rather than just reacting to public health crises. When implemented, the rule is intended to proactively manage risk for different sets of food producers.

Other rules promulgated under the FSMA with possible impacts on large-scale organic producers include the:

- Preventive Controls for Human Food Rule
- Foreign Supplier Verification Program for importers
- Sanitary Transportation rule and the
- Rule to Mitigate Food Against Intentional Adulteration.

The *Preventive Controls for Human Food Rule* updates requirements for registered facilities that manufacture, process, pack, or hold food for human consumption.[20] The Rule does not apply to farm stands at a farmer's market, off-farm community sustainable agriculture drop offs, or restaurants that sell directly to consumers because there is no requirement for producers marketing their products in these cases to register with the FDA. The regulation was drafted to protect small-scale community sustained agriculture from additional regulatory burden. In contrast to the Produce Safety Rule, which covers farms with as little revenue as $25,000 annually, the *Preventive Controls Rule* focuses on relatively large processors and handlers of food. A "very small business" is defined under the *Preventive Controls Rule* as a business with fewer than 500 full-time workers and less than one million USD of revenue. To assist entities with understanding whether the rule applies to their business or not, the FDA issues guidance.[21]

[20] 21 C.F.R. Part 117

[21] The FDA website provides a number of guidance documents. For example, when questions were raised by regulated parties about how to define an employee for purposes of the *Preventive Controls Rule*, the FDA published in 2019 a guidance document "Guidance for Industry: Determining the Number of Employees for Purposes of the "Small Business" Definition in Parts 117 and 507" which explains how the FDA will exercise its discretion in implementing the regulation. A copy of this guidance available at https://www.fda.gov/media/111951/download

Most of these guidance documents are issued in response to confusion by regulated parties over how a regulation will be implemented. Other examples of FSMA guidance rules available at the FDA website with applicability to the implementation of the Preventive Rule include: "Determination of Status as a Qualified Facility Under Part 117: Current Good Manufacturing Practice, Hazard Analysis, and Risk-Based Preventive Controls for Human Food AND Part 507: Current Good Manufacturing Practice, Hazard Analysis, and Risk-Based Preventive Controls for Food for Animals" (September 2018) and "Considerations for Determining Whether a Measure Provides the Same Level of Public Health Protection as the Corresponding Requirement in 21 CFR part 112 or the Preventive Controls Requirements in part 117 or 507" (January 2018)

A business that is covered is required to develop a Food Safety Plan based on "Hazards Analysis and Risk-Based Preventive Controls" to be overseen by a Preventive Controls-Qualified Individual (PCQI). This "individual" can be an existing employee, a third-party consultant, or a committee that has completed training under the FDA or has equivalent training. A plan consists of (1) hazard analysis (including biological, chemical and physical), (2) preventive controls, (3) a supply chain program for products received from other vendors, (4) a written recall plan, (5) procedures for monitoring the implementation of the preventive controls, (6) procedures for taking corrective actions, and (7) a verification procedure. In terms of ensuring the safety of the supply chain, a processor of human food who receives as part of its supply chain a product that may introduce hazards into the food product must either conduct an audit of the supplier, or review documentation of an audit indicating that the supplier recently passed.

The regulation on *Foreign Supplier Verification Program for Importers* under the FSMA requires importers (who make $1 million of food sales) to identify known or reasonably foreseeable hazards of imported food, analyze the risks associated with given suppliers, and evaluate a foreign supplier's performance to be sure that food is produced that meets safety criteria.[22] The Verification Program does not apply to fish or fishery products, juice, certain alcoholic products, certain egg and meat products, or low-acid canned food.[23] Parties that are in compliance with *Preventive Controls for Human Food Rule* are deemed to also be in compliance with the Verification Program.

The *Sanitary Transportation Rule* applies to shippers, receivers, loaders and carriers who transport food in the United States by road or rail.[24] It does not apply to those transporting food by plane or boat. It also does not apply to transport of live animals (except mollusks) or goods transported by a farm.[25] It requires regulated entities to adopt food safety practices during transportation that will contribute to proper refrigeration, adequate cleaning of vehicles between loads, and other measures (protection against allergens and other cross-contamination). The law does not apply to parties that hold permits under the National Conference on Interstate Milk Shipments' Grade "A" Milk Safety Program, the Interstate Shellfish Sanitation Conference's National Shellfish Sanitation Program, or food establishments supplying directly to consumers.

The Rule to protect food against intentional adulteration[26] applies to large food production firms that are regulated by the FDA (usually at least $10 million of business per year as calculated over a 3 year average) and not to farms. Every regulated facility is required to prepare and implement a food defense plan that identifies

[22] 21 C.F.R. Subpart L

[23] 21 C.F.R. 1.501 and 1.502; There are additional exception for farms not covered by the *Produce Safety Rule* and for small egg farms.

[24] 21 C.F.R. Subpart O

[25] Food transported by a farm is covered more broadly by the Food Drug and Cosmetic Act prohibiting transport of food in an unsanitary condition.

[26] 21 C.F.R. Subpart B

vulnerabilities within a firm where food could be potentially adulterated and pose a serious public health risk. As part of the plan, firms must provide a set of mitigation strategies, monitoring protocols, processes for correcting problems, and verification activities. A firm will be re-evaluated every 3 years if the FDA identifies problems or risks in manufacturing, processing, packing, or storage.

13.3 Select Challenges Under U.S. Law for Ensuring a Safe Organic Industry

With the rules promulgated under the FSMA being the primary regulatory framework for protecting consumers of organic products, questions remain about whether these rules will be sufficient to protect American consumers. In an improvement over previous industry practices, the *Produce Safety Rule* provides specific measures that farmers who are under the jurisdiction of the rule are expected to implement on-farm including animal proximity practices and composting practices. Having specific measures provides a harmonization across the industry but questions remain, however, about whether the measures reflect best business practices rather than the best science. For example, some researchers indicate researchers suggest that a 7 month wait time may be necessary for certain types of crops that are treated with manure to avoid contamination concerns.[27] If organic producers were required to wait a longer wait to harvest than the current standards, the Organic Trade Association indicated that a longer wait period would be too long for its members to remain profitable.[28] For some critics, there are questions of whether the *Produce Safety Rule* should be applied to more farms. Presently, the rule exempts a broad number of small-scale produce sellers who sell directly to consumers. These small-scale producers may still contribute to food-related illnesses.[29] Whether additional federal regulation would make a difference is unclear but the exemption under the FSMA for these types of producers is broad.

A second challenge arises under the current FSMA as regulators seek to ensure the effectiveness of the Foreign Supplier Verification Program. Although food accounts for one-third of the 42 million products imported into the United States each year, as of May 2019, fewer than 30% of overseas processors in a sample of individuals were receiving FDA inspections after 2 years of implementation of the

[27] M. Islam et al. supra note 6.

[28] Organic Trade Association, 'FDA Food Safety Modernization Act', https://ota.com/advocacy/food-safety/fda-food-safety-modernization-act

[29] Marc Bellemare and Ngoc Nguyen (2018) 'Farmers' Markets and Food-Borne Illness" *American Journal of Agricultural Economics*, 100(3): 676–690 (Finding that farmers markets in the United States contribute to measurable numbers of cases of food-borne illness, norovirus, and campylobacter.)

program.[30] This has substantial implications for the organic industry as at least some of the products being labeled as "organic" in the United States supply chain – coming from over 100 countries – are not actually organic products when evaluated under the standards set by the National Organic Board. The United States Department of Agriculture released a report in 2017 indicating that conventional agricultural substances prohibited under national organic standards were being used to fumigate "organic" products and mandatory audits to ensure proper labeling were not being performed.[31] A major concern is the inability of the regulators to reconcile differences between United States and foreign organic standards. The organic industry has responded to allegations of "organic fraud" against the industry with a quality assurance program for its members called "Organic Fraud Prevention Solutions" calling on members to ensure that food meets U.S. organic standards.[32]

Third, the legal right for processors to use certain non-organic processing and handling additives in food labeled as organic raises a question about what constitutes "organic". Although the additives may not be unsafe in the sense of contaminating a product, they raise questions for many organic consumers about what residue they might be ingesting that was used in the processing and handling of an "organic" product. In *Harvey v. Veneman*,[33] an organic blueberry farmer sued the US Department of Agriculture on a set of regulations that permitted "synthetic substances" to be used to process food that would eventually be labeled as organic food. The Appellate Court disagreed with the USDA that adding synthetic substances was consistent with the existing language of the Organic Food Production Act and ordered the Secretary of Agriculture to prepare new rules. U.S. Congress responded to the First Circuit opinion and amended the Act to provide explicitly for the use of certain non-organic substances in processing. The plaintiff farmer brought a section case to challenge Congress' amendment but the First Circuit found the Congressional amendments were valid.[34]

Finally, federalism poses unique challenges for the United States in food safety. For products that are not in interstate commerce, States and localities assert jurisdiction over matters involving public health and agriculture. In some cases, federal and state approaches coincide with, for example, state Meat and Poultry inspection units implementing requirements "at least equal to" those imposed under the Federal Meat Inspection Act and the Poultry Products Inspection Act. In other cases, jurisdiction on products involving food safety diverge. For example, the federal Pasteurized Milk Ordinance only permits the interstate sale of Grade A pasteurized milk to consumers and prohibits the sale of raw milk to consumers.[35] A number of

---

[30] Bob Ferguson, "What Industry and FDA Are Thinking about FSMA Implementation – Part 1", *Food Safety Magazine* (April/May 2019) https://www.foodsafetymagazine.com/magazine-archive1/aprilmay-2019/what-industry-and-fda-are-thinking-about-fsma-implementation-part-1/

[31] National Organic International Trade Arrangements and Agreements, Office of Inspector General, September 2017, https://www.usda.gov/oig/webdocs/01601-0001-21.pdf

[32] Organic Trade Association, https://ota.com/OrganicFraudPrevention

[33] 396 F.3d 28 (1st Cir.2005)

[34] *Harvey v. Johanns* 494 F.3d 237 (1st Cir. 2007) (Name change from *Harvey v. Veneman*)

[35] 21 C.F.R. 1240.61

States permit the sale of raw milk through statute though most of these States have adopted safety rules based on the federal Pasteurized Milk Ordinance.

13.4 Conclusion

The organic food industry is not immune to food safety challenges including contamination. The same food safety rules apply largely to both organic producers and conventional food producers. At the federal level, the Food Safety Modernization Act and its implementing rules governs organic food safety. While most of the implementation challenges under the Act are shared with conventional food products, under the Foreign Supplier Verification Program, there are growing concerns associated with organic products as certain products identified as "organic" have failed to meet U.S. national organic standards.

In thinking about organic food safety, it is important to recall that almost all of the relatively recent U.S. federal laws and regulations are targeted at large farms or processors. Large agricultural firms have been at the epicenter of many of the contamination outbreaks. State and local laws apply their public health regulations to small-scale organic producers. Whether these laws, which depend on the input of local lobbies of citizens and producers are, in fact, sufficient to protect public health remains to be seen.

Reference for Further Reading

Bellemare M, Nguyen N (2018) 'Farmers' markets and food-borne illness. Am J Agric Econ 100(3):676–690

Card M (2016) The paradox of clean Food and the Safety Modernization Act: understanding the FDA'S preventive controls for human food rule. SMU Sci Technol Law Rev 19:381–395

Ferguson B (2019) What industry and FDA are thinking about FSMA implementation – part 1. Food Safety Magazine (April/May 2019) https://www.foodsafetymagazine.com/magazine-archive1/aprilmay-2019/what-industry-and-fda-are-thinking-about-fsma-implementation-part-1/

Food, Drug, and Cosmetic Act, 21 U.S.C. §§ 301-399

Food Safety Modernization Act, Pub. L. No. 111-353, 124 Stat. 3885

Fortin N (2016) Food regulation: law, science, policy, and practice second edition. Wiley, Hoboken

Grüter R et al (2019) Long-term organic matter application reduces cadmium but not zinc concentrations in wheat. Sci Total Environ 669:608–620

Harper G, Makatouni A (2002) Consumer perception of organic food production and farm animal welfare. Br Food J 104(3/4/5):287–229

Harvey R, Zakhour CM, Hannah Gould L (2016) Foodborne disease outbreaks associated with organic foods in the United States. J Food Prot 79(11):1953–1958

Hutt P, Merill R, Grossman L (2013) Food and drug law. Foundation Press

Islam M et al (2005) Survival of Escherichia coli O157:H7 in soil and on carrots and onions grown in fields treated with contaminated manure composts or irrigation water. Microbiology 22:63–70

Lytton T (2019) Outbreak. Press, University of Chicago

Mie A et al (2017) Human health implications of organic food and organic agriculture: a comprehensive review. Environ Health 16:111–133

Pina K, Pines W (eds) (2017) A practical guide to FDA's food and drug law and regulation, 6th edn. Food and Drug Law Institute

Scallan E et al (2011) Foodborne illness acquired in the United States-unspecified pathogens. Emerg Infect Dis 17(1):7–15

Solomon EB, Yaron S, Matthews KR (2002) Transmission of Escherichia coli O157:H7 from contaminated manure and irrigation water to lettuce plant tissue and its subsequent internalization. Appl Environ Microbiol 68:397–400

Tai S (2015) Whole foods: the FSMA and the challenges of defragmenting food safety regulation. Am J Law Med 41:447–458

United States Office of Inspector General, National Organic International Trade Arrangements and Agreements (2017, September). https://www.usda.gov/oig/webdocs/01601-0001-21.pdf

Anastasia Telesetsky is a tenured Professor at the University of Idaho College of Law, United States, where she is a core faculty member for the Natural Resources and Environmental Law Programme. She has published numerous article and chapters on matters involving marine living resources and food law and coauthored and edited books on ecological restoration and disaster relief and planning law. She is a former Co-chair of the International Environmental Law Committee for the American Bar Association's Section of International Law and a current member of the IUCN World Commission on Environmental Law. She is a former Fulbright Scholar (the Philippines) and Axford Fellow (New Zealand). She has taught food law at the University of Auckland in New Zealand and China University of Political Science and Law in Beijing, China.

Chapter 14
Regulation of the New Zealand Organics Sector

Tracey Epps and Danae Wheeler

Abstract New Zealand's organics sector has grown significantly in recent years, with production reaching NZD $600 million in 2018 (Organics Aotearoa New Zealand (OANZ) OANZ Market Report 2018 at 16). In a 2017 survey, 72% of New Zealanders indicated that they buy organic products some, most, or all of the time (Colmar Brunton "Sustainable Development Goals" (2017) at 26). Despite this growth, New Zealand does not have a mandatory standard for organics. Instead, organics must comply with a number of overarching laws that apply to non-organic products, including the Food Act 2014, the Fair Trading Act 1986, the Wine Act 2003 and the Animal Products Act 1999. In addition, there are a variety of voluntary organic standards as well as private certifiers. Hence, it has been reported that only 7% of the consumers who purchase organic foods fortnightly are able to identify all aspects of an organic product (OANZ Market Report 2018 at 9).

Growing concerns from consumers, trading partners and industry highlighted the lack of a mandatory standard as a challenge facing New Zealand's organic sector. Recognising the impediment this presents to the growth of New Zealand's organic market, the New Zealand Government initiated a public consultation process in 2018. Following the consultation process, the Government announced that it would develop a national standard for organic production.

This chapter examines the history and development of organic food production in New Zealand. It begins by briefly examining early developments before turning to the regulation of food production in New Zealand and its application to organic products. The chapter then explores issues surrounding aspects of the voluntary regime, such as the involvement of private certifiers and application of private standards. It concludes by discussing the key factors behind the Government's decision to consult on the introduction of a mandatory domestic organics standard.

Keywords Organic regulation · New Zealand's organic sector · New Zealand's organic production · Organics in New Zealand · New Zealand's organic exports

T. Epps (✉) · D. Wheeler
Chapman Tripp, Wellington, New Zealand
e-mail: tracey.epps@chapmantripp.com; danae.wheeler@chapmantripp.com

GOH Bee Chen, R. Price (eds.), *Regulatory Issues in Organic Food Safety in the Asia Pacific*, https://doi.org/10.1007/978-981-15-3580-2_14

Glossary of Acronyms

DDT	Dichloro-diphenyl-trichloroethane
FSANZ	Food Safety Australia and New Zealand
MPI	Ministry for Primary Industries
OMARS	Overseas Market Access Requirements
CPTPP	The Comprehensive and Progressive Agreement for Trans-Pacific Partnership
TPA	Third Party Agency

14.1 Introduction

New Zealand's organic food sector has grown rapidly in recent years. Between 2015 and 2018 the total organics sector grew by 10% a year, with production reaching NZD $600 million in 2018.[1] Domestic retail sales totalled NZD $245 million in 2018 and grew at twice the rate of conventional products with a 2017 survey indicating that 72% of New Zealand consumers buy organic foods/beverages most or all of the time.[2] Yet while New Zealand's organic exports increased tenfold over the past decade and totalled NZD $355 million in 2018,[3] this remains small in relation to a global organic food sector worth €85 billion[4] and growing at over 10% per year.

Despite this market growth, it was only in December 2018 that the Government formally decided to develop a mandatory national standard for organic products produced and available in New Zealand.[5] Until this standard is developed, organic products must comply with a number of overarching laws that apply to non-organic products, including the Food Act 2014, the Animal Products Act 1999, the Wine Act 2003 and the Fair Trading Act 1986. In addition, there are a variety of private organic standards.

The focus of this chapter is on the history and development of organic food production in New Zealand. It begins by briefly examining early developments before turning to the regulation of food production in New Zealand and its application to organic products. The chapter then explores various aspects of the voluntary regime, such as the involvement of private certifiers and application of private standards. It concludes by discussing the key factors behind the Government's decision to consult on the introduction of a mandatory domestic organics standard.

[1] OANZ Market Report 2018 at 16.

[2] Colmar Brunton "Sustainable Development Goals" (2017) at 26.

[3] OANZ Market Report 2018 at 17.

[4] Ibid., at 5.

[5] The consultation was concerned with primary and processed products, including food and beverages, animal and plant products, live animals and wool.

14.2 What Do We Mean by "Organics"

This chapter discusses "organics" with reference to agricultural production and food. Definitions of "organic" differ internationally so the standard definition from the Codex Alimentarius is adopted which states that:

> Organic agriculture is a holistic production management system which promotes and enhances agro-ecosystem health, including biodiversity, biological cycles, and soil biological activity. It emphasises the use of management practices in preference to the use of off-farm inputs, taking into account that regional conditions require locally adapted systems. This is accomplished by using, where possible, agronomic, biological, and mechanical methods, as opposed to using synthetic materials, to fulfil any specific function within the system.[6]

Organic farming combines techniques such as on-farm nutrient recycling, composting, mulching and biological pest control. The use of synthetic fertilizers and synthetic pesticides, herbicides, plant growth regulators, livestock feed additives, and genetically modified organisms is excluded. Emphasis is put on animal welfare (for example, animals must be able to access pastures), and ensuring that products are not made with the use of genetically-modified organisms. In practice, many organic methods are derived from traditional conservation-based farming techniques, such as the use of crop diversity and rotations.

Organic production can also include adherence to certain ethical principles, for example those relating to health, ecology, fairness and care.[7] In New Zealand, traditional kaupapa Māori growing practices represent a distinct indigenous approach to organic production, which is where the overview begins.

14.3 History and Development of the Organic Industry

Agricultural production in New Zealand was de facto organic before the twentieth century. Practising organic agriculture was the only option for farmers before the advent of chemically synthesised fertilisers, biocides, medicines, mechanisation and fossil fuels needed for industrial scale agriculture. Hence farmers had little choice but to adapt to local biological and ecological conditions.[8]

Prior to the arrival of Europeans, Māori had a staple diet of seafood and birds for protein, and cultivated crops that their Polynesian ancestors brought with them.[9] Successfully introducing Polynesian root crops required skills in plant husbandry

---

[6] FAO/WHO Codex Alimentarius Commission, 1999.

[7] https://www.ifoam.bio/en/organic-landmarks/principles-organic-agriculture.

[8] Paul Kristiansen "Overview of organic agriculture" in Paul Kristiansen, Acram Taji and John Reganold (eds) *Organic Agriculture: A Global Perspective* (CSIRO, 2006) at 4.

[9] G T Alley and D O W Hall *The Farmer in New Zealand* (Department of Internal Affairs, 1951, Wellington) at 1. These include: kumara; taro; yam; gourd; ti pore; and aute.

and modifying soils to improve conditions for plant growth, particularly to account for New Zealand's seasonal temperature range, as different to the tropical Pacific.[10]

A shift to more intensive horticulture and pastoral production began when domesticated animals, crops and iron tools arrived in New Zealand in the late 18th and early nineteenth century from visiting explorers and traders. For example, when James Cook came to New Zealand in 1769 he brought cabbage, turnips and potatoes to Ngāti Porou in Ūawa (Tolaga Bay).[11] Later that year, Jean François Marie de Surville brought wheat, rice and peas to Doubtless Bay. Four years later, Cook revisited Ūawa and dropped off pigs and potatoes. By the early 1800s Māori were trading in potatoes, pigs, maize and other foodstuffs.[12]

Missionaries also assisted, with Samuel Marsden introducing horses and cattle in 1814 and John Butler bringing the plough 6 years later.[13] Initially, European settlers were reliant on Māori for food supplies as they were unfamiliar with the soils and climate.[14] Sir George Grey was the first governor to actively encourage the development of Māori agriculture along European lines, which he believed would be key to keeping peace between New Zealand's inhabitants at the time.[15] In doing so, he initiated the "sugar and flour policy" which involved providing loans for flour mills, agricultural implements and ships, as well as providing pensions to chiefs.[16] Occasionally the Government would even pay for the services of a full-time agricultural instructor. In a despatch to the Colonial Office in 1847, Sir George Grey wrote that when Māori were actively engaged in farming and owned quantities of produce, implements and mills, their property would "be too valuable to permit them to engage in war".[17] Within 30 years of the arrival of the plough, Māori had moved rapidly from subsistence gardening to highly successful commercial farming. In fact, the first dairy farmers were Māori.[18]

By 1856 the *New Zealander* referred to Māori as "landholders, farmers, graziers, seamen, ship owners, labourers and artisans".[19] Overall, Māori played a crucial role in New Zealand's agricultural exports to the Australian market, making a significant contribution to the Government's customs revenue.[20]

---

[10] Louise Furey "Maori gardening" Science & Technical Publishing 2006 at 10.

[11] Te Ara "Story: Ahuwhenua – Māori land and agriculture" https://teara.govt.nz/en/ahuwhenua-maori-land-and-agriculture/page-1

[12] Te Ara "Story: Māori foods – kai Māori" https://teara.govt.nz/en/maori-foods-kai-maori

[13] Ibid.

[14] Ibid.

[15] R.P. Hargreaves "The Māori Agriculture of the Auckland Province in the Mid-Nineteenth Century" *The Journal of the Polynesian Society* 68 (Brunton, 2017) 61 at 62–63 (1959).

[16] State Services Commission "The Story of the Treaty Part 2" (2005) at 6.

[17] Hargreaves above n 15.

[18] Geoff Kira "A utopia of food systems" New Zealand Land & Food Annual (2017) Vol 2 at 69.

[19] Quoted in G. T. Alley and D. O. W. Hall, The farmer in New Zealand. Wellington: Department of Internal Affairs, 1941 at 22.

[20] State Services Commission "The Story of the Treaty Part 2" (2005) at 6.

14.4 Conventional Farming

Conventional farming increased in prominence as farmers sought to increase yields and make farming more efficient. At the same time, most New Zealand soils lacked the phosphate needed to support the high levels of clover required for intensive grazing.

Successive governments played an increasing role in raising agricultural production. In 1881 the New Zealand Government offered up to £1500 to any firm or individual that could produce 50 tons of sulphuric acid a year.[21] Farmers were encouraged to topdress their pastures with superphosphate (with the base rock phosphate imported from Nauru). This helped usher in the "second grasslands revolution" and confirmed New Zealand's position as Britain's "far-distant stock farm".[22] The exchange of agricultural research and domestic farming techniques with other countries was encouraged.[23]

Farmer prosperity increased rapidly after the Second World War as Britain's demand for food increased,[24] and was further boosted by the wool boom during the Korean War. Sheep numbers soared from 32 million in 1949 to 70 million by 1980 and cattle numbers doubled to around 8 million.[25] As Professor Hugh Campbell writes, "it was the combination of these factors – new technologies, institutional structures to deliver them, state policies in support of increased production and powerful legitimising ideologies – which formed the post-war productivist regime in agriculture".[26] In the 1950s, New Zealand's per-capita income was 88% that of the United States (US). However, this prosperity was almost entirely based on agriculture, which in turn depended on access to British markets.

Following the Second World War, with a surplus of aircraft and trained pilots,[27] aerial topdressing was used to help increase agricultural production, and was often subsidised by the state.[28] By the 1950s and early 1960s New Zealand not only used superphosphate but also organophosphates and organochlorides (Dichloro-diphenyl-trichloroethane (DDT) family) for insect control in crops and pastures.[29] DDT was

[21] The Encyclopedia of New Zealand "Superphosphate" https://teara.govt.nz/en/superphosphate/page-3

[22] Tom Brooking "Past, present and future sustainability challenges for farming" RMupdate Issue 18 (April 2006) at 2.

[23] The following were established: the Department of Scientific and Industrial Research in 1926; Massey Agricultural College in 1927, along with an upgrading of Lincoln University, founded in 1878).

[24] Ibid.

[25] Brooking above n 22 at 2.

[26] Hugh Campbell "Recent Developments in Organic Food Production in New Zealand" Research Report No.1 Department of Anthropology, University of Otago (1998) at 11.

[27] Nick Kim and Matthew Taylor "A tale of two metals" New Zealand Land & Food Annual (2017) Vol 2 at 102.

[28] Campbell above n 26 at 10.

[29] Ibid., at 11.

originally produced in New Zealand to control grass grub and porina moth and 500 tons were applied, mainly on farms, in 1959.[30]

With the increased application of fertilisers, damaging side-effects including soil loss and degradation, water pollution and the loss of native plants and animals were emerging. Questions also surfaced about the use of pesticides. Overseas findings that 80% of DDT remains intact in the soil after 2 years led to investigations into insecticides by the Royal Society of New Zealand. While soil naturally contains cadmium and fluorine, it has a much lower concentration than that found in phosphate rock or the fertiliser made from it. An undesirable side effect of adding superphosphate to soil is that containment is retained in the soil layers and absorbed by organic matter and specific inorganic minerals. After superphosphate is applied, nearly 90% of the cadmium, which is toxic at high levels and can cause kidney and liver damage,[31] becomes bound to surface soils.[32]

As world prices for agricultural products fell and New Zealand developed a large trade deficit in the 1950s, questions about pesticide and chemicals were debated. The Conservation Committee in 1959 asked the Royal Society of New Zealand to investigate the increased use of insecticides, particularly those applied from the air, in response to an incident where a large number of bees were killed following spraying.[33]

Awareness of damage caused by agricultural chemicals to humans and animals was heightened following the release of the book, *Silent Spring*, with its message that "the pollution is for the most part irrecoverable".[34] In 1970 the use of DDT was prohibited on New Zealand farmland, and its sale for other purposes (e.g. borer bombs) was banned in 1989.[35]

14.5 Organic Renaissance in the 1970s

New Zealand organic organisations such as the Humic Compost Society and the Soil & Health Association were formed in the 1940s. The Humic Compost Society educated gardeners on composting and other aspects of organic gardening. Membership was around 2000 in 1944, and peaked at 6000 during the 1950s, before the movement began to lose traction as the government reinforced the message that

---

[30] J.T. Salmon "Report of Conservation Committee to the Royal Society of New Zealand on the Use and Effects of Modern Insecticides" Transactions and Proceedings of the Royal Society of New Zealand 1868–1961 (1959) at 11.

[31] Terry L Roberts "Cadmium and Phosphorous Fertilizers: The Issues and the Science" Procedia Engineering 83 (2014) 52 at 54.

[32] Kim and Taylor above n 27 at 104.

[33] Salmon above n 30 at 11.

[34] Rachel Carson *Silent Spring* (Mariner Book, 40th Anniversary Edition, New York, 2002) at 6.

[35] Ibid.

mass agricultural production was the key to increasing the nation's economic livelihood.[36]

Nevertheless, contact between the Humic Compost Society and the Soil & Health Association in New Zealand and similar groups in Europe[37] helped foster the domestic organic agriculture movement. Many individuals were influenced by debates about chemical usage in Europe and these views were often reinforced by the arrival of post-war European immigrants in New Zealand from countries that had intensively used chemicals and fertilisers for some time.

Interest in issues relating to organic production was raised when the Soil & Health Association published the first of their special issues on Organic Farming in 1963, followed by a second in 1968.[38] Organic gardening at home began to increase in the 1970s in response to growing consumer demand for nature-friendly food, free from toxic and polluting chemicals.

In 1983 New Zealand's first domestic organic certifying agency, BioGro, was founded by the New Zealand Biological Producers and Consumers Society to help ensure high production standards were maintained.

14.6 Organic Sales Today

Since the 1970s, organic food production, domestic consumption and exports have greatly increased. In 1990, Wattie Frozen Foods Ltd. (now known as Heinz Wattie) and the New Zealand Kiwifruit Marketing Board (now Zespri International) began exporting organically grown produce to Japan, the US and Europe. Europe, North America and Australia are the sector's primary export destinations, taking 69% of all exports.[39] Recognising this growth, more than 50% of producers polled in 2018 were interested in transitioning to or learning more about organic production.[40] Demand for organic produce has increased such that one of New Zealand's largest supermarket retailers, Countdown, has switched from conventional to organic Braeburn apples.[41] Yet while sales in both domestic and export sectors have grown rapidly (averaging 10% annually in recent years), New Zealand's exports are small relative to the global demand of around €85 billion.[42]

In 2016, New Zealand and China signed the Mutual Recognition Arrangement for Certified Organic Products. Under the agreement, labelling of all organic food

[36] Campbell above n 26 at 11.

[37] Ibid., at 12.

[38] Ibid.

[39] OANZ Market Report 2018 at 5.

[40] Ibid.

[41] The National Business Review "From hippies to millionaire-organic's rise to popularity" https://www.nbr.co.nz/story/hippies-millionaires-organic-s-rise-fame

[42] OANZ Market Report 2018 at 5.

would be compliant with Chinese certification requirement meaning that both countries mutually recognise each other's certifications for organic food imports. It is expected that exports to China will grow with the reduction in compliance costs, and with the certainty that the arrangement provides New Zealand exporters of organic products.[43] As of February 2020, this agreement is not yet operational. In addition, in February 2020, New Zealand and Taipei reached an agreement on mutual recognition of their regulatory systems governing organic products. This agreement builds on the free trade agreement entered into between the parties in 2013.

It is also expected that organic sales will increase with the signing of the Comprehensive and Progressive Agreement for Trans-Pacific Partnership (CPTPP) which provides that a party must enforce any requirement that it develops relating to the production, processing or labelling of products as organic.[44]

14.7 Regulation of Food Production in New Zealand

New Zealand's food regulatory regime is tightly bound up with international law and, in particular, operates in close concert with Australia. This section describes the regulatory approach taken by New Zealand, with a specific focus on food safety legislation and joint regulation with Australia.

The main legislation governing food production in New Zealand includes the Food Act 2014, the Animal Products Act 1999 and the Fair Trading Act 1986. Other relevant legislation includes the Wine Act 2003.

14.8 Food Act 2014

The Food Act 2014 replaced the Food Act 1981. The early legislation prohibited the sale of food that did not comply with prescribed standards, that was unsound, unfit for human consumption, contaminated, or contained anything or was in contact with anything injurious to health or harmful or offensive. It also prohibited misleading labelling and packaging. The current legislation also focuses on food safety, but is more detailed and takes a *risk-based* approach.[45] There is no mention in the Act of organics. It requires a person who trades in food to ensure that it is "safe and suitable".[46] As such, most provisions relate to food recalls and enforcement. While production is defined to include anything used from which food is derived (includ-

[43] MPI "New Zealand's organic food industry to benefit from new arrangement with China" https://www.mpi.govt.nz/news-and-resources/media-releases/new-zealands-organic-food-industry-to-benefit-from-new-arrangement-with-china/.

[44] CPTPP, Chapter 8 Technical Barriers to Trade, Annex A-G "Organic Products" at 8–38.

[45] Louise Delany *Health Law- A to Z of New Zealand Law* (Thomson Reuters, 2015) at 28.4.4.

[46] Section 14.

ing farming, raising, growing, harvesting, extracting and gathering), no provisions specify how production should be carried out.

Part 2 of the Act contains provisions relating to risk-based measures. The specific requirements under this Part vary depending on the risk classification accorded to the food sector. Under Schedules 1 to 3 of the Act, food sectors are classified into three risk-based classes depending on the risk to the public (high, medium to low, or low risk).

Part 3 of the Act deals with food imported for the purpose of sale, with further requirements set out in Part 5 of the Food Regulations 2015. Requirements for imports are also risk-based. For instance, importers of food for the purpose of sale must be registered or imported through a registered agent. Imported food that is categorised as either of "high regulatory interest" or "increased regulatory interest" requires a food safety clearance before entry and is monitored for specific hazards. The Act also strengthens the government's enforcement powers by introducing new offences and increasing penalties for breaching the Act.[47]

The Food Regulations 2015 implements New Zealand's obligations under the Australia New Zealand Joint Food Standards Agreement by giving lawful effect to joint food standards. The Minister may adopt a joint food standard for food manufactured or prepared for sale or sold in New Zealand, or imported into, or exported from, New Zealand.[48]

The Minister also has power under the Act to issue domestic food standards if standards have been or are being developed under the Australia New Zealand Food Standards Agreement for inclusion in the Australia New Zealand Food Standards Code, but New Zealand has chosen to opt out of the current standards.[49]

14.9 Other Legislation

Other legislation relevant to the regulation of food in New Zealand includes the Animal Products Act 1999 and the Wine Act 2003.

The Animal Products Act 1999 regulates the "production and processing of animal material" and animal products in order to manage associated risks and facilitate overseas market access. The Animal Products Act requires all animal products traded and used (including those exported from New Zealand) to be "fit for intended purpose", meaning that they must meet New Zealand animal product standards contained in Part 1 of the Animal Products Regulations 2000. The legislation manages physical, biological and chemical hazards that might present a risk, irrespective of where in the production or processing chain they occur (from production, processing and distribution to market). Risk-factors include risks from hazards to animal or

[47] The maximum penalties are now NZD $100,000 for individuals and NZD $500,000 for companies.

[48] Section 397(Kristiansen, 2006).

[49] Section 404(Organics Aotearoa New Zealand: New Zealand organic report 2018, 2018).

human health, risks from false or misleading labelling and risks to the wholesomeness of animal material or product.[50]

The Wine Act 2003 sets standards for "identity, truthfulness in labelling, and safety of wine", and provides for "minimising and managing risks to human health arising from the making of wine".

The Food Act 2014 recognises the general equivalence of food safety under the Animal Products Act 1999 and the Wine Act 2003. The Food Act 2014 applies to an animal product that is processed under the Animal Products Act 1999 if that animal product is also a food, and wine that is made under the Wine Act 2003, because wine is defined as a food under the Food Act. Similarly to the Food Act, neither the Animal Products Act nor the Wine Act make specific mention of organic production.

In addition to food safety legislation, New Zealand has various pieces of consumer protection legislation that are relevant to the sale of food. The Fair Trading Act 1986 requires that claims and labels must be true, not misleading, and able to be substantiated.[51] This has relevance to organics as it makes it an offence to make misleading claims as to the organic nature of a product. This means that if products are labelled or marketed as "organic", there needs to be information to substantiate that it is. Similarly, if products are "certified organic", it is important that certification is available if required; in 2006, a butcher was prosecuted for making such a false claim.[52] Overall though, this is of limited effectiveness without clear and agreed standards as to what constitutes "organic".

14.10 Trans-Tasman Food Safety

An agreement between New Zealand and Australia for a Joint Food Standards System came into force in 1996 and was updated in 2000. The agreement's objectives include reduced barriers to trade, and the adoption of a joint system for developing and promulgating food standards.

Food Safety Australia and New Zealand (FSANZ) is responsible for developing the joint food standards, known as the Australia New Zealand Food Standards Code (the Code) and operates under the Food Standards Australia New Zealand Act 1991. These standards are enforced by Australian state and territory governments, the

[50] The Animal Products Act 1999, s 4 (risk-factors).

[51] Section 10 of the Fair Trading Act 1986 says that no person shall, in trade, engage in conduct that is liable to mislead the public as to the nature, manufacturing process, characteristics, suitability for a purpose, or quantity of goods. Section 13(a) of the Fair Trading Act relates to false or misleading representations that goods are of a particular kind, standard, quality, grade, quantity, composition, style or model, or have had a particular history or particular previous use.

[52] New Zealand Commerce Commission "Fake organics land butcher with meaty $10,000 fine" https://comcom.govt.nz/news-and-media/media-releases/archive/fake-organics-land-butcher-with-meaty-$10,000-fine

Commonwealth Department of Agriculture (for imported food), and the New Zealand Government.

Parts one and two of the Code apply in New Zealand and regulate the use of processing aids, colourings, additives, vitamins and minerals. They also address the composition of some foods and include standards for genetically modified foods. There are no specific standards for organic production.

In Australia, FSANZ also sets primary production and processing standards, and maximum residue limits for agricultural and veterinary chemicals. In New Zealand, the Ministry for Primary Industries (MPI) is responsible for food policy and legislation.

The FSANZ Board decides whether or not to approve changes to the Food Standards Code and then notifies the Australian and New Zealand ministers responsible for food regulation (the Australia and New Zealand Ministerial Forum on Food Regulation). The forum can adopt, make changes to or reject standards and can ask FSANZ to review its decisions.

14.11 Labelling

Food labelling requirements in New Zealand are contained in the Australia New Zealand Food Standards Code. FSANZ is responsible for the labelling of packaged and unpackaged food, including mandating warnings or advisory levels. As for any food, organic products must be labelled to comply with the Code.

14.12 Laws That Govern Organics

There are no laws in New Zealand that govern the use of the term "organic" or impose specific standards for organic production. As noted above, such products must meet the same food safety and consumer standards that apply to all food for sale in New Zealand.

This contrasts with regulated markets where the term "organic" cannot be used unless production methods and labelling comply with the national requirements. For example, in regulated markets, claims such as "100% organic", "organic", or "made with organic ingredients" reflect strict composition and production requirements. In a similar manner, the use of national organic logos is also strongly regulated in such markets. In the US, for example, only products with more than 95% organic ingredients may use the organic seal.[53] In 2018, 93 countries had mandatory

[53] MPI Discussion Paper "Would New Zealand benefit from new organic regulation?" (August 2018) at 5.

requirements for organic production and a further 16 countries were in the process of drafting mandatory requirements.[54]

As discussed below, in December 2018 the New Zealand Government decided to develop a national standard for organic production. The next section focuses on the voluntary regime in place as at July 2019.

14.12.1 Voluntary Standards – Government (New Zealand Standard 8410)

A domestic standard (New Zealand Standard 8410 – Organic Production) was developed by Standards New Zealand in consultation with industry representatives in 2003. It outlines minimum requirements for the production, handling, processing and labelling of organic products.

This standard had not had significant uptake. In a 2018 discussion paper[55] MPI noted that it was not aware of any businesses or certification bodies currently using the standard, and suggested that reasons for the lack of uptake may include:

- the standard is not accredited by the International Federation of Organic Agricultural Movements (IFOAM) unlike other organic standards available in New Zealand (being accredited to IFOAM helps an organic standard to have international recognition);
- certification bodies are not currently accredited to assess organic products against the standard. Certification bodies incur fees to become accredited, and without market demand it is likely they did not consider this accreditation economically worthwhile; and
- the standard is not recognised as providing equivalent outcomes by New Zealand's trading partners (such as Europe, the US and Canada).

14.12.2 Voluntary Certifiers – Private

Several private organisations assess compliance with private organic standards, including their own standards, and provide certification services to overseas standards (e.g. Japanese Agricultural Standard, Pacific Organic Standard, Canadian Organic Regime and USDA Organic Regulations). In some cases, businesses choose to 'self-certify', which means they assess their own compliance with a private

---

[54] Beate Huber and others *The World of Organic Agriculture. Statistics and Emerging Trends* 2019 in Helga Willer and Julia Lernoud (eds) (Bonn Research Institute of Organic Agriculture (FiBL), Frick, and IFOAM – Organics International) at 32.

[55] MPI above n 53 at 6.

standard.[56] The main private certifying agencies are: BioGro Organic Standards; AsureQuality Organic Standard; Demeter; Hua Parakore and OrganicFarmNZ. New Zealand's two largest certifiers, BioGro and AsureQuality Limited, assist producers to meet international organic regulations in overseas markets.

BioGro Organic Standards BioGro is New Zealand's largest certifier for organic products and is owned by the Soil and Health Association. Use of the BioGro logo provides assurance that a product is made without animal testing, genetic modification or pesticides. The BioGro certification is internationally recognised and accredited. BioGro also assists producers to meet international organic regulations in overseas markets such as in Europe, Canada, the US and parts of Asia.

AsureQuality Organic Standard AsureQuality is a state-owned enterprise owned by the New Zealand Government. It was formed in 2007 from the merger of ASURE New Zealand Limited and AgriQuality Limited, companies originally formed in 1998 out of the service delivery arm of the Ministry of Agriculture and Forestry, MAF Quality Management.[57] The AsureQuality Organic Standard provides a minimum requirement for producers to comply with in order to gain organic certification for the production of, labelling of, and claims for organically produced foods.[58] AsureQuality also certifies to overseas regulated organic markets, as well as non-regulated organic export markets.

Demeter Demeter is an international certification system operated by the Bio Dynamic Farming and Gardening Association that registered Demeter as a certification trademark in 1984. Demeter developed its own standard which combines biodynamic methods and key organic principles.[59]

Hua Parakore Hua Parakore is an indigenous food verification system based on tikanga Māori. It has been described as a system that "verifies food and products from a kaupapa Māori perspective" to ensure that the food "enhances Papatuanuku

[56] Ibid., at 7.

[57] In 1998 MAF was restructured for the seventh time in a decade, two new State Owned Enterprises being spun off – AgriQuality NZ Ltd. and Assure NZ Ltd. AgriQuality had jurisdiction over quality assurance services for a wide range of foods while Assure was charged with auditing compliance.

[58] The AsureQuality Organic Standard can certify products to gain access to certain regulated and non-regulated organic export markets. Regulated markets include certification to the National Standard of Canada, whereas non-regulated organic export markets include Australia and Fiji.

[59] Inspectors visit the operators (farmers and processors) annually to collect information about their methods. In preparing for the inspection, the farmer needs to supply a farm management plan, an annual report, declarations from merchants and other farmers from whom supplies have been obtained and a farm profile detailing the farm physical characteristics such as soil type. A committee of assessors, most of whom are experienced biodynamic farmers and/or processors, then decides whether to grant certification.

(mother earth) and doesn't deplete her".[60] Hua Parakore ensures that the product purity and integrity is congruent with Māori cultural practices, such that it is free from genetic modification, chemicals and pesticides. Hua Parakore is administered by Te Waka Kai Ora, the national Māori organic organisation.

OrganicFarmNZ OrganicFarmNZ runs an organic certification scheme which is designed to be a low-cost option for local growers supplying the New Zealand market. It has around 130 producer members, with the average producer's holding being ten hectares in size. OrganicFarmNZ is best known for its "pod" system, where a group of between 3 and 5 growers review one another. Once the peer review process is complete, an independent auditor will review the paperwork of each producer. The auditor will audit/inspect one property in every four pod members each year.[61]

14.12.3 Export of Organic Products

The Official Organic Assurance Programme (OOAP) facilitates the export of organic products to Europe, Japan, Switzerland, Taiwan and the US. Other countries, such as Liechtenstein, Norway and Iceland take assurances from MPI based on New Zealand having agreements with the European Union (EU).

Organic products usually covered by the OOAP include:

- processed plant and animal products for human or animal consumption (for example, dairy products, wine and fruit juices); and
- unprocessed plant products, animal products and animals (for example, fruit and vegetables, seeds and unprocessed raw wool).

In order to export organic products, or supply them for export, a producer must fulfil various requirements, including: registering as an organic operator with an MPI-recognised third party agency (TPA); developing an organic management plan and having it verified by the TPA; ensuring all operators involved in producing the product are covered by the MPI organic standard OP3; and applying to MPI for an official organic assurance for their consignments.

TPAs are an organic certifying bodies recognised by MPI to audit and verify that: organic products exported under the OOAP are produced in accordance with the Overseas Market Access Requirements (OMARs) of the destination country; as well as MPI organic standard OP3 and the associated Technical Rules for Organic Production (Technical Rules). TPAs can also issue export certificates for markets where they have the authority to do so. MPI recognises AsureQuality and BioGro New Zealand Limited as TPAs that audit organic operators.

[60] "Tikanga Māori food verification system extends to other cultures", 27 November 2017, https://www.maoritelevision.com/news/education/tikanga-maori-food-verification-system-extends-other-cultures

[61] OrganicFarmNZ "OFNZ Farmers" https://www.organicfarm.org.nz/ofnz-farmers

Standard OP3[62] sets out the responsibilities and contains the outcomes for registered operators participating in the OOAP in order to receive and maintain registration, and outlines criteria for Organic Management Plans (OMPs). Standard OP3 applies in conjunction with the Technical Rules[63]; Standard OP1 ("Accreditation, Recognition, and Performance Measure Criteria for Third Party Agencies and their Personnel – Organic Products")[64]; and Standard OP2 ("Third Party Agency Responsibilities – Organic Products").[65]

All exports from New Zealand adopt the Technical Rules, which means those rules are seen as the main reference point for what is required for organic production. In the absence of a mandatory New Zealand standard, the Technical Rules are the benchmark against which TPAs undertake verification and make recommendations on official assurances on organic products intended for export. The Technical Rules were originally developed in 2000 on the basis of the relevant EU Regulation, and were subsequently amended to incorporate United States National Organic Standard requirements.

The OOAP scheme is reported to be very successful. For the year ending June 2017, the value of exports facilitated through OOAP was NZD $151.4 million, an increase of 19% from the previous year. Exports to the US were NZD $86.4 million and NZD $60 million went to the EU.[66]

For other markets, exporters often work with in-market affiliates of the Organic Exporters Association to ensure they meet the organic regulatory requirements of the destination country.

14.13 Decision to Consult on Law Reform

"With no government oversight of organic standards, and no national standard to protect the word 'organic', the integrity of the organic industry as a whole is compromised".[67]

---

[62] Registration and Performance Measurement Criteria for Operators – Organic Producers (NZFSA Standard OP3), August 2003 Version Two, available at: https://www.mpi.govt.nz/dmsdocument/593-ooap-standard-op3-registration-and-performance-measurement-criteria-for-operators

[63] Technical Rules for Organic Production, MAF Standard OP3, Appendix Two, Version 7.1, availableat:https://www.mpi.govt.nz/dmsdocument/596-ooap-standard-op3-appendix-2-technical-rules-for-organic-production

[64] NZFSA Standard OP1, August 2005, Version Two, available at: https://www.mpi.govt.nz/dmsdocument/120-ooap-standard-op1-accreditation-recognition-and-performance-measurement-criteria-for-third-party-agencies-and-their-personnel

[65] NZFSA Standard OP2, August 2005, Version Two, available at: https://www.mpi.govt.nz/dmsdocument/121-ooap-standard-op2-third-party-agency-responsibilities

[66] OANZ Market Report 2018 at 21.

[67] Soil & Health Submission on MPI Discussion Paper NO: 2018/19 "Submission on an organic standard in New Zealand" 11 June 2018.

New Zealand has an international reputation for being "clean and green".[68] However, New Zealand is one of only two countries in the world's top 25 organic markets (by value) that have not developed mandatory domestic organic standards (the other is Australia). Growing concerns from industry highlighted the lack of a mandatory standard as a challenge facing New Zealand's organic sector.[69] Recognising the impediment this presents to the growth of New Zealand's organic market, the government is now in the process of legislating mandatory requirements.

There are several disadvantages in not having a universally regulated organic certification standard for New Zealand.[70] These include: misuse of the term 'organic' to describe non-organic products; issues with equivalency negotiations in international trade; lack of consumer confidence in the domestic and international organic markets; confusion arising from multiple certifications as there are several different organic certification agencies in New Zealand with their own standards and labels[71]; and multiple, expensive export levies required to export organic products.

For example, nearly 80% of New Zealanders report that they purchase organic foods fortnightly, yet in a recent sample only 7% were able to identify all aspects of an organic product.[72] Consumers are uncertain as to which assurances are reliable, and only 5% of organic purchasers rely on a certification mark to determine if a product is organic.[73] Instead, 46% either simply read the label, or read the packaging to determine whether the product is organic.[74]

A universally regulated standard in New Zealand would bring several benefits. There are three key areas to reform. First, domestic consumer protection, as it is important to ensure that consumers are able to identify genuine organic products. Second, business certainty, as it is also important that New Zealand businesses are not overburdened with regulation. While some markets recognise MPI's oversight role, they require MPI to apply their standard as opposed to the New Zealand technical rules. In practice, this often requires certification of the "whole process" such that even the grower is audited.[75] Third, a universally regulated standard would put New Zealand in a better position to negotiate market access. Overseas trading partners expect New Zealand to have a comparable organic regulatory regime therefore mandating a requirement for organics would increase the credibility of New Zealand's organic produce.[76] Accordingly, in May 2018 MPI published a submission paper to gather feedback on a proposal to regulate organic production in New Zealand.

---

[68] Ministry for the Environment "Valuing New Zealand's clean green image" August 2001.

[69] Primary Production Committee "Briefing from Organics Aotearoa New Zealand" (April 2016).

[70] New Zealand House of Representatives Report of the Primary Production Committee "Briefing from organics Aotearoa New Zealand" (31 March 2016) at 2.

[71] Soil & Health Submission on MPI Discussion Paper above n 53.

[72] OANZ Market Report 2018 at 9.

[73] Ibid., at 10.

[74] Ibid.

[75] Organic Agriculture, http://www.organicag.co.nz/certification/

[76] New Zealand House of Representatives above n 70.

MPI stated that New Zealand's regime for organic products should meet the following objectives: increase customer confidence about the production of organic goods; provide businesses with regulatory certainty to invest in organic products; endeavour to enable effective trade in organic products; increase flexibility in the regulatory regime while also ensuring that it is effectively understood and administered; and ensure that the costs to business and consumers are proportionate to the overall benefits.

MPI proposed a mandatory standard for all relevant businesses. Ongoing verification would apply to most businesses, although a limited number would not be required to be verified, as is common in overseas regimes. These businesses would still need to meet the standard, but the exception would allow a reduction in disproportionate compliance costs for a small category of businesses. Enforcement action could still be taken in respect of these small businesses, if needed, when they claim that their products are organic.

In response to the discussion paper, MPI received 28 submissions.[77] Eighty-five percent of respondents supported a change in the way organics are regulated. The key factors cited were the opportunity to increase consumer confidence and support trade. Seventy-six percent of submitters supported a mandatory standard.[78]

On 27 February 2020 an Organic Products Bill was introduced to the New Zealand Parliament. This Bill will allow the development and introduction of National Organic Standards and methods to ensure that these requirements are met.

14.14 Conclusion

While the focus of this chapter has been on organic food production, there is growing global demand for non-food products that conform to organic farming practices. This includes the organic and natural beauty products industry, and the organic textile and organic aquaculture industries. Key factors driving this demand are increasing consumer awareness, rising disposable incomes, and the expansion in distribution channels that has also been facilitated by the growing e-commerce industry.[79]

Prospects for the continued rapid growth of the organic food industry look promising. A key element in realising this potential will be the opportunities that are expected to open up when New Zealand adopts mandatory national organic standards and the mechanisms to ensure their successful implementation.

[77] MPI "Would New Zealand benefit from new organic regulation?" Summary of submissions MPI Information Paper No: 2018/05 at 2.

[78] Ibid.

[79] PRNewswire "Organic Skin Care 2018: Global Industry Forecasts to 2024 - CAGR Expected to Grow at 10%" 25 May 2018 https://www.prnewswire.com/news-releases/organic-skin-care-2018-global-industry-forecasts-to-2024%2D%2D-cagr-expected-to-grow-at-10-300654926.html

References

Alley GT, Hall DOW (1951) The Farmer in New Zealand. Department of Internal Affairs, Wellington
Brooking T (2006, April) Past, present and future sustainability challenges for farming RMupdate issue 18
Brunton C (2017) Sustainable development goals
Campbell H (1998) Recent developments in organic food production in New Zealand research, Report No.1, Department of anthropology, University of Otago
Carson R (2002) Silent spring. Mariner Book, New York. 40th Anniversary Edition
Delany L (2015) Health law – A to Z of New Zealand law. Thomson Reuters
FAO/WHO Codex Alimentarius Commission, 1999 (1999)
Furey L (2006) Māori gardening science and technical publishing
Hargreaves RP (1959) The Māori agriculture of the Auckland province in the mid-nineteenth century. J Polynesian Soc 68(2):61
Huber B (2019) The world of organic agriculture statistics and emerging trends 2019. Bonn Research Institute of Organic Agriculture (FiBL), Frick, and IFOAM – Organics International, Frick
Kim N, Taylor M (2017) A tale of two metals. New Zealand Land & Food Annual 2
Kira G (2017) A utopia of food systems New Zealand Land & Food Annual 2
Kristiansen P (2006) Overview of organic agriculture in Paul Kristiansen. Acram Taji and John Reganold (eds) Organic agriculture: a global perspective. CSIRO, Victoria
Māori Television. Tikanga Māori food verification system extends to other cultures., https://www.maoritelevision.com/news/education/tikanga-maori-food-verification-system-extends-other-cultures. Accessed 10 February 2019
Ministry for Primary Industries. New Zealand's organic food industry to benefit from new arrangement with China. https://www.mpi.govt.nz/news-and-resources/media-releases/new-zealands-organic-food-industry-to-benefit-from-new-arrangement-with-china/ Accessed 5 Feb 2019
Ministry for Primary Industries Discussion Paper (2018) Would New Zealand benefit from new organic regulation?
Ministry for the Environment (2001) Valuing New Zealand's clean green image August 2001
New Zealand Commerce Commission. Fake organics land butcher with meaty $10,000 fine https://comcom.govt.nz/news-and-media/media-releases/archive/fake-organics-land-butcher-with-meaty-$10,000-fine. Accessed 10 May 2019
New Zealand House of Representatives Report of the Primary Production Committee: Briefing from organics Aotearoa New Zealand (2016)
NZFSA Standard OP1, August 2005, Version Two (2005a)
NZFSA Standard OP2, August 2005, Version Two (2005b)
Organic Agriculture. http://www.organicag.co.nz/certification/. Accessed 20 May 2019
OrganicFarmNZ. OFNZ Farmers. https://www.organicfarm.org.nz/ofnz-farmers. Accessed 20 February 2019
Organics Aotearoa New Zealand (2018) New Zealand organic report 2018
Primary Production Committee (2016) Briefing from Organics Aotearoa New Zealand
PRNewswire. Organic skin care 2018: global industry forecasts to 2024 – CAGR expected to grow at 10% https://www.prnewswire.com/news-releases/organic-skin-care-2018-global-industry-forecasts-to-2024%2D%2D-cagr-expected-to-grow-at-10-300654926.html. Accessed 2 May 2019
Registration and Performance Measurement Criteria for Operators – Organic Producers (NZFSA Standard OP3), August 2003 Version Two (2003)
Roberts T (2014) Cadmium and phosphorous fertilizers: the issues and the science. Procedia Eng 83:52–59
Salmon JT (1959) Report of conservation committee to the Royal Society of New Zealand on the use and effects of modern insecticides transactions and proceedings of the Royal Society of New Zealand 1868–1961

Soil & Health (2018) Soil & Health Submission on MPI Discussion Paper NO: 2018/19 "Submission on an organic standard in New Zealand" 11 June 2018

State Services Commission 2005 The story of the treaty part 2

Te Ara. Story: Ahuwhenua – Māori land and agriculture https://teara.govt.nz/en/ahuwhenua-maori-land-and-agriculture/page-1. Accessed 10 Mar 2019

Te Ara. Story: Māori foods – kai Māori. https://teara.govt.nz/en/maori-foods-kai-maori. Accessed 10 Mar 2019

Technical Rules for Organic Production (n.d.) MAF Standard OP3, Appendix Two, Version 7.1

The Comprehensive and Progressive Agreement for Trans-Pacific Partnership (n.d.) Chapter 8 Technical Barriers to Trade, Annex A-G "Organic Products"

The Encyclopedia of New Zealand "Superphosphate". https://teara.govt.nz/en/superphosphate/page-3. Accessed 10 Mar 2019

The National Business Review. "From hippies to millionaire-organic's rise to popularity." https://www.nbr.co.nz/story/hippies-millionaires-organic-s-rise-fame. Accessed 20 May 2019

Tracey Epps is a Trade Law Consultant at Chapman Tripp in Wellington, New Zealand. Previously, she was a Senior Advisor in the New Zealand Ministry of Foreign Affairs (*MFAT*) Legal Division. At Chapman Tripp, she advises government and private sector clients on all aspects of international trade and investment law. She has authored and edited a number of books and articles on topics related to international economic law, including climate change and international trade, subsidies, sanitary and phytosanitary measures and technical barriers to trade. She teaches International Trade Regulation and International Investment Law in the Faculty of Law at the University of Otago. She is Chair of the International Trade Committee in the Inter-Pacific Bar Association and regularly attends international conferences in the field of international economic law.

Danae Wheeler is a solicitor at Chapman Tripp in New Zealand. She has experience advising clients on arbitration and litigation proceedings and international trade and investment law. Having lived and studied in six countries over the past 8 years, she has a strong interest in how organisations operate in different cultural and legal settings. Before joining Chapman Tripp, she completed honours degrees at The Chinese University of Hong Kong (LLM in Chinese Business Law), Melbourne Law School (JD), and New York University (BA). In 2018, she was awarded the AMINZ-AUT Arbitration Young Practitioner Scholarship. She wrote her master's thesis on the enforceability of arbitral awards under the New York Convention in the People's Republic of China. Her undergraduate thesis explored the relationship between the rule of law and economic development, for which she was awarded the European and Mediterranean Studies Award at New York University. She speaks Spanish, Mandarin and Cantonese.

Chapter 15
Canadian Organics: Enhancing Food Safety and Trade Growth Through Regulatory Harmonization and International Collaboration

Alexander C. Phillips

Abstract This chapter outlines what *organic* means in Canada and how domestic and international consumers can recognize and trust that organic food, seed and animal feed products both within and from Canada are healthy and safe.

Specifically, it discusses the regulatory requirements that producers must meet for their products to be certified as *organic* by Canada. This includes recognized principles of sound organic farming and supply systems covering production, processing, storage, transportation, labelling and marketing. It also discusses how importers of organics can meet Canadian regulatory standards by accessing accredited certification bodies recognized by Canada or by utilizing an existing mutual recognition organic *equivalency agreement* with Canada.

The chapter then discusses the growing demand for Canadian organics domestically and internationally. Specifically, it assesses the growing demand for certain organic products from Canada and the reasons for it. In short, the chapter outlines Canada's efforts to enhance its capacity to enhance organic agri-food trade opportunities by promoting long-term relationships that include reducing trade impediments through reciprocity, harmonizing food safety regulatory regimes, instituting transparent organic certification standards and investing in world-class cold supply chain distribution systems.

By doing so, the chapter explores the relationship between consumer confidence and food security from a Canadian perspective as an open and trustworthy trading nation.

I would like to sincerely thank Mehrnaz Ashrafi, Sustainability Analyst with the Van Horne Institute for her research summaries, insights and commitment to this collaborative work. Research funding gratefully acknowledged from VersaCold Logistics Services, one of Canada's leading cold supply chain distribution companies.

A. C. Phillips (✉)
Van Horne Institute, Calgary, AB, Canada
e-mail: aphillips@vanhorneinstitute.com

GOH Bee Chen, R. Price (eds.), *Regulatory Issues in Organic Food Safety in the Asia Pacific*, https://doi.org/10.1007/978-981-15-3580-2_15

Keywords Canada · Organic logo · Standards · Trade collaboration

Glossary of Acronyms

BCCOP	British Columbia Certified Organic Program
CARTV	Conseil des appellations réservées et des termes valorisants (Quebec)
CB	Certification Body
CFIA	Canadian Food Inspection Agency
CGSB	Canadian General Standards Board
COABC	Certified Organic Association of British Columbia
COO	Canada Organic Office
COR	Canada Organic Regime
COS	Canadian Organic Standards
COTA	Canada Organic Trade Association
CVB	Conformity Verification Body
ISED	Department of Innovation, Science and Economic Development, Canada
OCO	Organic Council of Ontario
OFC	Organic Federation of Canada
OSIC	Organic Standards Interpretation Committee
OTA	Organic Trade Association (United States)
SCC	Standards Council of Canada
SFCA	*Safe Food for Canadians Act*, S.C. 2012, c.24
SFCR	*Safe Food for Canadians Regulations*, SOR/2018-108
TAC	Technical Advisory Committee (TAC)

15.1 Introduction

Canada's large and fertile land base, combined with its cooler climate, makes it an ideal place for producing organic food, seed and animal feed. Canada is also diversifying its organic products and is expanding its global markets. Both product diversification and market expansion have led to increasing annual growth in the production and sales of organic food.

As consumer demand for organic food grows, the need for a consistent system of certifying farming and processing operations that are compliant with recognized and trusted regulatory standards becomes more pressing to ensure that reliable, safe and healthy food meets customer expectations. Organic certification is, indeed, a consumer's guarantee that all food products that use the term *organic* are, in fact, produced and processed in a healthy and safe way.

In order for a food product to be certified by Canada as *organic*, all producers, processors and supply chain handlers must meet the legislative requirements set out in the *Canadian Organic Standards* (COS), must apply to a *Certification Body* (CB) accredited by the *Canadian Food Inspection Agency* (CFIA), must be able to show complete traceability of their products and must be open to inspection by an independent third party. Once a farmer or business's product is so certified, that organic operator is then permitted to use both the term *organic* and the *Canada organic logo* in its product's labelling and marketing. The *Canada organic logo* can be recognized as follows:

All organic operators are also inspected annually, by a verification officer appointed by a CFIA-accredited CB, to ensure that they continue to meet the COS.

Long-considered a *niche* industry, the Canadian organics sector is currently expanding at a much faster rate than conventional agricultural and agri-food industries and it is attracting increasing domestic and international investment. This growth is expected to continue well into the future as the organics sector in Canada becomes more organized, trade impediments are reduced or eliminated, and both domestic and international consumer demand increases due to greater public confidence in reliably safe and healthier *organic* food choices being available in and from Canada due to continuing efforts to implement *mandatory certification* to consistent *minimum standards* of food safety.

With this emerging trend in mind, an overview of Canada's organic agri-food regulatory regime is summarized below, in terms of organic principles and definition used, organic regulation and standards followed (including *equivalency agreements*) and organic food labelling required. In addition, the demand potential for enhanced organic agri-food trade with China is then discussed.

15.2 Organic Definition and Production Principles in Canada

The term *organic* refers to the way agricultural products are produced (i.e. grown) and processed.

Organic farming is a production system that focuses on sustaining a healthy ecosystem, including soils, plants, animal, air and water, as well as human well-being, and that has an added constraint – a prohibition on the use of synthetic chemicals. As defined in the COS pursuant to its General Principles and Management Standards, *organic production* is a holistic system designed to optimize the productivity and fitness of diverse communities within the agro-ecosystem, including soil organisms, plants, livestock and people (Government of Canada 2013a).

In addition, although organic production may look different depending on the ecosystem where the organic farming is located, as well as how it is being managed, *organic production* should also be based on the following *four principles*:

Health – Organic agriculture should sustain and enhance the health of soil, plants, animals, humans and the planet as one and indivisible.
Ecology – Organic agriculture should be based on living ecological systems and cycles, work with them, emulate them and help sustain them.
Fairness – Organic agriculture should build on relationships that ensure fairness with regard to the common environment and life opportunities.
Care – Organic agriculture should be managed in a precautionary and responsible manner to protect the health and well-being of current and future generations and the environment (Government of Canada 2013a).

These organic principles seek to increase the quality and the durability of the environment through specific management and production methods that are sustainable and harmonious with the environment. These principles are also ethically interconnected and serve as the roots from which organic farming should grow.

In Canada, an organic product is also legislatively defined as an agriculture or aquaculture product that has been *certified as organic* under Part 13 of the *Safe Food for Canadians Regulations,* SOR/2018-108 (SFCR).

These are the food safety regulations that require Canada's organic products be produced without the use of synthetic pesticides, chemical fertilizers, genetic engineering or irradiation. In addition, animals in the organic food chain must not be fed animal by-products, antibiotics or other growth hormones. Moreover, organically raised animals are required to be raised under specified conditions that promote animal health and welfare, such as providing plenty of outdoor exercise to socialize with other animals.

At the processing level, Canada's organic products must also be free of chemical preservatives and synthetic additives, like colourings and waxes.

Organic farmers, consumers and regulators therefore strive for continuous improvement by adhering to all four of the above organic principles and by strictly complying with Canada's food safety legislation and its national standards outlined in the COS governing the production and processing of organic products.

15.3 Canada Organic Regime (COR)

15.3.1 Organic Food Safety Legislation

The Government of Canada recently strengthened its regulatory oversight of food commodities being traded domestically or internationally by enacting the *Safe Food for Canadians Act,* S.C. 2012, c.24 (SFCA) (Justice Laws 2019) to harmonize and modernize its food safety systems. Regulations covering *organic* food, seed and animal feed were also specifically included in this process. The SFCA, assented to in 2012, fully came into force on January 15, 2019, after the *Safe Food for Canadians Regulations*, SOR/2018-108 (SFCR) were established under this Act.

The SFCR outlines Canada's *organic* agri-food products *certification requirements* in Part 13. Part 13 is designed to protect consumers against false and misleading organic claims and to govern the use of the *Canada organic logo*. Part 13 or the SFCR applies to the inter-provincial and international trade of organic products. However, organic agri-food products sold *intra*-provincially in Canada (i.e. solely within a Province) are not within the scope of Part 13 of the SFCR, unless the product bears the *Canada organic logo*. With respect to the later, provincial organic product certification regulations do exist in some Provinces, including Quebec, New Brunswick, Nova Scotia, Manitoba and British Columbia. As a result, regulatory enforcement can also be a shared jurisdiction between these specific Provinces and the federal *Canadian Food Inspection Agency* (CFIA).

Under the SFCR, any food, seed, or animal feed that is labelled *organic* is federally regulated by the CFIA. The CFIA is responsible for administering Canada's food safety regulations and policy initiatives. The *Canada Organic Regime* (COR) is Canada's national regulated system for organic agricultural products, implemented by the CFIAs *Canada Organic Office* (COO).

The purpose of the COR is to regulate all parties involved in the certification of organic products and to verify that all applicable regulatory requirements, standards and guidelines are being met.

The CFIA is the federal regulatory agency that oversees, monitors and enforces the requirements of the COR using a third-party service delivery model that includes *Conformity Verification Bodies* (CVBs), *Certification Bodies* (CBs), and organic operators working collaboratively.

Specifically, under this model, each of these entities perform the following functions:

First, the CFIA sets the federal rules for the COR; designates and audits CVBs; accredits CBs; and can also establish organic *equivalency agreements* with other countries (see below).

Second, the CVBs assess applications from CBs and submit recommendations for accreditation of CBs to the CFIA. The CVBs also monitor the accredited CBs. There are three CVBs designated by the CFIA.

Third, the CBs verify that organic operators produce organic products in compliance with the *Canada Organic Standards* (COS – see below); issue certificates for organic products, as well as certificates for the activities of packaging and labelling that meet the required standards.

Lastly, the organic operators (who are producing or processing organic products) must comply with both the COS and with any other relevant Canadian legislation, including the *Food and Drugs Act*, RSC 1985, c.F-27, the *Consumer Packaging and Labelling Act*, RSC 1985, c.C-38 and any other commodity-specific requirements, such as those for dairy, eggs and meat. Those looking to be *certified* as an *organic* producer or processor must apply to a CB and be approved before they can use the term *organic* or the *Canada organic logo*.

In all, there are 17 CFIA-accredited CBs within Canada (CFIA 2019a) and over 150 CFIA-accredited CBs outside Canada (CFIA 2019b) that certify organic products under Part 13 of the SFCR.

15.3.2 Canadian Organic Standards (COS)

The CFIA, in partnership with the *Organic Federation of Canada* (OFC), has also created and relies on the *Organic Standards Interpretation Committee* (OSIC) to provide its COR team with legislative interpretations and regulatory guidance on issues related to the COS. Although these national standards are prescribed by the *Canadian General Standards Board* (CGSB), the federal government organization that oversees national standards and their review processes, the COS are written in collaboration with organic industry stakeholders (renewed every 5 years), and are coordinated by the OFC (OCO n.d.).

The COS consist of three national standards, developed by the Standards Council of Canada (SCC). The SCC is Canada's national accreditation body, a federal Crown corporation within the portfolio of Innovation, Science and Economic Development (ISED) Canada which reports to Parliament through the Minister of Industry Canada. With the goal of enhancing Canada's economic competitiveness and social well-being, the SCC oversees Canada's national standards system, leads and facilitates the development and use of national and international standards, coordinates Canadian participation in standards development, and identifies strategies to advance Canadian standardization efforts.

The three national standards in the COS governing organic production, processing and handling systems in Canada include:

15.3.2.1 General Principles and Management Standard

The General Principles and Management Standard (Government of Canada 2013a) outlines the principles and management standards required for farming systems engaged in organic food production. The principles govern the production of high-quality organic food using sustainable management practices that avoid damage to the environment and ensure the ethical treatment of livestock. The standards outline the specific criteria that must be met when food products and other inputs used in organic food production are described as *organic*. These requirements are applied to the production, processing and handling of organically produced foods, and must be met or exceeded in order to meet this standard for organic agriculture.

15.3.2.2 Permitted Substances Lists

The Permitted Substances Lists (Government of Canada 2013b) provides lists of substances that are allowed to be used in organic production systems. This includes crop production (such as fertilizers, plant foods, soil amendments, crop production aids and materials, and weed management); livestock production (such as feed, feed additives and supplements, and health care products and production aids); and processing and sanitation (such as organic ingredients, non-organic ingredients and with organic ingredients, processing aids, cleaners, disinfectants and sanitizers, and pest control substances). Prohibited substances are also listed in the General Principles and Management Standards.

While agriculture products must be certified in accordance with both the General Principles and Management Standard and Permitted Substances Lists, aquaculture products must be certified in accordance with the Aquaculture – General Principles (below).

15.3.2.3 Aquaculture – General Principles, Management Standards and Permitted Substances Lists

The Aquaculture – General Principles, Management Standards and Permitted Substances Lists (Government of Canada 2013c) applies to the following organic products:

(a) unprocessed crop and crop products, livestock and livestock products, to the extent that the principles of production and specific verification rules for them are described in the standard;
(b) processed crop and livestock products intended for human consumption or use and derived from the items mentioned in a);
(c) livestock feed;
(d) processed crop and livestock products intended for animal consumption or use and derived from the items mentioned in a).

However, organic aquaculture products are not currently required to be certified under the SFCR for a 24-month period that will end January 15, 2021, and therefore, are not within the scope of Canada's current equivalency agreements with other countries for organic products. This delay will provide for less market disruption while new requirements are being determined.

15.3.3 Equivalency Agreements with Canada

In order to enhance and facilitate international trade in organic products, and to verify that the organics requirements of an importing country are equivalent to and in compliance with the COR in Canada, an equivalency determination between Canada and that country can be undertaken by the CFIA.

An *equivalency agreement* is a trade arrangement that deems both the foreign country's organic conformity assessment system, as well as its standards, are equivalent to the requirements of Canada's organics regulations and standards. An equivalency determination therefore allows two differing standards, regulations or procedures to remain as is (in each country) but treats them as if they are the same, (i.e. as long as they achieve the same results and policy objectives, regardless of means). Equivalency is determined by assessing and comparing the two regulatory systems, including the standards, to determine whether the principles and outcomes achieved are equivalent. Elements that are assessed by the CFIA include: regulations, accreditation criteria, certification criteria, standards, and monitoring and enforcement activities (CFIA 2019c).

Canada currently has five organics equivalency agreements. The first equivalency agreement was signed in 2009 with the United States – Canada's largest organics trading partner. Subsequent agreements were signed with the European Union (2011), Switzerland (2012), Costa Rica (2013), and Japan (2015) (CFIA 2019c).

The process to reach an equivalency agreement with another country is outlined in the *CFIA Procedure for Organic Equivalency Determination and On-going Monitoring of the Existing Arrangements* (CFIA 2019d). The CFIA verifies the effectiveness of the organic equivalency arrangements by conducting peer reviews, as outlined in the *CFIA Peer Review Procedure Related to the COR* (CFIA 2019e).

A body known as the *Technical Advisory Committee* (TAC), under the COR, is responsible for providing a comparative analysis to the CFIA on Canadian organic production standards and those of foreign countries. As the expertise needed might change depending on what country an equivalency agreement may be negotiated with, TAC members can be selected based on the product areas involved and the expertise required. The TAC therefore serves as a stakeholder forum to ensure that the diverse viewpoints of the organic industry are taken into account in the CFIA's organic equivalency determination process (COTA n.d.-a).

15.3.4 Organic Food Labelling

The use of the word *organic* or the *Canada organic logo* is voluntary. However, whenever it is used, the requirements of the SFCR must be complied with. Specifically, compliance with Part 13 of the SFCR is required if the product: (i) has an *organic* claim on the product label and is sold between provinces or is imported from another country; or (ii) displays the *Canada organic logo* on the label and is sold within or outside of Canada.

Under Part 13 of the SFCR, products must be certified as *organic* according to the COS. An *organic* claim is only permitted on products with 95% or more *organic content* that have been *certified* according to the requirements of the COR.

Terms such as *organically grown*, *organically raised*, or *organically produced*, or similar words, abbreviations of, symbols for and phonetic renderings of these words are considered the same as *organic* claims and must meet the requirements of the SFCR (CFIA 2019f). Terms such as *100% organic* or *100% organic (product name)* is not permitted in Canada. An organic product always bears the name of the CB that has certified it by the SFCR, and thus, the claim *certified organic* is considered misleading as it implies to consumers that products not bearing this claim are not certified. However, terms such as *certified by* or *certified organic by* immediately followed by the name of the CB, or as part of the CB logo, is acceptable.

Furthermore, the claims such as *made with organic ingredients* or *made with organic (naming the ingredient)* are not acceptable, as it is not clear how much of the product is made with organic ingredients (CFIA 2019g).

Multi-ingredient products with 70–95% organic content, may carry the declaration *contains x% organic ingredients*, with the percentage rounded down to the nearest whole number. Unlike in the 95% or greater organic category, non-organic ingredients may be used even if there is a commercially available organic alternative. However, the use of organic and non-organic versions of the same ingredient is not permitted. Multi-ingredient products with less than 70% organic content may not use the organic logo nor use the claims *organic* or *contains x% organic ingredients*, but only carry organic claims in the product's ingredient list.

There are no restrictions as to size or location of the organic logo on the product label. The name of the CBs must appear on the label, anywhere on the product, except the bottom of the container.

All *organic* claimed products must have either been *certified* by an accredited CB or by one which works under a country system which the CFIA has deemed equivalent (i.e. recognized under an organic *equivalency agreement* with a foreign competent authority under the SCFR). Regardless of which standard a product is certified to, organic products must meet domestic labelling requirements (the labelling requirements of the country in which they are sold), such as language requirements, unique nutritional labeling, and different product grades.

The CFIA approves the Conformity Verification Bodies (CVBs) which assess the ability of CBs both inside and outside of Canada to conduct inspection and certification of farms and processors involved in the production of organic products destined to be sold on the Canadian market. CVBs perform assessment of CBs according to ISO/IEC 17065 and the relevant requirements set out in the COR Operating Manual (CFIA 2019h), and then make a recommendation for accreditation to the CFIA, which is the ultimate decision maker. The COR Operating Manual provides an overview of the procedures and policies governing the relationship between the CFIA and CVBs, as well as the procedure to accredit CBs and to certify organic products.

Products labelled as *organic* which are sold *intra*-provincially can be certified to either federal or provincial standards. Among Canada's ten provinces, two provinces British Columbia (BC) and Quebec have organic logos for products that are produced and sold within each province. In BC, the *Checkmark* logo is the official mark/symbol of the *British Columbia Certified Organic Program* (BCCOP). It, along with the phrase *British Columbia Certified Organic*, can be used to identify products produced in BC containing 95–100% certified organic ingredients, and are certified by the *Certified Organic Association of British Columbia* (COABC).

Organic producers may use an organic label with respect to agricultural products if the products are certified under the federal COR or the provincial BCCOP. Products that are certified organic under the COR by a CFIA-accredited CB may display the *Canada organic logo* if desired. These products may also carry the BCCOP *Checkmark* logo if certified by a COABC-accredited CB. Organic products certified under the BC provincial program for organic products may only display the BCCOP *Checkmark* logo (COABC n.d.).

In Quebec, the *Bio Québec* logo is a so-called *compliance logo*, an official trademark whose use is strictly regulated. The activities of businesses that use this logo on their products are monitored by a CB accredited *Conseil des appellations réservées et des termes valorisants* (CARTV). Organic producers can apply to the CARTV-managed logo program to use the *Bio Québec* logo to show their commitment to providing genuine organic products prepared in Quebec. Similar to the BC provincial organic logo, the Quebec provincial organic logo can be used on products that are certified organic under the COR by a CFIA-accredited CB if desired. In this case, the products would carry both provincial and federal organic logos (CARTV n.d.).

15.4 Organic Food Demand and Trade

The Canadian organic food industry is currently thriving from increasing growth. For example, the total value of the organic products market in Canada in 2012 was $3.5 billion CAD. By 2017, it had grown to $5.4 billion, with organic retail food purchases accounting for 81% of that total or $4.4 billion CAD. In addition, the market share of organic food and beverages sold by the retail food industry in Canada grew from 1.7% in 2012 to 2.6% by 2017. Canada's organic food market size is also ranked seventh largest in the world by value (OTA n.d.).

According to the latest Canada Organic Trade Association's (COTA) comprehensive analysis of Canada's organics market in 2017 (COTA n.d.-b), Ontario has the largest organics market, while British Columbia (BC) continues to have higher organics sales per capita. Two-thirds of Canadian grocery shoppers are purchasing organics weekly. Among all Canadian provinces, Alberta has the highest percentage of organics sales with 74% of Albertans buying organic food each week. Of the Canadian organic grocery products, fresh fruit and vegetables remains the most purchased category.

According to COTA's Organic Food Service Study 2018 (COTA 2018), consumer demand remains high, continuing to push the industry to offer more and more organic food options. *Millennial* buyers are key drivers of this demand with 83% purchasing organic food and beverages – the highest of any generation.

To meet this growing demand, the production and supply of organic food across Canada has increased. Specifically, there are now almost 5000 certified organic farms, processors and handlers in Canada. And Canada's organically managed acreage reached 3.15 million acres in 2017, an increase of 130,000 acres from 2016, and up over 700,000 acres from 2015, according to the latest figures from COTA (COTA n.d.-b).

Nevertheless, the current demand for organic products at present still exceeds domestic supply in Canada. This has resulted in an increased reliance on organic food imports (Kamchen 2019). Canada currently tracks 65 organic imports and 17 organic exports. Although grain is by far Canada's largest organic export commodity, Canada's organics products industry is also becoming more diversified.

Specifically, Canada also produces and exports a full range of fresh, frozen and processed organic foods, including everything from fruit and fruit juices to peas, lentils and other pulses, and from maple syrup to breakfast cereals, fruit smoothies, nut butters, and a wide range of meat and seafood products.

The top organics imports are unroasted coffee, bananas and olive oil, totaling approximately $216 million (of $637 million in total imports) for 2017. Green and red lentils, and red spring wheat account for approximately two thirds of organic exports, contributing to the total value of organic exports reaching $607 million by the end of 2017, according to *Canadian Organic Market: Trends and Opportunities 2017* (COTA n.d.-b).

Canadians consumers are demanding sustainably produced foods that avoid preservatives, synthetic fertilizers, artificial additives and genetically engineered ingredients (GMOs). Consumers want to trust the food they consume and understand the environmental impacts of both organic food choices and organic farm production systems. Canadian consumers' increasing health consciousness and a greater focus on social and environmental responsibility are driving the current organic food sales growth trend.

This trend is expected to grow as the concept of healthy lifestyle becomes increasingly more popular in parallel with the diversification of new organic products and greater investment in organic production, processing and handling.

15.4.1 Enhancing Organic Products Trade with China

In China, the demand for organics is growing at an even faster rate than in Canada. Specifically, the sales of organic products in China was recently forecasted to increase at a compound annual growth rate (CAGR) of 16% from 2016 to 2021 (Agriculture and Agri-Food Canada 2019).

Foreign food imports to China have also increased 15% annually in recent years, and the value of food imports has more than quadrupled in 10 years from 2005 to 2015 to 480 billion RMB (i.e. $77 billion USD). Edible vegetable oils, cereal and milk products account for roughly half of all food imports to China. A rough estimate for the market value of *organics* imports to China by 2021 will be around $32 million USD, with typical consumers being middle and upper class Chinese citizens with children [i.e. 2015 $20 M x 16% CAGR]. In a consumer study carried out on 204 Chinese organic consumers in Beijing and Shanghai, 71% were ready to pay a premium of 20–50% for organic products. The top five reasons for buying organic were all related to food quality assurance (Wright 2015).

China therefore poses a significant trade opportunity for Canadian organics producers. One Canadian agri-product export to China that has grown significantly is cherries. Specifically, in 2017, China imported 102,000 tons of cherries from eight countries valued at $770 million USD. Chile and the United States' market share by volume was 67.7% and 26.5% respectively. But Canada's market share was the third largest at 4.2% (or 4289 tons) or $26.35 million USD, representing an increase of 125% in volume and a 75% increase in value over 2016 (Zang 2018).

Canada's other recognized food trade priority opportunities with China include: barley, durum wheat, canola, linseed (flax), dried peas, salmon, shrimp, clam products, crab products, lobster products, maple syrup and pork (Zang 2018).

Given the potential growth opportunity for exporting *organic* products as part of these food trade priorities to a growing middle class in China, the negotiation of an organic *equivalency agreement* with China by Canada would enhance organic products trade, improve food safety and strengthen Canada's long-term economic relationship with China. Canada has long prided itself as being the "food basket" to the world and its continued respect by other nations will depend on it remaining a trusted trading partner that is more than willing to unselfishly share its food, safe food practices and reliable food distribution systems through greater trade collaboration, regulatory harmonization and mutual investment.

In short, Canadian organic farmers want to be *trusted friends* of China providing its citizens with reliable sources of imported safe and healthy food for its growing population. In return, Canada should be open to greater *collaboration* on supply chain investment and developing trade information networks where China has the proven capacity to be a technology and infrastructure development leader.

15.5 Conclusion

Organic production is a system of farm management and food production that intends no harm either to humans or the environment, including plant or animal health. The principles governing organic food production in Canada are Health, Ecology, Fairness, and Care.

These principles have also been codified into rules of sound ecological practices that must be implemented by organic operators. The *Safe Food for Canadians Regulations* (SFCR), outline the legal requirements in Canada governing organic production, processing and labelling. The SFCR incorporates and references the *Canadian Organic Standards* (COS) – the backbone of the Canadian organics industry – and assures consumers that products labelled *organic* were produced in compliance with these national standards. The Canadian organic standards outline and describe the practices that are allowed, as well as the substances that can be used by organic operators.

One of the strengths of Canada's organic standards is that they recognize the climatic and ecological biodiversity of the country, providing flexibility and compatibility. Yet, there is room for further improvement, by continually reforming regulations, polices, and programs to promote optimal land use practices by farmers based on the ecology of their region. Moreover, while some provinces have their own organic standards, it is critical for all provinces and territories to adopt organic regulations that are equivalent to, or more robust than, the federal organic regulatory regime.

Investing in local organic marketing, promoting international opportunities for organic exporters, particularly with China, and supporting increased capacity in the organic production supply chain, are all important policies to be recommended to encourage continued organic farming growth in Canada.

Lastly, the pursuit of greater regulatory *harmonization* and international *collaboration* through the negotiation of more organics *equivalency agreements* with more countries will enhance both minimum food safety standards and greater trade market access for organic food products. These goals and outcomes will be necessary to meet the rapidly growing demand by both domestic and international consumers for identifiably safe and healthy organic food products from trusted and reliable suppliers.

References

Agriculture and Agri-Food Canada (2019) Outline of opportunities in China. Retrieved from http://agr.gc.ca/eng/industry-markets-and-trade/international-agri-food-market-intelligence/reports/outline-of-opportunities-in-china/?id=1513879312343

Canada Organic Trade Association (COTA) (2018) Organics reach highest-ever market share and production capacity. Retrieved from https://www.canada-organic.ca/en/news/organics-reach-highest-ever-market-share-and-production-capacity

Canada Organic Trade Association (COTA) (n.d.-a) Equivalency arrangements. Retrieved from https://www.canada-organic.ca/en/what-we-do/market-access/equivalency-arrangements. ON, Canada

Canada Organic Trade Association (COTA) (n.d.-b) Canadian organic market report: trends and opportunities, 2017. Retrieved from https://canada-organic.myshopify.com/collections/canadian-organic-market-report-2017

Canadian Food Inspection Agency (CFIA) (2019a) Certification bodies accredited by the CFIA in Canada. Retrieved from https://www.inspection.gc.ca/food/requirements-and-guidance/organic-products/certification-bodies/in-canada/eng/1327861534754/1327861629954

Canadian Food Inspection Agency (CFIA) (2019b) Certification bodies accredited by the CFIA outside Canada. Retrieved from https://www.inspection.gc.ca/food/requirements-and-guidance/organic-products/certification-bodies/outside-canada/eng/1327862136347/1327862268980

Canadian Food Inspection Agency (CFIA) (2019c) Organic equivalency arrangements with other countries. Retrieved from https://www.inspection.gc.ca/food/requirements-and-guidance/organic-products/equivalence-arrangements/eng/1311987562418/1311987760268

Canadian Food Inspection Agency (CFIA) (2019d) CFIA procedure for organic equivalency determination and on-going monitoring of the existing arrangements. Retrieved from https://www.inspection.gc.ca/food/requirements-and-guidance/organic-products/equivalence-arrangements/equivalency-determination/eng/1328067000985/1328067099099

Canadian Food Inspection Agency (CFIA) (2019e) CFIA peer review procedure related to the Canada organic regime (COR). Retrieved from https://www.inspection.gc.ca/food/requirements-and-guidance/organic-products/equivalence-arrangements/cfia-peer-review-procedure/eng/1328067544284/1328067663737

Canadian Food Inspection Agency (CFIA) (2019f) Permitted claims. Retrieved from https://www.inspection.gc.ca/food/requirements-and-guidance/labelling/industry/organic-claims/eng/1389725994094/1389726052482?chap=3

Canadian Food Inspection Agency (CFIA) (2019g) Non permitted claims. Retrieved from https://www.inspection.gc.ca/food/requirements-and-guidance/labelling/industry/organic-claims/eng/1389725994094/1389726052482?chap=4

Canadian Food Inspection Agency (CFIA) (2019h) Canada organic regime operating manual general information. Retrieved from https://www.inspection.gc.ca/food/requirements-and-guidance/organic-products/operating-manual/eng/1389199079075/1554143470958?chap=2

Certified Organic Associations of BC (COABC) (n.d.) Organic certification in BC. Retrieved from https://www.certifiedorganic.bc.ca/cb/certification.php. BC, Canada

Conseil des appellations réservées et des termes valorisants (CARTV) (n.d.) 'Bio Québec' Logo. Retrieved from https://www.cartv.gouv.qc.ca/en/logo-bio-quebec. Quebec, Canada

Government of Canada Publications (2013a) Organic production systems: general principles and management standards – CAN/CGSB-32.310 – Published in 2006, revised in 2018. Retrieved from http://www.publications.gc.ca/site/eng/9.854643/publication.html

Government of Canada Publications (2013b) Organic production systems: permitted substances lists. CAN/CGSB-32.311 – Published in 2006, revised in 2018. Retrieved from http://www.publications.gc.ca/site/eng/9.854645/publication.html

Government of Canada Publications (2013c) Organic production systems: aquaculture – general principles, management standards and permitted substances lists. CAN/CGSB 32.312 – Published in 2012, revised in 2018. Retrieved from http://www.publications.gc.ca/site/eng/9.851011/publication.html

Justice Laws Website (2019) Safe food for Canadian regulations. Retrieved from https://laws-lois.justice.gc.ca/eng/regulations/SOR-2018-108/index.html

Kamchen R (2019) Room to grow organic. Retrieved from https://www.country-guide.ca/guide-business/canadian-farmers-increasing-organic-output-with-plenty-of-room-to-grow/

Organic Council of Ontario (OCO) (n.d.) Canada's organic sector: an overview. Retrieved from https://www.organiccouncil.ca/canadas-organic-sector-an-overview/. ON, Canada

Organic Trade Association (OTA) (n.d.) Organic Canada. Retrieved from https://globalorganictrade.com/country/canada. DC, USA

Wright K (2015) Going organic: investing in China's growing health foods market. China business review. Retrieved from https://www.chinabusinessreview.com/going-organic-investing-in-chinas-growing-health-foods-market/

Zang J (2018) 2017 year in review: China's cherry market. Producer report. Retrieved from https://www.producerreport.com//article/2017-year-in-review-china%E2%80%99s-cherry-market

Alexander C. Phillips is the President and CEO of the Van Horne Institute and a trade and transportation lawyer based in Calgary, Alberta, Canada. The Van Horne Institute is a national 'think tank' focused on growth issues related to trade, transportation, infrastructure and sustainable resource development. In his 35-year career, he has provided law, economics and regulatory policy advice to carriers, shippers, forwarders, airports, ports, electricity grids and related supply chain stakeholders. He completed his BComm (Hons) and LLB degrees from the University of Manitoba; his Master of Laws degree in Aviation Deregulation from the London School of Economics, United Kingdom; and his doctorate (SJD degree) in Infrastructure Privatization, Trade Access Rights and Dispute Resolution from Bond University, Australia.

Chapter 16
Concluding Observations: Perspectives and Prospects for the Regulatory Environment of Organic Food Safety in Asia Pacific

Rohan Price and GOH Bee Chen

Abstract The issues encountered by twenty-first century Asian nations in the field of organics regulation and food safety are not new. In striking a balance between agricultural productivism and crop production based on organic principles, nearly all the nations of Asia Pacific play out an instalment of a history close to hand. In several respects, the dilemma encountered in Europe after the World War Two has merely been reanimated in modern Asia as its nations rise to ever-greater prosperity. Policy steers a course, or floats adrift, between a desire to feed millions and revering romantic ideas about producing food. Yet in other aspects, perhaps more important ones, including the rise of marketing and its focus on consumer perception and verification of food origin, the debate over what role the state can and should play in the production of clean, unpolluted farm produce free of synthetic inputs has entered a new territory. It seeks only to defend what consumers *believe* about food notionally, at least, until the science is in and sensible accommodation of light touch conventional and eco-purist approaches is found. This concluding chapter conducts a broad survey of this book's chapter contributions in a theme of the more things change the more they stay the same. But it also flags the discussions in this book which, although not entirely outside the scope of conventional historical or political frameworks, do give pause for thought or suspicion about what the future holds.

Keywords & Phrases Organic Agriculture · Organic Food Safety · Regulatory Framework · Asia Pacific · History of Organics

R. Price
School of Law and Justice, Southern Cross University, Gold Coast, QLD, Australia
e-mail: rohan.price@scu.edu.au

GOH Bee Chen (✉)
School of Law and Justice, Southern Cross University, Gold Coast, QLD, Australia

Faculty of Law, Chiang Mai University, Chiang Mai, Thailand
e-mail: beechen.goh@scu.edu.au

GOH Bee Chen, R. Price (eds.), *Regulatory Issues in Organic Food Safety in the Asia Pacific*, https://doi.org/10.1007/978-981-15-3580-2_16

16.1 Introduction

Spirits have been running high in the preparation of this edited volume, because its publication is timely. The book has derived valuable insights from contributing authors with diverse interdisciplinary and geographical backgrounds. By way of serendipity, the field of organics has witnessed another important piece of work with the 2018 publication of Venus Bivar's *Organic Resistance: The Struggle Over Industrial Farming in Postwar France.*[1] This chapter will draw upon various parallels from Bivar's publication for comparative purposes.

In this book, it has appeared that Asia-Pacific's leading lights on organic regulation had told a story that, although seemingly fragmenting the organic dream along the lines of national and cultural reception, was conformable in some respects to the post-war policy torment that might be thought characteristic of France or Europe. What has occurred in the historical continuum of Asia-Pacific has simply happened later in time – an adjustment of thinking entirely consistent with its economic reassertion. In Bivar's framing: were farmers to produce "high quality, nutritious" food or adopt mass farming to produce "safe" food?[2] In a telling way, most of the chapters of this volume have addressed one or other variant of this tension and anxiety.

16.2 International Dimensions

This volume has ventured from the macro insights delivered from regional perspectives and trading blocks toward the finer observations of national polity, both large and small, across the far-flung Asia Pacific. **Francis Snyder** (Chap. 2) opened the book's contributions with a consideration that 'ecological food' in China was comprised of a troika of standards which each maintained a different shade of promised health in their guarantee to the food-consuming public. He argued that, although the regulation of ecological food was beset by inconsistency and public suspicion, it would provide a template for future regulation i.e. the basic categories of 'pollution free', 'green' and 'organic food' would tighten up and hold good in the future. Snyder places faith in the growth of legal pluralism in China and, in particular, contends that it will improve food quality by increasing participating guarantee schemes including RFID tracking and block-chain technology. Snyder's investigation of China's food regulation notes that the government has concentrated on 'green food' while development of IFOAM organic standards has taken a parallel path. This appears in an emblematic light in his conclusion that there is an "uneasy fit between agricultural

[1] Venus Bivar, *Organic Resistance: The Struggle Over Industrial Farming in Postwar France* (Chapell Hill: University of North Carolina Press: 2018).

[2] Ibid, 3.

productivism, environmental protection and food safety which remains a central part of Chinese rural and food policy". Snyder's contention that Chinese consumers do not trust government or third party eco-food labelling because of its deservedly poor reputation finds an echo in Bivar's work which dismissed historical governmental allusions to food quality in the Eurosphere as "a version of quality" that was "vacuous" and "largely an invention of marketing".[3] Snyder rightly baulks at glorification of the China food system but his contribution comes out nevertheless up-beat for his remit: "...standards, whether legally binding or not, tend to favour certain interests, by harmonising rules, determining market access and even creating markets".

The tone of cautious optimism for pluralism and the determinacy of markets characterising Synder's assessment of China's regulatory example in eco-food distribution paves the way of other investigations in the book. **GOH Bee Chen**'s detailed treatment of organic food labelling in Thailand and China (Chap. 3) argued that country-of-origin labelling induced higher public confidence in those countries and, in a broader arena, provided an outstanding opportunity for seasoned players in organic exports including Australia, New Zealand and Canada. Goh's contention is that the organic food movement of the Asia Pacific is a return to the agricultural past of India and China prompted by middle class concerns about the safety of post-industrial broad acre farming. This argument of revivalism, given example by Thailand, prompts the reader to ask, if government promotion of worthy organic agriculture is quite the same idea as government marketing of eco-foods through a certification process. In this light, Goh highlights the role of certification in "satisfaction" of "credence claims" in maintaining "the perceived inherent health value of organic foods". Certification gives grounds to perception. After setting down the stringencies of Australian organic food labelling, Goh turns to the question of country of origin marketing as part of the credence claim.

Australia, it turns out, is part of decided drift of mature organics economies from voluntary to compulsory labelling declaring the origin of agricultural food. As Bivar pointed out, the spurious "quality" marketing of French agricultural produce after World War Two relied on convincing consumers that "French products were somehow superior to their British or American counterparts".[4] Goh shows how poorly such queries have dated through a discursion into the topic of trust in food labelling. Goh argues that "perceptual bias" that mature organic producing economies vet their food exports in a reliable way is clearly to the trade advantage of such countries. i.e. an American leek is not better than a French leek but a consumer would instinctively trust either in preference to a leek produced in a country with no track record of regulating according to an organic standard. Hence, Goh finds that when a mutual recognition standard is established between mature organics producers with a satisfied claim to credence, "the perception that each has achieved the high organics standards will undoubtedly be present in the minds of consumers willing to pay a pre-

[3] Bivar, *Organic Resistance*, 3–4.

[4] Bivar, *Organic Resistance*, 4.

mium price for trusted organic foods". The nationalism pervading discussions of food quality in the 1950s has become, in Goh's reading, a form of old club internationalism.

Ying Shen tackled two goliaths (Chap. 4): climate change mitigation and the potential contribution to it of organic food production practices in China. Having noted the liberalisation of Chinese agriculture in the late 1970s in the direction of personal and household ownership and benefit from rural produce production, Shen traces the rise of the Green Food Program in the 1980s as 'a halfway house' between conventional and organic food production and goes on to note the meteoric rise of organic food practices since the year 2000. After detailing the plethora of regulation and certification measures in China in aim of food safety, Shen's chapter describes the role of organics in adaptation to climate and its potential in carbon capture. And so, China moves from one great social aim for agriculture to a new one. As Eisenman pointed out, the 'decollectivisation dividend' in the late 1970s was aimed at disbursement of commune capital and reducing the extractive powers of the commune dramatically.[5] These days, the great march of the nation is toward what Shen describes as increasing the 'production capability' of China's rice farmers in disrupted and unprecedentedly variable climatic conditions. Also part of the response to climate change is the verification of positive values of organic farming practices, including moderate water usage, continued commitment to preservation of seed and crop diversity. The key to this is consumer trust in the Food Safety Law and its role in promoting organic an agriculture that is transparent and credible in order to make organic food consumption increase to a point where it is transformative in environmental terms. Echoing both Goh and Snyder, Shen's references to public trust in governmental processes continued a familiar theme of the book. Shen's chapter confirmed that, in some ways, China's vastly felt social movements and dramatic lurches toward economic reform pushed put its legacy into an internalised, domestic category unto itself. Nevertheless, the prospect of ever-increasing consumer demand for more sustainably produced food from the Motherland and abroad, stands a strong chance of contributing positively to climate change abatement through wholesale review of agricultural practices.

16.3 Historical Influences and National Responses

Satoshi Kodera sought to examine the organic food production of Japan, one of Asia's economic powerhouses and a world-leading trend-setter in the fields of food consumption and production (Chap. 5). Kodera reviewed the two decades of operation enjoyed by the Japanese Agricultural Standard (JAS). Kodera reports that initially, in the boom decades after the Second World War, Japanese

[5] Joshua Eisenman, *Red China's Green Revolution: Technological Evolution, Institutional Change, and Economic Development Under the Commune* (New York: Columbia University Press, 2018) 246.

consumers established a DIY system whereby their concern about polluted food produced by industrial agriculture led them to form private buying groups which dealt directly with small, trustworthy agricultural producers. This was occurring elsewhere in the world in the 1960s; Bivar noted a trend of small-scale French farmers concentrating on bee hive products and goats' cheese dealing directly with discerning members of the public who were concerned with the degradation of the environment by mass farming practices and who were prepared to pay for expensive food sourced somewhat eccentrically.[6] Kodera opens a fascinating window: that when dealings over food are face-to-face with partner farmers there is little need for regulation and that, perhaps, the convenience of delivered or supermarket-bought food has devalued our discernment of quality. Kodera reports the development of a post-war 'organic bubble' in which food misdescription became rife. Law was needed to set it straight, culminating in the 1990s in a set of detailed laws on food production and processing standards. Moreover, the specificity of Japanese aims in the production of organic plants lays down a blueprint for any advanced nation considering what it means by certification of organic standards. In particular, noteworthy attention is given to recycling processes within agriculture, productivity from developing soil properties without artificial chemicals and reducing the impact of agriculture on the wider environment containing it.

Depending on how one looks at it, China's organics project is decades or millennia in the making. **Rohan Price** canvasses the twentieth century past to produce an encounter with the China history problem with organic characteristics (Chap. 6). His work does not yield to pretensions of humanism, or revere operative trading hierarchies, that mystically clarify or dissemble when we begin to talk about a particular nationality or economic interest. Price acknowledges the catastrophic mistakes in Chinese agricultural policy after 1949 but frames them as the hardest lessons best learned. In this undertaking, Price invokes a futurist argument: considering that only 14% of China's land mass is arable, its old tragedies and new money culminate in a determination to pursue quality agricultural produce imported from elsewhere on a scale that will be deep and wide in its impacts on the nations of the West. Due to the country's shift from quantity of food to its quality, this growing national obsession will change the world's economic balance. China's history of struggle since 1949 to produce palatable food for hundreds of millions of people has resolved into a period of mass prosperity leading to outsourcing as an ultimate solution to national want. Its increasingly wealthy middle class seek clean food from elsewhere, not necessarily a sour reminiscence of the nation's highly variable historical balance of necessary agricultural output, environmental cost, climatic conditions, and externally dictated balances of trade. But this history is not a dead letter. Concerns about the old failures in food security have evolved into consumer indignation, a kind of civil rights of minimum health, requiring government accountability for food safety and quality. China cannot address these problems alone, but it

[6] Bivar, *Organic Resistance*, 161.

now has the economic muscle to do something about them. The distrusted terrain of bilateralism and its hidden cargoes has produced a fascinating twist to the post-1949 Chinese mass line: when the country's leadership rhetorically asks why wealthy Chinese consumers still cannot get access to the best food, it appeals to both individualism and nationalism. Organics takes a starkly political form.

Natalie Wong took a case study on gutter oil in the Chinese mainland restaurant trade that left its readers asking profound questions about the limits of regulation in cases of mass food contamination (Chap. 7). Undoubtedly, this case study highlighted the serious concerns pertaining to food safety. Considering that in China participation in restaurant dining remains a generally affordable mass activity, Wong's chapter also asked question about the primacy of human health in a context where even the most basic safeguards against toxic food ingredients are routinely flouted. Most sobering about Wong's contribution is that the conversion of inedible food into notionally edible food is known at high levels of government to be a serious national issue but, in regulatory terms, it continues to be handled as a minor or emerging problem through guidelines recommending what should be done. Wong has produced more than an example of China's long history of adulterating or repurposing food. The chapter sounds as a warning about the health-adverse state of affairs in China and how political will must actively follow consumer affluence, not rely on neglectful state habits waiting for a scandal to force change.

Li Hui built on work of Francis Snyder and Ying Sheng in detailing the perplexing bureaucratic frameworks regulating organic food production and processing in China (Chap. 8). Li, like several of our contributors, highlighted the effect of regulatory failures on consumer trust and confidence in certification standards and the practical meaning of labelling as a quality reassurance. Li identified low consumer awareness of organic food, low trust in certification, unreasonable price premiums for organic produce, poorly coordinated and less than transparent supervision of organic standards, and a weak penalty regime for transgressive practices concerning organic food, as the key challenges facing the Chinese organic food industry despite it recording huge year-to-year strides in the market place. Li, in a similar vein to Price, also tracked the development of government policy from quality increase to quality improvement as its prime objective but noted that China has a long way to go before its certification standards are backed by public trust.

Suhail Nathani and **Mehfuz Mollah** drew our gaze to the organic food system of a consumption giant, the Indian nation (Chap. 9). In recent past, India has emerged as a prominent organic food producer, accounting for about 30% of the total organic producers of the world.[7] These co-authors delineate the organic laws in India covering aspects as diverse as labelling, proof of providence, logistics, as well as retailing and importing organic food. India, Pakistan and the Philippines led the developed world in the green revolution of the 1960s and 1970s and such nations were well in advance of China. Although the rural landscape was green with high

[7]FiBL Survey 2019, reproduced in Helga Willer and Julia Lernoud (eds), *The World of Organic Agriculture: Statistics and Emerging Trends 2019* (Frick and Bonn: Research Institute of Organic Agriculture (FiBL), and IFOAM Organics International, Bonn: 2019), 24.

yield varieties of rice and wheat, and the middlemen and export traders lined their wallets with much 'green', it came at the cost of land degradation and overuse of chemical inputs that were anything but 'green'. Nathani and Mollah report that government policy has been instrumental in the rising availability of organic food since the bad old days but that the regulatory framework had to wait until 2017 for a comprehensive overhaul to an essentially piecemeal system.

The co-authors recite an incredibly detailed regime for food safety in India ranging for the penalties for various types of food contamination to exacting certification processes. Similar to Kodera in the Japanese context, and Bivar in a French example, allusion is made by the chapter co-authors to a small producer guarantee system in India that sits alongside government sanctioned third party certification according to a single standard. Under the Participatory Guarantee System for India (PGS-India) small-holder producers assess, inspect and verify production practices to enable an organic certification process. Thus, producers and consumers of food work together in a "foundation of trust, social networks and knowledge exchange". Such systems have challenges, but must be seen as a vital part of the organic puzzle because they do not shift certification costs back on to the small producer in developing nations and they emphasise community involvement in tracing the quality of food back to the growing environment of agricultural foodstuffs.

Our South East Asian coverage began with a study of the Malaysian legislative organic food regime in a jointly written chapter by **Mohammad Aziz Bin Abdul Aziz**, **Muhanad Shakrin Bin Mispan** and **Febri Doni** (Chap. 10). As with most, if not all, our contributors, the three joint authors reported a considerable gap between government policy interventions in the form of law and the degree of trust among the consuming public that the 'organic claim' made about food was credible. They point out that there is a lack of hard evidence proving the health-giving benefit of organic over conventional methods of production: "the safety of the products from both origins are indecisively dependent on good agriculture practices and maintaining hygienic conditions from farm to the plate". This raised a question whether governments should be in the business of promoting a purist pursuit of organic farming practices when practices which limit bio-contamination of all food crops, and have only the most requisite pesticide and fertiliser inputs, could produce safe and nutritious food.

Thararat Chitov supplied a chapter from a scientific perspective on the Thailand organic enterprise which, in turns, alarmed and warned (Chap. 11). The work detailed extremely high pesticide residues found on a wide variety of fruit and vegetables available in Thai supermarkets. The findings of this chapter will give pause for thought for holidaymakers in the Kingdom when consuming chillies, pakchoi and basil. Chitov gave a detailed overview of the network of laws on food safety in Thailand and also deeply considered the literature on bio-contamination hazards noting that both modern technology and traditional practices have a role to play in the production of safe food. Chitov's main contribution is to demonstrate how apparently ideal conditions for the organic production of milk in Thailand nevertheless yielded unacceptable bacterial counts in milk. As a scientist in the front line of food safety, Chitov knows the standards and the practice. This prompted reflection on the marketing of food cleanliness and the everyday, inadvertent risks to it that make a lie of the promise. Food really can look too good. Ultimately, food safety remains the prime consideration.

Van Kien Nguyen is our organics contributor from Vietnam (Chap. 12). The chapter showed that Vietnam is fast joining Indonesia and the Philippines as the rising star of organic agriculture in South East Asia. Nguyen reported that, despite ever-larger tracts of Vietnamese soil being given over to organic agriculture, the domination of small land holders is a major challenge to continued growth of the sector because of the financial impost of organic requirements. If the governments of South East Asia truly wish to transform agriculture to organic ways that are sustainable and premium-bearing, there nevertheless seems to be a debate about the role of subsidy and ultimate economic benefit that is not occurring. Nguyen also recorded allegations that consumers were being cheated by masquerading organic produce, that there were problems with the trustworthiness of certification and that the non-accreditation of Vietnamese produce as organic makes export-oriented market development difficult. Nguyen also reports that Vietnamese universities show little enthusiasm for sponsoring research on how chemical inputs can be replaced with more sustainable agricultural measures. Taken alongside governmental reluctance to sponsor farmer transition to organic farming methods, investment failure in research must be regarded as a problem extending beyond Vietnam's borders to its near neighbours and regional colleagues.

Anastasia Telesetsky provided a chapter on the regulatory constellation of the world's leading bench marker of organic produce, the United States of America (Chap. 13). In a similar vein to Chitov, Telesetsky detailed the significant water and soil-borne bacterial risks posed by improperly composted manure used in production of fruit and vegetables, as well as milk. After detailing an extensive web of national and local laws, Telesetsky refers to an intriguing initiative in California whereby produce handlers agree to only deal in produce from farms that regularly test for pathogen risks, especially *E. Coli*. This is another example of the self-help measures used in some organic industries to ensure that contamination-free produce reaches consumers. It speaks less of the failure of government and law than it does of the importance of industry agents to work actively and cooperatively on a manageable scale to mitigate risk. The voluntary approach has also replicated the regulatory approach by adopting rules on animal proximity, time minima on manure breakdown before harvest, and the like, to improve food hygiene. As Bivar pointed out in her study of French agriculture, until the 1960s 'quality' of milk or other farm produce was most often rated by the absence of illness-causing pathogens and, as time has gone by, 'quality' has developed diffuse meanings.[8] It may be rather a stretch to contend that marketing qualities on a basis of what is perceived as healthy by a consumer has come at the cost of scientific vigilance over farming practices. Nevertheless, the organic claim to healthfulness of organic leaf vegetables and milk

[8] Bivar, *Organic Resistance*, 3.

in the United States has, on occasion, been reprimanded by mass hospitalisations through consumption of pathogenic produce prompting a less nuanced and more basically accountable concept of quality.

Tracey Epps and **Danae Wheeler** gave us an update on all things organic from Aotearoa, New Zealand (Chap. 14). In their national report, New Zealand's clean, green image is a good deal more than marketing a perception, but is backed by a significant commitment to locally sourced, contamination-free food that rejects use of synthetic chemicals. The organic food industry has continued to grow at an impressive and enviable rate in New Zealand. The government of the country has been active in the development of a mandatory organics standard; one which supplements a range of private standards and a strong legislative commitment to ensuring the safety of food across the board, including non-organic products. New Zealand followed the conventional path in the post-war years in liberally applying to farmland superphosphates, organophosphates and chemicals in the DDT family leading to soil degradation, environmental damage to plants and wildlife. From the 1970s onward, the country underwent an organic renaissance and consumer demand for nature-friendly, unpolluted food increased through media popularisation and the adoption of a voluntary organic standard in 1983. Perhaps the most path breaking aspect of the New Zealand approach to organics has been to sign a mutual recognition agreement of organic certification with China and a trans-pacific arrangement requiring member states to properly enforce organic standards. These achievements place the country at the forefront of globalising organic standards through trade. At this juncture, the reader needs to wait eagerly for what lies ahead in legislative reform for organics in New Zealand.

Rounding out our Pacific coverage, **Alexander Phillips** followed a similar line to our New Zealand colleagues by making trade and organics certification a focus of his chapter which took a Canadian lens to the organic movement (Chap. 15). Phillips outlines the core component principles of Health, Ecology, Fairness and Care at the heart of Canadian philosophy underpinning its legislated regime establishing the standards that growers and handlers of organic food must achieve. The previous chapter by Epps and Wheeler illustrated a similar constitutional approach to defining organic food as promoting agro-ecosystem health, use of management systems rather than off-farm inputs, mulching and biological control in preference to synthetic chemicals, and so on. Whether based on farm practices or social outcomes, the few remaining Western nations including Australia and New Zealand that have not yet developed their own national mandatory organic standard, would benefit from the approach of Canada in starting with constitutional goals and developing their standard from them, rather than simply take U.S. and EU standards as the beginning and end of what certification can achieve. Phillips reviews the intricate latticework of laws and regulations in force in Canada to comprise the Canadian Organic Standard (COS) and proceeds to make a telling point: "[Canadian] consumers want to trust the food they consume and understand the environmental impacts of both organic food choices and organic farm production systems". From this, it is conceivable that a significant reason of a mature and well-informed organics consuming society for opting for organic food is that its imprint on the environment it is produced in is more in harmony with the needs of a broad biosphere than conventional farming. Therefore, any extraordinary health benefits accruing from organic food consumption can be regarded a bonus rather than a dominant benefit.

16.4 Future Challenges

It has been keenly observed that, globally, organic food sales have edged towards US$100 billion.[9] This is a staggering figure indeed. On the one hand, the prospects for organic food consumption and trade are undoubtedly very bright. On the other hand, the associated challenges can hardly be ignored. We have seen throughout the various chapter expositions in this volume that the global rise of organic food consumption has been phenomenal. Health and well-being factors, climate activism and animal welfare concerns are considered the chief attributing factors to organic food safety.

An immediate challenge can be discerned from the regulatory environment in organics. No matter how well-perceived the idea of organic food is to the everyday consumer, the stakeholders involved know far too well that the picture is not always rosy, nor the path smooth or risk-free. Take, for instance, the subject of certification of organic food production. The time, expense and bureaucracy involved can easily act as deterrents to those farmers wishing to engage in organic farming, particularly for small-scale farmers or family-owned farming enterprises. When costs are prohibitive in order for organic farmers to achieve compliance with standards, and organic trade returns are not as lucrative as initially perceived, organic farming becomes an ideal rather than a practical reality. Furthermore, the regulatory environment for organics is complex and a minefield, where standards can vary from jurisdiction to jurisdiction.[10] Added to this is the ongoing challenging task of regulatory enforcement. There arises confusion or misunderstanding as consumers tend to overlook the fact that compliance with organic standards such as in the area of organic food labelling may be mandatory or voluntary, depending on jurisdictions.

Allied to food labelling is the variety in the marketplace pertaining to labelling of health foods. Organic labels exist alongside other competing labels or eco-brands. It has been remarked that "the food industry now has over 200 labels that represent some ethical/sustainable attributes".[11] For the health conscious consumer,

---

[9] Amarjit Sahota, "The Global Market for Organic Food and Drink" in Helga Willer and Julia Lernoud (eds), *The World of Organic Agriculture: Statistics and Emerging Trends 2019* (Frick and Bonn: Research Institute of Organic Agriculture (FiBL), and IFOAM Organics International, Bonn: 2019), 148.

[10] Ibid.

[11] Ibid, 149.

wading through the variety of health labels can be quite daunting. The adverse impact for organic food trade caused by confusing or competing labelling is that consumers have more choices in healthy eating than just being limited to eating only organic food.

Another significant challenge lies in the prospect of supply and demand of organic foods. On a global scale, there is growing disparity between available organic agricultural land for production and organic food consumption. Furthermore, there is also a lopsided concentration of organic food sales in the Western world so much so that organic consumption is regarded as a "luxury" good.[12]

With the next decade of the twenty-first Century afoot, the uncertainty caused by Brexit as well as bilateral trade wars between economic powers such as the United States and China can pose serious threats to the world of organic food trade.[13] Organic farmers and traders need to be able to act with some certainty with regard to the trade and regulatory environment. Otherwise, the economic risks can be too high to warrant commencing or continuing with organic food production. In the end, consumers will end up paying higher prices than are already commanded by the premium pricing of organic foods.

16.5 Conclusion

The findings contained in this edited volume have both shown the common historical grounds as well as shared perspectives of organic food safety, and teased the reader about its prospects. It has demonstrated that the legacy of the Western past and the exciting present of Asia Pacific crisscross each other in a variety of ways.

Perhaps the last word should be given to Eisenman and China: "many experts and most laymen, Chinese and non-Chinese alike, trace China's sustained economic growth to the expansion of rural markets and material incentives beginning in 1979".[14] In a world where those matters called 'rural' are deemed insignificant or uneducated or lacking because they are cast as pre-industrial, how refreshing it is to have a date for the greatest national ascent in human history and to know that its cause is 'rural'. How great it is to know that when the rural world is perfected, it is moderated or joined by science rather than dominated by it to the earth's peril. What enormous consolation there is that the laws of humans must address the discontents of most humans, for if they do not, all there can be is holding down of the many, their routine abuse or hoodwink, and the purposeful ignorance of a few about the needs of many. Such is the remedial promise of organics studies, beyond the immediate excitements of trade and middle class health.

12 Ibid, 148.

13 Ibid, 149.

14 Eisenman, *Red China's Green Revolution*, 1.

References

Bivar V (2018) Organic resistance: the struggle over industrial farming in postwar France. University of North Carolina Press, Chapel Hill

Eisenman J (2018) Red China's green revolution: technological evolution, institutional change, and economic development under the commune. Columbia University Press, New York

Sahota A (2019) The global market for organic food and drink. In: Willer H, Lernoud J (eds) The world of organic agriculture: statistics and emerging trends 2019. Frick and Bonn: Research Institute of Organic Agriculture (FiBL), and IFOAM Organics International, Bonn, pp 146–150

Willer H, Lernoud J (eds) (2019) the world of organic agriculture: statistics and emerging trends 2019. Frick and Bonn: Research Institute of Organic Agriculture (FiBL), and IFOAM Organics International, Bonn

Rohan Price A lecturer in the School of Law and Justice, Southern Cross University, Dr. Rohan Price has established a reputation as a foremost historian of the role of nationalism in the British/Chinese colonial encounter. He was awarded a Doctor of Philosophy from the University of New England (Australia) for his thesis on the use of property law to encourage civic loyalty to colonial Hong Kong between the world wars. His books, including *Reading Colonies: Property and Control of the British Far East* and *Resistance in Colonial and Communist China* (1950–1963), are based on extensive archival and digital repository research. His books have been described in reviews as "passionate", containing "argumentative strength and forthright originality" and "enormous attention to historic, theoretical and political detail". He has enjoyed lengthy stints as a Visiting Professor in three Chinese universities over the last decade, teaching in fields including common law history, maritime law and the law of trusts. His interest on Chinese food safety issues was prompted by a friend in Hong Kong who, in 2008, casually mentioned he should not buy the same brand of noodles every time he went to the supermarket. He has finished the year 2019 with a new monograph published by City University Press, Hong Kong, *Violence and Emancipation in Colonial Ideology: Hong Kong and British Malaya.* It has been described in an academic review as "a major contribution to the literature".

GOH Bee Chen A Former Malaysian Rhodes Scholar, Professor GOH Bee Chen is Professor of Law and Director of the Judge-in-Residence Programme, School of Law and Justice, Southern Cross University, Australia. She has recently been a Visiting Professor in the Faculty of Law, Chiang Mai University, Thailand. She is a Director and Fellow of the Australian Academy of Law and Fellow of Cambridge Commonwealth Society and of the Society for Advanced Legal Studies in London. Besides organic regulatory framework, her current collaborative research project deals with Law and Theatre. Her scholarly interests include mediation and ADR, especially on cross-cultural (Sino-Western) dispute resolution, and international law of peace. Her publications include *Negotiating with the Chinese* (Dartmouth/Routledge, 1996), *Law Without Lawyers, Justice Without Courts: On Traditional Chinese Mediation* (Ashgate/Routledge, 2002), *Activating Human Rights and Peace: Theories, Practices and Contexts* (Ashgate/Routledge, 2012, with Offord and Garbutt); and *Scholarship, Practice and Education in Comparative Law: A Festschrift in Honour of Mary Hiscock* (Springer, 2019, with Farrar and Lo).

Index

GOH Bee Chen, R. Price (eds.), *Regulatory Issues in Organic Food Safety in the Asia Pacific*, https://doi.org/10.1007/978-981-15-3580-2

The manufacturer's authorised representative in the EU is Springer Nature Customer Service Centre GmbH, Europaplatz 3, 69115 Heidelberg, Germany. If you have any concerns regarding our products, please contact ProductSafety@springernature.com

Printed and bound by CPI Group (UK) Ltd, Croydon, CR0 4YY
30/09/2026
02225325-0011